STARSHIP CENTURY

Edited by

James Benford & Gregory Benford

Based on the 2011 100 Year Starship Symposium

Proceeds from this book will be donated to research on interstellar travel.

under license
agreement with

under license agreement with

16 JUNE

Starship Century
Copyright 2013 by Microwave Sciences
All rights reserved

Cover Art: NASA image
Cover Design: James Benford

Published by Microwave Sciences
10 9 8 7 6 5 4 3 2 1
First Edition

Table of Contents

(Fiction in italics)

THE STARSHIP ERA

AUTHORS, ARTISTS, ORGANIZATIONS & FURTHER READING

DEDICATION

This volume is dedicated to two who have labored
to make the future happen:

Pete Worden
David Neyland

Nothing truly worth doing is easy....

ACKNOWLEDGEMENTS

For much needed assistance creating and assembling this book we thank Paul Gilster, Jill Tarter, Stewart Brand, Trish Morrisey, Kelvin Long, Pete Worden, Stephen Hawking, Neal Stephenson, Kathryn Cramer, David Neyland, Stan Borowski, Dominic Benford, Harold White, Harry Jones, Jack McDevitt, Steve Postrel, Bart Kosko and, of course, Hilary Benford and Elisabeth Malartre.

Interstellar Ramjet leaves orbit of planet around a red dwarf star. (Don Dixon)

Starships: Reaching for the Highest Bar

Come, my friends.
'Tis not too late to seek a newer world.
Push off, and sitting well in order smite
The sounding furrows; for my purpose holds
To sail beyond the sunset, and the baths
Of all the western stars, until I die.

~Tennyson, "Ulysses"

Is this the century when we begin to build starships? The preliminary work was done in the 20th century. Several new organizations have starships as their goal. For some starship concepts, we know the physics and the engineering appears at least possible.

We assembled this volume to address *how* to build starships, *why* to build them, and *what* the implications for the long-term future will be if we do. The key question is how to energize civilization for a really long-term objective: the stars.

So will we mount the effort? And why should we?

Why go to the stars?

Because we are the descendents of those primates who chose to look over the next hill.

Because we won't survive here indefinitely.

Because the stars are there, beckoning with fresh horizons.

As this book was being created, word came that an Earth-sized planet had been discovered circling Alpha Centauri B, one member of the three-star system closest to us. Alpha Centauri Bb is locked facing the star, a hellish half-world indeed, with a frigid, dark shadowed hemisphere. This rocky planet around our nearest neighbor has ignited interest in sending probes there, as a first venture into interstellar space. Moreover, based on results from NASA's Kepler spacecraft—looking for planets around stars in a segment of the sky—we now know that multiple-planet systems are common, especially when small rocky planets are found close to the central star.

Plausibly, Alpha Centauri A—the bigger partner of Centauri B, and closer to it than Pluto is to our sun—has a good chance to also host planets. Observations and calculations indicate that massive gas giants are ruled out in the Alpha Centauri system, but rocky planets are now ruled in, as we now know that at least one exists. The dimmer third star of the Centauri system, Proxima, lies a fractional light-year away from the other two, but it is of interest for exploration.

So we have nearby the best situation: three stars that may have Earthlike planets, but can be explored together. Intensive observations of the Centauri system are underway and in a few years we will know much more about it. This will doubtless fuel further interest in building telescopes to observe the planets in the Centauri system and probes to voyage there. This is what we have done with the outer planet explorations of the last thirty years in our solar system. This process continues; this adventure will continue.

We can now begin to develop the technologies to reach out to the nearby stars, for our observations are bringing them into the realm of what we might call Known Space.

Whatever the reason, there is a growing interest and effort toward the greatest challenge we can imagine. To begin exploring other stars within a century opens up grand panoramas. Scientific, technical, biological, social, and economic challenges abound. But we can now see how to meet those challenges. Despite all the difficulties, interstellar travel *is* possible.

A potentially pivotal event took place in October 2011: the 100 Year Starship Symposium. Its goal was first to set a bar high enough and hard enough to seriously challenge the next generations. Second, to inspire by showing the scope of the problems and possible solutions.

The Symposium felt much like a science fiction convention with solid content and a zest seldom seen. DARPA intends to spur research and select an organization that will sustain and develop interplanetary resources over the next century, culminating in an interstellar launch. David Neyland, the director of tactical technology at DARPA, spoke of *"creating a culture* centered on human expansion into the solar system, and onward to the stars." A science fictional staple, yes—so it needed SF writers, who appeared on several panels and gave papers.

Still, "Vision without execution is daydreaming," as Neyland dryly noted.

How High the Bar?

The distances are enormous. If you made a scale model of the distance between Earth and the moon inside a large sixty-foot-long meeting room, on that scale the nearest star (Alpha Centauri) would be as far away as the actual Moon.

The speed needed for interstellar flight is sobering. In thousands of years, humans have progressed from a stroll (4 km/hr) to an astronaut's return from the moon aboard the Apollo spacecraft (40,000 km/hr). That's a factor of 10,000. Reaching nearby stars in reasonable time, say decades, would require a velocity jump of another factor of 10,000—close to the speed of light. To do that in a century means increasing the velocity 100 times faster than we did over the last ten millennia.

By comparison, the fastest rocket we've ever sent into space would cover the distance to the Alpha Centauri system in seventy-four thousand years! As detailed in this book, lightweight sails driven by beams of microwaves or lasers might promise to cover this immense distance in less than a thousand years. For you and me, there isn't much difference between a thousand years and seventy-four thousand years. But in the lifetime of civilizations, the difference between these numbers is significant.

Any interstellar transport must require a steady buildup toward high velocity vehicles. Many expected to see this after the Apollo landings, but politics did not work that way.

Congress came to see NASA primarily as a jobs program, not an exploratory agency. Slowly, NASA complied with the post-1972 vision—safety-obsessed, with few big goals for manned flight beyond low Earth orbit. Very little science got done in the Station. NASA never did the experiments needed for a genuine interplanetary expedition—centrifugal gravity to avoid bodily harm, and a true closed biosphere. We're not doing that. The Station was not about living in space, but camping in space. This echoed the earlier Russian Mir station, where crews got a weekly vodka, cognac and cigarette ration to pass the time.

"We had the Shuttle to reach the Station, and the Station to give the Shuttle a destination," an old NASA hand remarked: "A school bus route writ large."

Starships in Science Fiction

The effect on science fiction (SF) writers mirrors this decline. Even before the first moon landing, English SF writer J.G. Ballard wrote nostalgic stories regarding the space program as a glorious folly of the 1960s. Barry Malzberg's 1972 "Beyond Apollo" depicted an astronaut driven insane by the experience of an expedition to Venus. American writers' optimism faded much more slowly. The 1970s saw a lot of space fiction, especially about living in habitats there.

Many of the best writers, such as Arthur Clarke, Isaac Asimov ("Foundation") and especially Robert A. Heinlein (*Stranger in a Strange Land, The Moon is a Harsh Mistress*) set epics in futures that assumed ready access to space. Over time, such settings became less common. The "conceptual blowout" and frontier imagery of interplanetary exploration is now more often tied specifically to interstellar travel, as in Poul Anderson's *Tau Zero* (1970), Vonda McIntyre's *Superluminal* (1984), and Allen Steele's popular *Coyote* series (2002).

Today, writers like Geoffrey Landis and Stephen Baxter both believe that pop SF such as *Star Trek, Star Wars,* and the like has distorted the difficulties of space. "Exploring space is so very easy," Landis says. "You just jump in your ship and go, and nobody every questions why or asks 'Who's funding this?' or even 'What's the energy source here?' In the real world, there's little margin for error. That guy who says 'It's a crazy idea, but it just might work!'—well, in the real world, ninety-nine times out of a hundred, it doesn't work."

Now, science fiction is becoming more firmly economically based, because the history of post-1972 NASA has sobered the writers. The writers in this book do know the realities, and most have PhDs. They see that in today's frugal climate, space is going commercial. A nuclear thermal rocket used only in space (a technology already largely developed), and resupplied with fuel by the launchers we have now, could open whole new industries. Some envision profitable businesses that could build a genuine interplanetary future repowering of satellites in geosynchronous orbits, big spinning wheel hotels, asteroid mining, and more. Getting people into orbit may well best be done not with risky rockets, but with two stages—an airplane carrying a rocket plane, so takeoff is not on a Roman candle but on a runway.

Despite the challenge, many now think our economic problems can be dealt with by using the resources of the solar system. Such a development path could also open up means to begin exploring other stars, all within a century. This book looks at how this can be done.

There is a deeper reason why thinking of starships now opens fresh horizons. By working through the technical steps for propulsion, habitat, motivations, and more, we gain some perspective on how interstellar flight looks to alien minds that face the same physical problems. We live in a sparse outer part of our galaxy, and our star is on average about a billion years younger than similar stars elsewhere. In the vast hub of stars farther toward the galactic center, distances between stars are shorter, and so travel is on average easier. Worlds there have had more time to develop space travel. Alien minds might have mastered the huge problems of such voyages long ago. This brings to mind an old question, due to Enrico Fermi: *Where are they?*

Clearly any society, human or alien, that sends starships into the deep dark must be both very rich and very long lived. Our human experience shows that extravagances seldom exceed 1 percent of a society's annual spending; the Apollo moon program grazed this limit. Interstellar ships will need vast energies and capital. Similarly, few human enterprises exceed a human life span to pay off.

As we have prospered, so have we lived longer. Our average life span doubled over the last two centuries. As Robert Zubrin argues in his piece here, economic growth can be rapid and better than linear in time, as it has now for several centuries. This has enormous cultural impact; fields such as archeology and environmental concern require a long enough perspective in time to exist at all.

So there is some hope that our future societies will see people who live well beyond 100 years and take the long view in their investments. Their societies would, too. Then they could see the point in such a vast dream as flight to the stars.

Suppose we invert this argument. Fermi's famous question—*Where are they?*—may imply that aliens do not live long and prosper, to quote *Star Trek*. If so, it may be up to us to do so, perhaps uniquely in our galaxy.

This book may help thinking about the Fermi Paradox. More, it will place our era in a larger perspective, as a galactic presence on the threshold of greatness.

How to Get There

The thrust generated by rocket-powered spacecraft is limited by the speed at which gases leave the exhaust nozzle. Rockets only go about twice the exhaust speed—far too slow for interstellar travel. Future starships will need new and powerful forms of propulsion because they'll be massive and need every possible efficiency.

Thrust can only be applied so long as there's fuel onboard, and rocket speed depends exponentially on the mass of fuel. Going three times the exhaust speed would require fuel twenty times the mass of the rest of the rocket (the "dry weight"). Chemical combustion of hydrogen and oxygen is simply too slow.

Nuclear thermal rockets—in which a fluid flows through a fission reaction core—would open up the solar system to large, fast spacecraft. So start with nuclear thermal rockets. The fuel—probably hydrogen because it's lightweight—passes through a fission reactor core, gets very hot, and exits through an exhaust nozzle. That's still not hot enough for interstellar flight, but can open the solar system to large spacecraft, harvesting its resources. Much work was done in the Cold War on nuclear thermal rockets. There are now new efforts in the United States, the best hope of getting high thrust at high speed soon.

To make interstellar missions possible requires using the nuclear reaction that drives the sun, fusion rockets. Eventually we can try for matter-antimatter rockets. Fusion rockets are the mainline of starship concept studies. They will be complex and large. In 1978, the British Interplanetary Society produced the most detailed starship concept, called Daedalus, using inertial fusion, then new but still undemonstrated today, to drive the rocket. Lasers crush a pellet of fusion fuel—isotopes of hydrogen—from all sides, compressing the material into a tiny volume. Pressure rises enough for the hydrogen nuclei to fuse, releasing energy. The hot mass thrusts out of the nozzle. Lasers fire rapidly, giving pulsed thrust. Daedalus was complex, as big as an aircraft carrier; it would dwarf the Saturn 5, our rocket to the moon.

That might work for nearby stars, but would still take centuries to get there. The group *Project Icarus* is revisiting this idea to see what a twenty-first century concept would be like. But fusion rockets must follow from achieving fusion here on Earth. Several big facilities are underway, but success has been elusive. So far, fusion hasn't been conquered; fusion has conquered us for sixty years.

A very different, genuinely twenty-first-century type of spacecraft to offer the promise of interstellar flight is the beamed energy sail, or *sailship*. It uses the ability of electromagnetic waves to move power through space and produce force at great distances.

The source of the beam (beam source plus antenna, the "beamer") projects powerful laser, millimeter wave, or microwave beam onto a large sail, a light screen. This reflects the beam, picking up momentum and moving away. The expensive part of this utility is the beamer. But the great feature of beaming energy is that the beamer, with all its mass and complexity, gets left behind. The far simpler, ultralight sail with its payload gets driven far away. The beamer can be used for many sailship missions. Like the nineteenth-century railroads, once the track is laid, the train itself is a much smaller added expense.

We've already demonstrated the physics of sailships. The unresolved question is engineering of large beamers and sails. Economic studies show them to be inefficient and expensive. Research underway will determine if the time for beam-driven sails may be here. James Benford describes this in his piece.

Science fictions such as *Star Trek* and *Star Wars* have routinely used "warp drives" and "wormholes" to get around the galaxy quickly. They're popular in science fiction because they enable the plot. But that means faster-than-light travel, doesn't it? But General Relativity says that's maybe possible, if there is such a thing as negative mass—matter of opposite sign to that of normal matter, for which there is no evidence. But quantum mechanics, the realm of the very small, does allow it. Can this be used for star travel? The physics jury is still out.

To do so means a warping of space-time by energy so dense it's far more than the energy density of any star. Black holes have high energy densities, but you can't go FTL by passing near a black hole. One idea is a distorted space-time, called a "warp bubble," is just large enough to hold a ship. Estimates for the energy to make a warp bubble used to be the equivalent of the mass of a galaxy. By making the walls of the bubble the right thickness, the energy has fallen greatly, but is still the equivalent of the mass of Jupiter. Note that the biggest H-bomb actually converted only a few kg of matter to energy. Warping of space-time on the scale needed for wormholes and warp bubbles are beyond mere twenty-first century science. For more, see John Cramer's piece in this book.

Habitats

For crewed missions, protection from the hazards of space, or even to live well, is challenging. To provide artificial gravity by centrifugal force, the craft must spin. For the crew, the essentials of air, food, and water drive the volume of space needed, meaning that private space will be compact, so public spaces will be emphasized.

What's needed is a closed recycled ecosystem, which is far from risk free and is far beyond our present capability. After survival challenges, there are the psychological and social issues—confinement, limited companionship, social dynamics. The principles of dealing with prolonged isolation and confinement were to stay busy, keep a steady day-night schedule, and watch your health. And in a crewed starship, there are going to be both sexes, especially if they are going to colonize, which brings interesting issues. Slowing down or ceasing the metabolism at low temperatures with reanimation at the end of the voyage would greatly lessen the need for supplies and eliminate the long-term psychological and social issues. Bears do it naturally, but we don't know how. But some progress is being made in cryogenics, the preservation of the dead, trying for survival to a time they might be revived. We know far too little now of this shortcut to either embrace or dismiss the prospect of suspended animation. These issues are dealt with by Richard Lovett, in *Living Large*.

Perspectives

Any dreams of interstellar flight must start somewhere. The next century could see an enormous expansion of the human prospect by opening up the resources of our solar system. Doing that demands an infrastructure meeting many requirements—high ship velocities, radiation shielding, robust crew and passenger health systems, training for long duration missions, robotics, and, certainly, long-range financial investment.

While doing all this, and so writing a fresh chapter in the human story, we can keep in the back of our minds the constant presence of the glimmering stars, beckoning above.

THE BIG
PERSPECTIVE

Here we assemble several of Stephen Hawking's many remarks on the long-range prospects of humanity. His hallmark in all such comments has been caution. He feels radiating powerful signals to announce our presence is an unnecessary risk, in case aliens might wish us ill. Risk aversion also features in his thoughts on subjects he has thought on a great deal, since his teenage years, when he was a prodigious science fiction fan.

Our Only Chance
Stephen Hawking

We are entering an increasingly dangerous period of our history. There have been a number of times in the past when survival has been a question of touch and go, like the Cuban Missile Crisis of 1963, and the frequency of such occasions is likely to increase in the future. We shall need great care and judgment to negotiate them all successfully.

Our population and our use of the finite resources of planet Earth are growing exponentially, along with our technical ability to change the environment for good or ill. But our genetic code still carries the selfish and aggressive instincts that were of survival advantage in the past. It will be difficult enough to avoid disaster in the next hundred years, let alone the next thousand or million.

Our only chance of long-term survival is not to remain lurking on planet Earth, but to spread out into space. This is why I favor "personed" space flight and encourage further study into how to make space colonization possible.

But I'm an optimist. If we can avoid disaster for the next two centuries, our species should be safe, as we spread into space. Once we establish independent colonies, our entire future should be safe.

In his 1960 paper "Search for Artificial Stellar Sources of Infra-Red Radiation," published in the journal Science, *Freeman Dyson famously argued that the long-term evolution of technological alien societies might lead to capturing the bulk of all their star's emissions, forming what came to be called by others Dyson Spheres. Dyson once said, "Science is my territory, but science fiction is the landscape of my dreams." Though he has never written science fiction, his scientific imagination has inspired a great deal of it. Here he looks again at the very long term, but for life far from stars. Still, his focus is on their energy needs.*

Noah's Ark Eggs and Viviparous Plants
Freeman Dyson

Science-fiction stories about starships usually depict the universe as a collection of stars and planetary systems separated by vast stretches of empty space. The space between stars is imagined to be filled with dilute interstellar gas and nothing else. The real universe is much more interesting. The real universe contains a multitude of objects of various sizes, giving interstellar travelers places to stop and visit friends and collect fresh supplies between the stars. We know almost nothing about these objects except for the fact that they exist. We know that in the space around our own planetary system there are two populations of comets, known as the Kuiper Belt and the Oort Cloud.

The Kuiper Belt is the source of short-period comets, and the Oort Cloud is the source of long-period comets. We know that they exist because the comets which we see coming close to the sun are visibly disintegrating and cannot survive for a long time. The tail which makes a comet beautiful is proof of its mortality. Meteor showers are the debris marking the graves of dying comets. To keep new comets appearing at the observed rate, the source populations must be large, of the order of billions of comets of each kind. A few of the biggest and closest objects in the Kuiper Belt population can be directly observed, orbiting the sun with orbits concentrated around the plane of the planets. The brightest and most famous of these objects is Pluto. The Oort Cloud is invisible from the Earth. It is a spherical population of objects at much greater distances from the sun, loosely attached to the sun by weak gravitational forces.

There is no reason to believe that the space between the Oort Cloud and the nearest stars is empty. We know that a large fraction of all stars are born with planetary systems. It is also likely that large numbers of planets are born unattached to stars. Furthermore, we know that the normal processes of formation and evolution of planetary systems result in ejection of planets and comets from the systems. As a result of these processes, the universe probably contains more unattached planets than stars, and billions of times more unattached comets. The space between our solar system and the nearest stars is probably infested with unattached planets and far more numerous unattached comets. In addition, there may be other objects of intermediate kinds which we have not yet observed, from snowballs to black-dwarf stars. It is conceivable that some of the intermediate objects might be alive, a population of mythological monsters making their home in space.

The existence of abundant way-stations between the stars is likely to have a decisive influence on the development of starships. We shall not jump in one huge step from planetary to interstellar voyages. We shall be exploring one group of objects after another, first the Kuiper Belt, then the Oort Cloud, then a string of farther-out oases in the desert of space, before we finally come to Proxima Centauri. In the history of mankind on this planet, there were two very different kinds of explorers who learned to navigate the oceans. There were the European navigators who sailed from fixed bases in Europe to destinations in America and Asia and came home to Europe with loot from their trade and conquest. Columbus was typical of these explorers, making three voyages back and forth across the Atlantic. But before the Europeans, there were Polynesian navigators who built canoes to sail long distances on the Pacific and populated the Pacific islands from Asia to Hawaii and New Zealand. The Polynesians did not have home bases in Asia and America, and they were not interested in sailing all the way across the ocean. They made their voyages from island to island, stopping to make a new home when they found a new island suitable for raising their crops and pigs and children.

The Polynesians were navigating the Pacific for a thousand years before the Europeans crossed the Atlantic. Island-hopping came first, intercontinental voyages later. It is likely that the future of our traveling beyond the solar system will follow the same pattern. The evolution of starships, like the evolution of Polynesian canoes and European galleons, will proceed by a process of trial and error. Unattached comets and planets will be like the islands in the Pacific Ocean. We will begin like the Polynesian navigators, modestly. Developing starships one step at a time, we can learn by trial and error how to do the job right. Perhaps, after a thousand years, we will be ready to build grand superhighways conveying traffic along non-stop routes from star to star.

Two things are needed to make starships fly, a place to go and a way to get there. The first problem is mainly a problem of biology, the second a problem of engineering. Let us look at biology first. To have a place to go, we must learn how to grow complete ecosystems at remote places in the universe. It is not enough to have hotels for humans. We must establish permanent ecological communities including microbes and plants and animals, all adapted to survive in the local environment. The populations of the various species must be balanced so as to take care of each others' needs as well as ours. Permanent human settlement away from Earth only makes sense if it is part of a bigger enterprise, the permanent expansion of life as a whole. The best way to build human habitats is to prepare the ground by building robust local ecologies. After life has established itself with grass and trees, herbivores and carnivores, bacteria and viruses, humans can arrive and build homes in a friendly environment. There is no future for humans tramping around in clumsy spacesuits on lifeless landscapes of dust and ice.

The recent revolution in molecular biology has given us new tools for seeding the universe with life. We have learned to read and write the language of the genome, to sequence the DNA that tells a microscopic egg how to grow into a chicken or a human, to synthesize the DNA that tells a bacterium how to stay alive. We have sequenced the genomes of several thousand species. The speed of sequencing and of synthesis of genomes is increasing rapidly, and the costs are decreasing equally rapidly. If the increase in speed and the decrease in costs continue, it will take only about twenty years for us to sequence genomes of all the species that exist on our planet. The genetic information describing the entire biosphere of the planet will be available for our use. The total quantity of this information is remarkably small. Measured in the units that are customary in computer engineering, the information content of the biosphere genome amounts to about one petabyte, or ten to the power of sixteen bits. This is a far smaller amount of information than the data-bases used by enterprises such as Google. The biosphere genome could be embodied in about a microgram of DNA, or in a small room full of computer memory disks.

Looking ahead fifty or a hundred years, we shall be learning how to use genetic information creatively. We shall then be in a position to design biosphere populations adapted to survive and prosper in various environments on various planets, satellites, asteroids, and comets. For each location we could design a biosphere genome, and for each biosphere genome we could design an egg out of which an entire biosphere could grow. The egg might weigh a few kilograms and look from the outside like an ostrich egg. It would be a miniature Noah's ark, containing thousands or millions of microscopic eggs programmed to grow into the various species of a biosphere. It would also contain nutrients and life-support to enable the growth of the biosphere to get started. The first species to emerge from a Noah's ark egg would be warm-blooded plants designed to collect energy from sunlight and keep themselves warm in a cold environment. Warm-blooded plants would then provide warmth and shelter for other creatures to enjoy. In this way, life could be seeded in great abundance and variety in all kinds of places, traveling on small spacecraft carrying payloads of a few kilograms. Since life is inherently an unpredictable phenomenon, many of the biospheres would fail and die. Those that survived would evolve in unpredictable ways. Their evolution would continue forever, with or without human intervention. We would be the midwives, bringing life to birth all over the universe, as far as our Noah's ark eggs could travel.

The second problem, the problem of engineering, is to build machines that can take us from here to there. To have space travel over long distances at reasonable prices, we must build a public highway system so that the costs of the initial investment can be shared by a multitude of users. A public highway system in space will require terminals using sunlight or starlight to generate high-energy beams along which spacecraft can fly. The beams may be laser-beams or microwave-beams or pellet streams. The massive energy-generating machinery at the terminals remains fixed. The spacecraft are small and light, and pick up energy from the beams as they fly along. Unlike chemical or nuclear rockets, they do not carry their own fuel. For the system to operate efficiently, the volume of traffic must be big enough to use up the energy of the beams. Spacecraft must be flying along the beams almost all the time. As with all public highway systems, the system can only grow as fast as the volume of traffic. The cost of travel will be high at the beginning and will become low when every terminal is crowded with passengers waiting for a launch.

In every public transport system, things work better if we build separate vehicles for passengers and freight. On the roads, cars for passengers and trucks for freight. On the railroads, fast short trains for passengers and slow long trains for freight. The Space Shuttle was a system designed to put passengers and freight on the same vehicle, and that was one of the reasons why it failed. It was supposed to be cheap and safe and reliable, with frequent flights and a high volume of traffic, and it turned out to be expensive and unsafe and unreliable. The public highways of the future will be like roads and railroads and not like the Shuttle. But the relation between passengers and freight in the future will be the opposite of what it was in the past. In the past, humans were small and light, freight was big and heavy. Cars were small and agile, trucks were big and clumsy. In space today, this relation between human passengers and freight is already inverted. Because of the miniaturization of instruments and communication systems, unmanned spacecraft have become smaller and lighter than manned spacecraft. Payloads of unmanned missions have remained roughly constant while their performance and capability have improved by leaps and bounds. Payloads of manned missions have remained larger while politicians fail to decide what they are supposed to do.

In the future, when missions go beyond the solar system, the difference between passengers and freight will become greater. Freight will no longer be bulk materials such as fuel and water. Freight will be information, embodied in ultralight computer memory or in DNA. Freight will be several orders of magnitude lighter than human passengers. Payloads of unmanned missions may be measured in grams, while payloads of manned missions will always be measured in tons. As a result, the public highway system will consist of two parts, a heavy-duty system transporting human passengers between a small number of metropolitan human habitats, and a light-freight system transporting packages of information along a wider network of routes to more distant destinations. A typical light-freight mission might be like the Starwisp proposed by Bob Forward. The Starwisp is an ultralight sail made of fine wire mesh, driven through space by a high-power beam of microwaves. The wire mesh is not only the vehicle but also the payload, carrying sensors to explore the environment and transmitters to send information collected by the sensors to humans far away. Starwisp could also be a vehicle for carrying Noah's ark eggs to bring life to remote places. It is likely that the travel-times of voyages will become longer than a human life-time. After life has spread that far, it will no longer make sense for humans to travel with it. Instead of imprisoning human travelers for a lifetime in a spacecraft, it would make more sense to load the spacecraft with a few human eggs, which could grow into humans at the destination. In the end, we would populate the galaxy by broadcasting the information required for growing humans, rather than by carrying deep-frozen human bodies for thousands of years.

When we are thinking about the spread of life into the universe, the most important fact to remember is that almost all the real estate in the universe is on small objects. Real estate means surface area. The universe contains objects of all sizes. Most of the mass and volume belong to big objects such as stars and planets. Most of the area belongs to small objects such as asteroids and comets. Most of the life will have to find its home on small objects. The majority of small objects have three qualities that make them unfriendly to life. They are far from the sun or other stars, they have no atmosphere, and they are cold. In spite of those disadvantages, they can be seeded with life. They can support biospheres as diverse and as beautiful as ours.

The key technology for bringing life to small cold objects in space is the cultivation of warm-blooded plants. Warm-blooded plants are more essential to the ecology of cold places than warm-blooded animals are to the ecology of our warm planet. Life on Earth might have evolved happily without birds and mammals, but life in a cold place could never get started without warm-blooded plants. Two external structures make warm-blooded plants possible, a greenhouse and a mirror. The greenhouse is an insulating shell protecting the warm interior from the cold outside, with a semitransparent window allowing sunlight or starlight to come in but preventing heat radiation from going out. The mirror is an optical reflector or system of reflectors in the cold region outside the greenhouse, concentrating sunlight or starlight from a wide area onto the window. Inside the greenhouse are the normal structures of a terrestrial plant, leaves using the energy of incoming light for photosynthesis, and roots reaching down into the icy ground to find nutrient minerals. Since there is no atmosphere to supply the plant with carbon dioxide, the roots must find mineral sources of carbon and oxygen to stay alive. We see in the light emitted from comets, as they come close to the sun, that these icy objects contain plenty of carbon and oxygen as well as nitrogen and other elements essential to life.

The embryonic warm-blooded plant must grow the greenhouse and the mirror around itself while still protected within the greenhouse of its parent. The seeds must develop into viable plants before they are dispersed into the cold environment. These plants must be viviparous as well as warm-blooded. It seems to be only an accident of evolution on our own planet that animals learned to be viviparous and warm-blooded while plants did not.

The optical concentration that the mirror must provide will depend on the distance of the plant from the sun or star providing the energy. Roughly speaking, the optical concentration must increase with the square of the distance from the source. For example, if the plant is on the surface of Enceladus, a satellite of Saturn at ten times Earth's distance from the sun, the intensity of sunlight is one hundredth of the intensity on Earth, and the optical concentration must be by a factor of a hundred. If the plant is in the Kuiper Belt at a hundred times Earth's distance, sunlight is reduced by a factor of ten thousand and the mirror must concentrate by a factor of ten thousand. Existing biological structures can do much better than that. The human eye is not an extreme example of optical precision, but it can concentrate incoming light onto a spot on the retina by a factor larger than a million. That is why staring at the sun is bad for the health of the eye. A mirror as precise as a human eye would be good enough to keep a plant warm at a distance ten times farther from the sun than the Kuiper Belt. Eagles and hawks have better eyes than we do, and a simple amateur telescope costing less than a hundred dollars is better still. There is no law of physics that would prevent a warm-blooded plant from growing a mirror to concentrate enough starlight to survive anywhere in our galaxy. The main difficulty in achieving a high concentration of starlight is that the mirror must track the source accurately as the object carrying the plant rotates. The plant must be like a sunflower, tracking the sun as it moves across the sky. If high accuracy is needed, the plant must grow an eye to see where it is pointing.

These speculations about viviparous plants and Noah's ark eggs and life spreading through the galaxy are my personal fantasies. They are only one possible way for the future to go. The real future is unpredictable. It will be rich in surprises that we have not imagined. All that we can say with some confidence is that biotechnology will dominate the future. The awesome power of nature, to evolve unlimited diversity of ways of living, will be in our hands. It is for us to choose how to use this power, for good or for evil.

As the most world's most prominent astronomer, Martin Rees is singularly well qualified to outline the implications of a simple observation—that on the scale of our galaxy, our own species' evolution is just beginning.

To the Ends of the Universe
Martin Rees

Astronomers like myself are professionally engaged in thinking about huge expanses of space and time. But this doesn't make us serene and relaxed about the future. Most of us worry as much as anyone about what happens next year, next week, or tomorrow. Nonetheless, our subject does offer a special perspective. We view our home planet in a cosmic context. We wonder whether there's life elsewhere in the cosmos. But, more significantly, we're mindful of the immense future that lies ahead.

The stupendous time spans of the evolutionary past are now part of common culture (but maybe not in the Bible Belt, nor in parts of the Islamic world). We're at ease with the idea that our present biosphere is the outcome of four billion years of Darwinian evolution. But the even longer time-horizons that stretch ahead—though familiar to every astronomer—haven't permeated our culture to the same extent. Our sun is less than halfway through its life. It formed 4.5 billion years ago, but it's got 6 billion more before the fuel runs out. It will then flare up, engulfing the inner planets and vaporising any life that might then remain on Earth. But even after the sun's demise, the expanding universe will continue—perhaps forever—destined to become ever colder, ever emptier. To quote Woody Allen, "eternity is very long, especially towards the end." That, at least, is the best long-range forecast that cosmologists can offer.

Any creatures witnessing the sun's demise 6 billion years hence won't be human—they'll be as different from us as we are from a bug. Posthuman evolution—here on Earth and far beyond—could be as prolonged as the Darwinian evolution that's led to us, and even more wonderful. Indeed this conclusion is strengthened when we realise that future evolution will proceed not on the million-year timescale characteristic of Darwinian selection, but at the much accelerated rate allowed by genetic modification and the advance of machine intelligence (and forced by the drastic environmental pressures that would confront any humans who were to construct habitats beyond Earth). Natural selection may have slowed: its rigours are tempered in civilised countries. But it will be replaced by "directed" evolution. Already, performance-enhancing drugs, genetic modification, and cyborg technology are changing human nature, and these are just precursors of more drastic changes.

Darwin himself realised that "No living species will preserve its unaltered likeness into a distant futurity." We now know that "futurity" extends much farther, and alterations can occur far faster than Darwin envisioned. And we know that the cosmos, through which life could spread, offers a far more extensive and varied habitat than he ever imagined. So humans are surely not the terminal branch of an evolutionary tree, but a species that emerged early in the overall roll-call of species, with special promise for diverse evolution—and perhaps of cosmic significance for jump-starting the transition to silicon-based (and potentially immortal) entities that can more readily transcend human limitations.

We humans are entitled to feel uniquely significant, as the first known species with the power to mold its own future. And we live at a crucial time. Our Earth has existed for 45 million centuries, and still more lie ahead. But this century may be a defining moment. It's the first in our planet's history where one species (ours) has Earth's future in its hands, and could jeopardise the immense potential stretching for billions of years. And, more central to the theme of this book, it's the century when the spread of life beyond Earth could begin.

A famous picture in the English edition of Newton's *Principia* shows cannon balls being fired from the top of a mountain. If they go fast enough, their trajectory curves downward no more steeply than Earth curved away underneath it—they go into orbit. This picture is still the neatest way to explain orbital flight. Newton calculated that, for a cannon-ball to achieve an orbital trajectory, its speed must be 18,000 miles/hour—far beyond what was then achievable.

Indeed, this speed wasn't achieved until 1957, when the Soviet Sputnik was launched. Four years later Yuri Gagarin was the first human to go into orbit. Eight years after that, and only sixty-six years after the Wright Brothers' first flight, Neil Armstrong made his "one small step." The Apollo programme was a heroic episode. And it was a long time ago—ancient history to today's young people. Those in England know that the Americans landed on the moon, just as they know the Egyptians built pyramids—but both enterprises seem driven by equally arcane goals.

Since 1972, humans have done no more than circle Earth in low orbit—more recently, in the International Space Station. This has proved neither very useful nor very inspiring. But space technology has burgeoned—for communication, environmental monitoring, satnav, and so forth. We depend on it every day. And unmanned probes to other planets have beamed back pictures of varied and distinctive worlds.

But it has been pictures of Earth itself, showing how its delicate biosphere contrasts with the sterile moonscape where the astronauts left their footprint, that have become iconic, especially for environmentalists. We've had these images for forty-five years. But suppose some aliens had been viewing such an image for our planet's entire history, what would they have seen? Over nearly all that immense time, 4.5 billion years, Earth's appearance would have altered very gradually. The continents drifted; the ice cover waxed and waned; successive species emerged, evolved and became extinct.

But in just a tiny sliver of Earth's history—the last one millionth part, a few thousand years—the patterns of vegetation altered much faster than before. This signaled the start of agriculture. The pace of change accelerated as human populations rose. Humanity's "footprint" got larger because our species became more demanding of resources—and also because of population growth.

Within fifty years—little more than one-hundredth of a millionth of Earth's age—the carbon dioxide in the atmosphere began to rise anomalously fast. And something else unprecedented happened: Rockets launched from the planet's surface escaped the biosphere completely. Some were propelled into orbits around Earth; some journeyed to the moon and planets.

If they understood astrophysics, the aliens could confidently predict that the biosphere would face doom in a few billion years when the sun flares up and dies. But could they have predicted this sudden "fever" halfway through Earth's life—these human-induced alterations occupying, overall, less than a millionth of Earth's elapsed lifetime and seemingly occurring with runaway speed?

If they continued to keep watch, what might they witness in the next hundred years? Will the spasm be followed by silence? Will the planet make a transition to sustainability? And, most important of all for the log-term future, will an armada of rockets leaving Earth have led to new communities elsewhere—on Mars and its moons, on asteroids, or freely floating in space?

The Next Century

Scientific forecasters have a dismal record. One of my predecessors as Astronomer Royal said, as late as the 1950s, that space travel was "utter bilge." Few in the mid-twentieth century envisaged the transformative impact of the silicon chip or the double helix. The iPhone would have seemed magical even twenty years ago. So, looking even a century ahead we must keep our minds open, or at least ajar, to what may now seem science fiction. Indeed some proponents of the "singularity"—the takeover of humanity by intelligent machines—claim this transition could happen within fifty years.

Had the momentum of the 1960s been maintained over the next 40 years, there would be footprints on Mars by now. But after Apollo the political impetus for manned spaceflight was lost. This was one of many instances of the widening gap between what could be achieved technologically, and what is actually done. As with many technical forecasts, we can be more confident of what could happen than of how soon it will happen. Development of supersonic airliners, for instance, has languished (Concorde having gone the way of the dinosaurs); in contrast, the sophistication and worldwide penetration of Internet and smartphones advanced much faster than most forecasters predicted.

I'd venture a confident forecast that during this century, the entire solar system   planets, moons, and asteroids—will be explored and mapped by flotillas of tiny robotic craft. The next step would be space mining and fabrication. (And fabrication in space will be a better use of materials mined from asteroids than bringing them back to Earth). The Hubble Telescope's successors, with huge gossamer-thin mirrors assembled under zero gravity, will further expand our vision of stars, galaxies, and the wider cosmos.

But what role will humans play? There's no denying that NASA's "Curiosity," now trundling across Martian craters, may miss startling discoveries that no human geologist could overlook. But robotic techniques are advancing fast, allowing ever more sophisticated unmanned probes; whereas the cost gap between manned and unmanned missions remains huge. The practical case for manned spaceflight gets ever-weaker with each advance in robots and miniaturisation—indeed as a scientist or practical man, I see little purpose in sending people into space at all. But as a human being, I'm an enthusiast for manned missions. I hope some people now living will walk on Mars—as an adventure, and as a step toward the stars. They may be Chinese: China has the resources, the dirigiste government, and maybe the willingness to undertake an Apollo-style programme. And China would need to aim at Mars, not just at the moon, if it wanted to assert its superpower status by a "space spectacular": a re-run of what the US achieved fifty years earlier would not proclaim parity.

NASA's manned programme, ever since Apollo, has been impeded by public and political pressure into being too risk-averse. The Space Shuttle failed twice in 135 launches. Astronauts or test pilots would willingly accept this risk level, but the Shuttle had, unwisely, been promoted as a safe vehicle for civilians. So each failure caused a national trauma and was followed by a hiatus while costly efforts were made (with very limited effect) to reduce the risk still further.

Unless motivated by pure prestige and bankrolled by superpowers, manned missions beyond the moon will need perforce to be cut-price ventures, accepting high risks—perhaps even "one-way tickets." These missions will be privately funded; no Western governmental agency would expose civilians to such hazards. There would, despite the risks, be many volunteers—driven by the same motives as early explorers, mountaineers, and the like. Private companies already offer orbital flights. Maybe within a decade adventurers will be able to sign up for a week-long trip round the far side of the moon—voyaging farther from Earth than anyone has been before (but avoiding the greater challenge of a Moon landing and blast-off). And by mid-century the most intrepid (and wealthy) will be going farther.

(The phrase "space tourism" should however be avoided. It lulls people into believing that such ventures are routine and low-risk. And if that's the perception, the inevitable accidents will be as traumatic as those of the Space Shuttle were. Instead, these cut-price ventures must be "sold" as dangerous sports, or intrepid exploration.)

But don't ever expect mass emigration. Nowhere in our solar system offers an environment even as clement as the Antarctic or the top of Everest. Space doesn't offer an escape from Earth's problems. Nonetheless, a century or two from now, there may be small groups of pioneers living independent from Earth—on Mars or on asteroids. Whatever ethical constraints we impose here on the ground, we should surely wish these adventurers good luck in genetically modifying their progeny to adapt to alien environments. This might be the first step towards divergence into a new species: the beginning of the post-human era. And machines of human intelligence could spread still farther. Whether the long-range future lies with organic post-humans or with intelligent machines is a matter for debate. Either way, dramatic cultural and technological evolution will continue not only here on Earth but far beyond.

The most crucial impediment to space flight, even in Earth's orbit and still more for those venturing farther, stems from the intrinsic inefficiency of chemical fuel, and the consequent requirement to carry a weight of fuel far exceeding that of the payload. (It's interesting to note, incidentally that this is a generic constraint, based on fundamental chemistry, on any organic intelligence that had evolved on another planet. If a planet's gravity is strong enough to retain an atmosphere, at a temperature where water doesn't freeze and metabolic reactions aren't to slow, the energy required to lift a molecule from it will require more than one molecule of chemical fuel.)

Launchers will get cheaper when they can be designed to be more fully reusable. It will then be feasible to assemble, in orbit, even larger artifacts than the International Space Station. But so long as we depend on chemical fuels, interplanetary travel will remain a challenge. Nuclear power could be transformative. By allowing much higher in-course speeds, it would drastically cut the transit times to Mars or the asteroids (reducing not only astronauts' boredom, but their exposure to damaging radiation). And it could transform manned spaceflight from high-precision to an almost unskilled operation. Driving a car would be a difficult enterprise if, as at present for space voyages, one had to program the entire journey beforehand, with minimal opportunities for steering on the way. If there were an abundance of fuel for mid-course corrections (and to brake and accelerate at will), then interplanetary navigation would be a doddle—indeed simpler than driving a car or ship, in that the destination is always in clear view.

But even with nuclear fuel, the transit time to nearby stars exceeds a human lifetime. Interstellar travel is therefore, in my view, an enterprise for post-humans, evolved from our species not via natural selection but by design. They could be silicon-based. Or they could be organic creatures who had won the battle with death, or perfected the techniques of hibernation or suspended animation. Even those of us who don't buy the idea of a singularity by mid-century would expect sustained, if not enhanced, rate of innovation in biotech, nanotech, and in information science. I think there will be entities with superhuman intellect within a few centuries. The first voyagers to the stars will not be human, and maybe not even organic. They will be creatures whose lifecycle is matched to the voyage the eons involved in traversing the galaxy are not daunting to immortal beings.

3000 and Beyond

By the end of the third millennium, travel to other stars could be technically feasible. Before setting out from Earth the voyagers would know what to expect at journey's end. Most importantly of all, they will know whether their destination is lifeless or inhabited, and robotic probes will have sought out already-existing biospheres, or planets that could be terraformed to render them habitable.

It could happen, but would there be sufficient motive? Would even the most intrepid leave the solar system? We can't predict what inscrutable goals might drive post-humans. But the motive would surely be weaker if it turned out that biospheres were rare. The European explorers in earlier centuries who ventured across the Pacific were going into the unknown to a far greater extent than any future explorers would be (and facing more terrifying dangers)—there were no precursor expeditions to make maps, as there surely would be for space ventures. Future spacefarers would always be able to communicate with Earth (albeit with a time lag). If precursor probes have revealed that there are indeed wonders to explore, there will be a compelling motive—just as Captain Cook was motivated by the biodiversity and beauty of the Pacific islands. But if there is nothing but sterility out there, the motive will be simply expansionist—in resources and energy—and that might be better left to robotic fabricators.

How bright are the prospects that there is life out there already? There may be simple organisms on Mars, or remnants of creatures that lived early in the planet's history; and there could be life, too, in the ice-covered oceans of Jupiter's moons Europa and Ganymede. But few would bet on it; and certainly nobody expects a complex biosphere in such locations. For that, we must look to the distant stars—far beyond the range of any probe we can now construct.

In the last twenty years (and especially in the last five) the night sky has become far more interesting, and far more enticing to explorers, than it was to our forbears. Astronomers have discovered that many stars—perhaps even most—*are* orbited by retinues of planets, just as the sun is. These planets are not detected directly. Instead, they reveal their presence by effects on their parent star that can be detected by precise measurements: small periodic motions in the star induced by an orbiting planet's gravity, and slight recurrent dimmings in a star's brightness when a planet transits in front of it, blocking out a small fraction of its light.

Data are accumulating fast, especially from NASA's Kepler spacecraft, which is monitoring the brightness of 150,000 stars with high enough precision to detect transits of planets no bigger than Earth. Some stars are known to be orbited by as many as seven planets, and it's already clear that planetary systems display a surprising variety—our own solar system may be far from typical. In some systems, planets as big as Jupiter are orbiting so close to their star that their "year" lasts only a few days. Some planets are on very eccentric orbits. One is orbiting a binary star which in turn orbits another binary star: it would have four "suns" in its sky. But there is special interest in possible "twins" of our Earth—planets the same size as ours, orbiting other sun-like stars, on orbits with temperatures such that water neither boils nor stays frozen. NASA's Kepler spacecraft has identified of hundreds of these.

But we'd really like to see these planets directly—not just their shadows. And that's hard. To realise just how hard, suppose an alien astronomer with a powerful telescope was viewing Earth from (say) thirty light-years away—the distance of a nearby star. Our planet would seem, in Carl Sagan's phrase, a "pale blue dot," very close to a star (our sun) that outshines it by many billions: a firefly next to a searchlight. But if the aliens could detect Earth at all, they could learn quite a bit about it. The shade of blue would be slightly different, depending on whether the Pacific ocean or the Eurasian land mass was facing them. They could infer the length of the "day," the seasons, whether there are oceans, the gross topography, and the climate. By analysing the faint light, they could infer that Earth had a biosphere.

Within twenty years, the unimaginatively named ELT ("Extremely Large Telescope") planned by European astronomers, with a mosaic mirror thirty-nine meters across, will be able to draw such inferences about planets the same size as our Earth, orbiting other sun-like stars. (And there are two somewhat smaller US telescopes in gestation too.)

But do we expect alien life on these extra-solar planets? We know too little about how life began on Earth to lay confident odds. What triggered the transition from complex molecules to entities that can metabolise and reproduce? It might have involved a fluke so rare that it happened only once in the entire galaxy—like shuffling a whole pack of cards into a perfect order. On the other hand, this crucial transition might have been almost inevitable given the "right" environment. We just don't know—nor do we know if the DNA/RNA chemistry of terrestrial life is the only possibility, or just one chemical basis among many options that could be realized elsewhere. Even if simple life is common, it is of course a separate question whether it's likely to evolve into anything we might recognize as intelligent or complex—whether Darwin's writ runs through the wider cosmos. Perhaps the cosmos teems with life; on the other hand, our Earth could be unique among the billions of planets that surely exist.

And it might be too anthropocentric to limit attention to Earthlike planets. Science-fiction writers have other ideas—balloon-like creatures floating in the dense atmospheres of Jupiter-like planets, swarms of intelligent insects, nano-scale robots, etc. Perhaps life can flourish even on a planet flung into the frozen darkness of interstellar space, whose main warmth comes from internal radioactivity (the process that heats Earth's core). There could be diffuse living structures, freely—floating in interstellar clouds; such entities would live (and, if intelligent, think) in slow motion, but nonetheless may come into their own in the long-range future.

No life will survive around on a planet whose central sun-like star became a giant and blew off its outer layers. Such considerations remind us of the transience of inhabited worlds (and life's imperative to escape their bonds eventually). We should also be mindful that seemingly artificial signals could come from super-intelligent (though not necessarily conscious) computers, created by a race of alien beings that had already died out. Maybe we will one day find ET. On the other hand, SETI searches may fail; Earth's intricate biosphere may be unique. But that would not render life a cosmic sideshow. Evolution is just beginning. Our solar system is barely middle aged and if humans avoid self-destruction, the post-human era beckons. Life from Earth could spread through the entire galaxy, evolving into a teeming complexity far beyond what we can even conceive. If so, our tiny planet—this pale blue dot floating in space—could be the most important place in the entire galaxy. The first interstellar voyagers from Earth would have a mission that would resonate through the entire galaxy and perhaps beyond.

"Fast-forward" To The End Of Time

In cosmological terms (or indeed in a Darwinian timeframe) a millennium is but an instant. So let us "fast forward" not for a few centuries, nor even for a few millennia, but for an "astronomical" timescale millions of times longer than that. The "ecology" of stellar births and deaths in our galaxy will proceed gradually more slowly, until jolted by the "environmental shock" of an impact with Andromeda, maybe four billion years hence. The debris of our galaxy, Andromeda, and their smaller companions within the local group will thereafter aggregate into one amorphous galaxy. If the cosmic acceleration continues, then, as Freeman Dyson and others have noted, the observable universe gets emptier and more lonely. Distant galaxies will not only move farther away, but recede faster and faster until they disappear—rather as objects falling onto a black hole encounter a horizon, beyond which they are lost from view and casual contact.

But the remnants of our Local Group could continue for far longer—time enough, perhaps for Kardashev Type III phenomenon to emerge as the culmination of the long-term trend for living systems to gain complexity and "negative entropy." All the atoms that were once in stars and gas could be transformed into structures as intricate as a living organism or a silicon chip but on a cosmic scale.

But even these speculations are in a sense conservative. I have assumed that the universe itself will expand, at a rate that no future entities have power to alter. And that everything is in principle understandable as a manifestation of the basic laws governing particles, space, and time that have been partly disclosed by twentieth-century science. Other chapters in this book envisage stellar-scale engineering to create black holes and wormholes—concepts far beyond any technological capability that we can envisage, but not in violation of these basic physical laws. But are there new "laws" awaiting discovery? And will the present "laws" be immutable, even to a Type III intelligence able to draw on galactic-scale resources?

We are well aware that our knowledge of space and time is incomplete. Einstein's relativity and the quantum principle are the two pillars of twentieth-century physics, but a theory that unifies them is unfinished business for twenty-first-century physicists. Current ideas suggest that there are mysteries even in what might seem the simplest entity of all—"mere" empty space. Space may have a rich structure, but on scales a trillion-trillion times smaller than an atom. According to string theory, each "point" in our ordinary space, if viewed with this magnification, would be revealed as a tightly wound origami in several extra dimensions. Such a theory will perhaps tell us why empty space can exert the "push" that causes the cosmic expansion to accelerate; and whether that "push" will indeed continue forever or could be reversed. It will also allow us to model the very beginning—an epoch where densities are so extreme that quantum fluctuations can shake the entire universe—and learn whether our big bang was the only one.

The same fundamental laws apply throughout the entire domain we can survey with our telescopes. Atoms in the most distant observable galaxies seem, from spectral evidence, identical to atoms studied in laboratories on Earth. But what we've traditionally called "the universe"—the aftermath of "our" big bang—may be just one island, just one patch of space of time, in a perhaps infinite archipelago. There may have been an infinity of big bangs, not just one. Each constituent of this "multiverse" cooled down differently, ending up governed by different laws. Just as Earth is a very special planet among zillions of others, so—on a far grander scale—our big bang was also a very special one. In this hugely expanded cosmic perspective, the laws of Einstein and the quantum could be mere parochial bylaws governing our cosmic patch. Space and time may have a structure as intricate as the fauna of a rich ecosystem, but on a scale far larger than the horizon of our observations. Our current concept of physical reality could be as constricted, in relation to the whole, as the perspective of Earth available to a plankton whose "universe" is a spoonful of water.

And that's not all—there is a final disconcerting twist.

Post-human intelligence (whether in organic form, or in autonomously-evolving artifacts) will develop hyper-computers with the processing power to simulate living things—even entire worlds. Perhaps advanced beings could use hyper-computers to surpass the best "special effects" in movies or computer games so vastly that they could simulate a universe fully as complex as the one we perceive ourselves to be in. Maybe these kinds of super-intelligences already exist elsewhere in the multiverse—in universes that are older than ours, or better tuned for the evolution of intelligence. What would these super-intelligences do with their hyper-computers? They could create virtual universes vastly outnumbering the "real" ones. So perhaps we are "artificial life" in a virtual universe. This concept opens up the possibility of a new kind of "virtual time travel," because the advanced beings creating the simulation can, in effect, rerun the past. It's not a time loop in a traditional sense: it's a reconstruction of the past, allowing advanced beings to explore their history.

Possibilities once in the realms of science fiction have shifted into serious scientific debate. From the very first moments of the big bang to the mind-blowing possibilities for alien life, parallel universes, and beyond, scientists are led to worlds even weirder than most fiction writers envisage. We have intimations of deeper links between life, consciousness, and physical reality. It is remarkable that our brains, which have changed little since our ancestors roamed the African savannah, have allowed us to understand the counterintuitive worlds of the quantum and the cosmos. But there is no reason to think that our comprehension is matched to an understanding of all key features of reality. Scientific frontiers are advancing fast, but we may sometime "hit the buffers." Some of these insights may have to await post-human intelligence. There may be phenomena, crucial to our long-term destiny, that we are not aware of, any more than a monkey comprehends the nature of stars and galaxies.

If our remote descendents reach the stars, they will surely far surpass us in insight as well as technology.

Peter Schwartz has made internationally known his methods of foreseeing the possible futures of companies, movements, and even ideas. Such a view is new to long-range thinking about exploration, and here he describes in detail what his methods imply about our long future. Most strikingly, he argues that star flight is nearly inevitable.

Starships and the Fates of Humankind
Peter Schwartz

Over the centuries ahead, the fate of humankind will be very different if the stars are open to us, rather than finding ourselves trapped in our solar system.

Will planets around distant stars provide the opportunity for many new variations of human civilizations, or even species? Or will we be forced to turn in on ourselves as the frozen vacuum of space proves an insurmountable barrier to the human potential? This concluding chapter explores several different answers to the questions of whether we can get to the stars and if so who, why, and how, and with what implications for life on Earth.

I will begin by framing the driving forces that are shaping the future, such as scientific and technological progress and human motivations. The driving forces yield varying key uncertainties, leading to an array of possible scenarios. I then frame richer descriptions of these scenarios. I conclude with the implications for today's choices of the full scenario range.

The Driving Forces

First and most fundamentally, is it even possible to voyage to the stars?

The most interesting and optimistic scenarios flow from a "yes" to that question, though we will explore what "no" means as well. If the answer is *yes* then *how* matters, because there may be multiple very different paths to the stars. Is it a voyage of many centuries or do the voyagers return within human lifetimes? Or are the voyagers even human in any ordinary sense? We also need to know something about conditions on Earth in the centuries ahead and how that motivates the explorations. Earthly scenarios will tell us why we are headed outward.

And finally, one wild card in the whole story that could profoundly influence the shape of the scenarios; do we discover evidence of life in space and most especially intelligent life in our neighborhood in space and time? If the answer to any form of that question is "yes," that could provide a profound motivation for star flight.

Will voyages to the stars ever become possible? Most serious scientists and engineers today would almost all say "probably not." A few might say "just maybe." And a very few would answer "almost certainly yes, eventually." Earlier chapters in this volume explored the various scientific and technical possibilities, so I won't go into any detail, but simply remind the reader of the key variations among the possible options for building a starship.

Skepticism has to do with the profound difficulties of propelling a starship and/or keeping human beings alive for a very long time. Almost any imaginable propulsion system will take huge amounts of energy, much greater than anything we can do today with chemically powered rockets. Unless we can go much faster-than-light, we will need some way to keep human beings alive for a very long time, at least years, but more likely centuries. Even at the speed of light the nearest stars are years away. Or we will need to do something extremely difficult, finding a way to put human consciousness into nearly eternal machines.

The problem of propulsion is basically an energy issue, moving any meaningful mass a vast distance at a reasonable speed. The biggest thing we have ever launched into space was the Saturn V that took Apollo to the moon. Imagine how much energy it would take to get it up to speeds that would get you to even a nearby star in less than a thousand years. Let's imagine the various possibilities. There are likely possibilities in new physics we have yet to understand, say capturing dark energy or using antimatter, but I will focus first on physics we know something about.

Getting There

Perhaps the most likely source of energy is some form of nuclear power that accelerates some form of reaction mass to very high energies. With a century or more of progress, it is not hard to believe that today's nuclear power plants—that only extract a tiny fraction of the energy from the nuclear fuel to boil water—will be surpassed by some method that directly captures the energy of the fuel. Using nuclear energy to generate a high-energy plasma for a very long time might achieve the sustained acceleration. Beyond the level of fission energy we may one day develop a means of creating and containing a nuclear fusion reaction. To get to a significant fraction of the speed of light, to reduce the trip time to decades, would take vast amounts of fuel and an engine running nonstop for many years. Among the ideas for fueling such reactions is capturing the hydrogen in deep space with a huge magnetic field and feeding it into the nuclear fusion reactor like an interstellar ram jet (better known as a Bussard rocket, named for its inventor).

A very different class of propulsion uses energy beamed from Earth or its vicinity. So imagine an immense solar array somewhere inside the orbit of Venus, capturing huge amounts of solar energy. That energy can be converted into some of form of beamable energy, say light or microwaves. At the far end that energy might be captured as physical momentum in a sail miles across, or captured as electromagnetic radiation and converted to plasma propulsion.

While new physics might lead to fuels like antimatter, the great hope, of course, is the discovery of some means of going faster than the speed of light. The nature of the universe as we understand it today prohibits mass from going at the speed of light or faster. At those speeds mass becomes energy. That is the meaning of Einstein's famous equation:

$$E=mc^2$$

For faster-than-light travel to become a possibility, we will have to reinvent the universe. Our model for this universe has become increasingly strange populated by unimaginably small multidimensional strings, dark matter, dark energy, and even multiple universes. This may be a hint that the model needs a fundamental reconsideration. In a recent conversation with Saul Perlmutter, my neighbor and winner of the Noble Prize for discovering dark energy, he agreed that just such a revolutionary reinvention was plausible, given his discovery. This era of maximum confusion echoes other eras in the past, particularly the end of the nineteenth century; the revolutionary theories of relativity and quantum mechanics grew from those fruitful confusions. A grand new theory requires a great imagination to see the possibilities. Perhaps a revolution in physics lies just over the visible horizon.

Maybe we will be able to make a black hole that warps space, so we can surf faster-than-light. Or perhaps we will be able to warp space so much that we can take shortcuts through both space and time. Maybe we will be able to create bubbles of alternate universes where faster-than-light travel, FTL, is possible. All of these and any other FTL possibilities are merely speculation based on no physics we understand today. So long before we build FTL starships we will need a great deal of theoretical and experimental progress in physics. We only need to reinvent the universe.

Staying Alive

The second great technical challenge is keeping people alive during an interstellar voyage. Most scenarios for star flight imply a voyage measured in many years or even centuries. There are several imaginable strategies for very long voyages. The first and least problematic would be an ordinary spaceship travelling at some meaningful fraction of the speed of light, say 1 percent. The nearest stars would be forty years away at that speed. An eighty-year round trip might be accomplished by extending human life spans by a few decades, a plausible goal of modern biology. A second way would be some form of suspended animation technology that would slow aging for decades and "freeze" the voyagers to minimize the life support burdens. Much longer voyages, say centuries, imply longer periods of "sleep," multigeneration populations or extremely long life spans.

If we are able to get up to a high fraction of the speed of light, another possibility might exist. According to the theory of relativity, voyagers travelling near the speed of light would experience a very different rate of time relative to people who remained on Earth. Voyagers would experience much slower time relative to Earth, and so could age much less relative to the home world. A hundred-year trip Earth time might be only a few years of relativistic time on the starship. Our voyagers might have aged only a few years while Earth moved on a century or more by the time they return.

Of course, if we can actually go faster-than-light, as they do in *Star Trek,* then ordinary human beings with normal life spans may become interstellar astronauts. But going faster-than-light might be impossible for normal human beings. Perhaps the accelerations, due to the space-time distortions that might be required, are so great they would pulp human flesh. Science fictions movies, like the 1950s classic *Forbidden Planet,* dealt with this problem by some form of protective conversion process for going trans-light. But there is a more exotic possibility beyond that.

Depending upon a very different kind of scientific advance, perhaps we will learn how to download human consciousness into some device that could be sent to the stars. Imagine that some point in the future you could create an exact duplicate of your consciousness in an electronic device. That device might be, in effect, immortal and physically invulnerable to all the stresses of interstellar travel. The original biological instantiation of you would remain on Earth while the electronic self would voyage to the stars. This more durable version of human beings could open up planets that would be uninhabitable by biological human beings, perhaps making interstellar travel even more likely.

After science and technology, the next big drivers of interstellar exploration mirror what happens here on Earth and who drives us forward toward the stars. It is one thing to look a few decades ahead, as many future studies have, but far harder to look centuries ahead. To make our probe a bit more rigorous, I have based scenarios on a recent study by the United Nations of very long-term population trends looking three centuries into the future, all the way to the year 2300.

The UN population scenarios come from a simple set of premises—what if fertility levels out roughly where they are headed today, vs. what if fertility rises or falls a bit. Over three centuries small changes can have large impacts. If fertility levels out, population will peak somewhere between eight billion to ten billion people and oscillate slowly around that level. If the fertility rate climbs for some reason by about half a point the population will also climb to over thirty billion. If it falls by a similar amount the number of people will fall to 2.4 billion. So far this is simply an exercise in mathematics. It says nothing about *why* and *how* these population dynamics might come about.

If nothing changes, that middle, stable scenario of roughly eight billion to ten billion people appears most likely. The fundamental question is how are they living? Are they rich and sustainable or deeply divided and despoiling the planet? But the really interesting question that takes a bit of imagination is, what could move the fertility trend meaningfully up or down?

There is an obvious answer to what might drive fertility up: religion. The communities that emphasize babies and growing more believers are the rigorous religions of Islam, Christianity, Orthodox Judaism, and Hinduism. The Jews are too small in number to matter much but the other three could come to dominate the planet and reverse the long-term decline in fertility. The rigorous religions could ban birth control and abortion and reward big families. After long periods of conflict and competition they are likely to carve up Earth geographically, roughly along current lines, with Christianity dominating the Americas, Europe, Russia and China, Islam the Middle East, Central Asia and Africa, and Hinduism South and South East Asia.

After many wars and ecological disasters, by 2300 most people live in vertical mega cities (100 million or more population) that are essentially one big internally self-sustaining building.

The object is to minimize the human footprint on the land. The religions enforce a discipline of necessary frugality as next to Godliness. So the first scenario is a high population world driven by and divided by religion, living on a highly urban planet.

There are two very different versions of low population scenarios. Populations can fall because of war, disease, famine, and ecological disasters. Or population might fall because we have grown rich and long-lived, with very few children. Both are plausible, but I am going to focus on the second possibility. So this is a world of relatively few people of great wealth scattered around the planet and living sustainably using advanced technology.

The third scenario of stable population will take the negative view of a world struggling with environmental, economic, and ideological conflict with enormous amounts of criminal activity. It is a dark future with the few wealthy living on islands of wealth amid oceans of despair.

These three different population forecasts lead us to three very futures for Earth. And those in turn lead to three very different visions of who it is that might lead the voyages to the stars. History tells us that missionary zeal is a powerful motivator of exploration and conquest. Bringing the word of your God to the heathens has driven much of the spread of civilizations around the world. So if we were to discover evidence of "convertible" beings living around other stars, it is not hard to imagine the great religions driving us into space to save the souls of these poor aliens.

Escaping miserable conditions has also been a motivator. So it is not hard to imagine the elites of some nations driving the conquest of interstellar space to assure the survival of their civilization, in the face threats to survival on Earth.

Today we are also seeing something new, the outward urge of the new super-rich, building companies like Space X and Virgin Galactic. Perhaps one day trillionaires will put their vast resources behind voyages to the stars.

An interesting and potentially significant wild card in these scenarios is the discovery of evidence of life or even intelligent life elsewhere in the universe. We might wish to either head out to meet them or hide if we fear them. But as noted above, they could easily become the target of Earthly religions missionary passions. If they live on inhospitable planets they might provide some impetus to develop the technology of downloading consciousness into machines.

Key Uncertainties

This array of driving forces shaping the future leads to a set of key uncertainties;

1. Form of starship
    a. None: The technological hurdles are too great and we are stuck in our solar system.
    b. Multigenerational: Modest speeds from, say, an atomic fission drive lead to a starship that requires more than one and perhaps many generations to get to its target star.
    c. Sleeper ship: Modest speeds and successful suspended animation technology lead to long voyages with ships full of "sleeping" crew.
    d. Time dilation (Tau) ship: Higher speeds, say from a fusion drive, lead to a meaningful slow-down of relativistic time and apparently shorter voyages.
    e. Download ship: Fundamental advances in brain science allow downloading a human consciousness into a machine that can be sent to the stars, slow or fast.

    f.   Faster-than-light ship: A fundamental breakthrough in physics allows us to travel faster than the speed of light, making the wider universe accessible.

2. World Scenario
    a.   High-population religious world: Thirty billion people in a world geographically divided into religious zones of Christianity, Islam, and Hinduism.
    b.   Mid-population high conflict world: Ten million people struggling to survive in a world of environmental and economic stagnation, and ubiquitous conflict of all sorts.
    c.   Low-population rich world: Rapid advances in technology and sustained economic growth lead to great wealth, long life, sustainable systems, and a long view.

3. Who leads the way to the stars
    a.   Religious Institutions: As with earlier explorations, religious institutions might try to demonstrate their moral superiority and even possibly to convert the newly found heathens.
    b.   Nations: Nationalism and national survival have driven exploration and colonialism throughout history.
    c.   The super wealthy: The new, vastly wealthy have taken an interest in space and their ultra-wealthy successors are likely to as well.

4. Aliens
    a.   No evidence of life elsewhere in space: Despite continuing efforts and many planets discovered, no evidence of life is found on any of them.
    b.   Evidence of life: We detect biochemical evidence of molecules that could have come from life on planets whose physics makes it plausible.
    c.   Evidence of intelligent life: We detect information signals of advanced intelligent life.

Scenarios

Each of these scenarios aims to be a coherent story of the future. The various forces need to reinforce each other and not fundamentally conflict. In principle, all the uncertainties noted above can be combined in 162 unique ways. But not all of them are very distinctive or interesting. Our criterion for interesting is that we can learn something of value from it. Second, the set of scenarios ought to span a wide range of possibilities. Finally, each scenario should be plausible. The reader might combine them in a different fashion and find some interesting scenarios that the author missed. But I think the four basic scenarios, one of which has two distinct variations (to give us five scenarios), are both sufficiently instructive, distinct, plausible, and cover a wide range of possibilities.

Stuck in the Mud imagines that we are trapped in our solar system, and is only interesting as a base for comparison with the others. *Escape from a Dying Planet* says it all; the motive is survival as Earth may no longer be survivable. *God's Galaxy* is all about saving the souls of distant aliens. And finally, *Interstellar Trillionaires* is about the pursuit of the unknown by the fabulously wealthy. This last scenario might have two variations, one based around faster-than-light travel and the other around downloading consciousness.

Stuck in the Mud

Over the next century or two, great scientists and engineers struggle with the extremely hard challenges of interstellar propulsion, long-term life support and even downloading consciousness—but with little to show for it. Human beings, at least for the foreseeable future, are trapped in their solar system and in their biological bodies. Dreams of deep space begin to fade. Perhaps it is possible that in this scenario Mars, or the moons of Jupiter or Saturn or even our own Moon, might be sufficiently interesting to invite modest colonies. That, of course, assumes that we lower launch costs by a lot. Perhaps we might move an asteroid fairly near Earth, hollow it out and create a space colony. But none of these possibilities is likely to fundamentally change the human condition. Indeed they are likely to reinforce the sense of limits.

These limits of human capability might be combined with any of the world scenarios. In the event of a high-population, religious scenario or stable-population and high-conflict scenario, the closing in of horizons might lead to a much higher likelihood of conflict, as we find no outlets for our aspirations in the wider universe. In the low-population, high-wealth future, human creativity and energy will have to find other outlets, say the oceans or the arts.

The great risk in all these scenarios is lowering the sights of human aspiration. Human progress has come from great dreamers doing great things. Stagnation and decline may be the ultimate result of the closing in of frontiers, even if great wealth makes almost anything possible. The result might be long lives filled with boredom?

Escape from a Dying Planet

What if the pessimists are right, and Earth can't sustain a population of nearly ten billion people at a high standard of living? What if the struggles for water, energy, food, and resources become the ubiquitous source of conflict? This is not a future of a grand apocalypse, a massive die off from war, or famine or disease or ecological collapse, or some combination of all of the above. It is misery by a thousand cuts. Nothing works right. Climate change disrupts agriculture and undermines infrastructure and little can be done about it. Governments are incompetent and corrupt, basing their legitimacy on fear of the neighbors. Corporations are narrowly self-serving and individuals are unwilling to sacrifice for the greater good. Over the decades the world comes to resemble the cities of immense urban squalor, with a very few living protected lives in vast wealth. It is the world as Mumbai or Lagos or Mexico City. The city of tomorrow is a slum and organized crime the most common career choice. Small private wars simmer.

Now suppose that the 100 Year Starship Project meets with some success. In this scenario the project develops a propulsion system powerful enough to drive a large starship to nearby stars in less than a few centuries. Suppose the project also develops the technology to put large numbers of people into suspended animation of some sort for very long periods. It would mean a vastly reduced life support problem, particularly if the passengers were only awakened upon arrival. In that case a sleepship on a voyage of decades or centuries becomes imaginable. The final key element of this scenario is the discovery of one or more inhabitable planets within reach of our sleepships.

As decades pass more and more of the potential of the world is consumed in friction, until by the middle of the twenty-second century a few nations come together to launch an audacious survival bid, the conquest of interstellar space. One can imagine in 2150 leaders of some of the remaining great nations reach the conclusion that Earth is really in trouble and that the only way to assure the survival of their civilization is colonizing the stars. And they had better do it soon while they still can. For example, the English speakers might come together, the Europeans, the Chinese, and a few more as well, in coalitions of nations rich enough in both financial and technical resources to build giant sleepships and launch them toward habitable planets orbiting distant stars. If there is only one habitable planet within reach then it will almost certainly become a race for survival and perhaps even a war.

Each group of nations would almost certainly launch several ships to assure that at least one succeeds. Each of the ships would carry large numbers in stasis of fully developed human beings of various ages. It would also carry a very large number of fertilized human eggs to assure adequate biological diversity and rapid scale-up of populations. Small robot probes to determine whether the target planets really are inhabitable would precede each of the starships. If not, the advance probe can redirect the starship to the next target planet and keep the passengers in deep sleep. So in this scenario escape from misery and the fight for survival drive human beings toward distant star systems.

God's Galaxy

Most scenarios of the future assume a leveling off of population growth because a world of very high population appears quite unsustainable. One also has to ask the question, what would lead fertility to rise again? Today nearly everywhere as wealth increases, as access to medicine increases, and women's rights improve, women are gaining control of their wombs and having many fewer babies. So something would have to change for population to start rising again in a big way. It leads one to the question, "Where do you see the urge to have lots of babies being the strongest?" And the answers, of course, are the conservative religions. The Catholics, the Evangelicals, the Muslims, very Orthodox Jews, and some Hindu sects all emphasize big families and creating more believers.

Only a few decades ago there was a widespread academic view that as education and science spread human beings would become more secular and less religious. *Time* magazine famously and wrongly declared *God Is Dead* in 1967—probably the biggest sociological error of the last fifty years. Rather than declining, religion is growing in numbers of adherents and impact, especially the most conservative sects. It turns out that most people in most places really want to believe in some higher level, to give purpose and direction to their lives. So one of the only ways one might envision a reversal of the decline in fertility is by religious compulsion "to be fruitful and multiply."

Perhaps religion also might be one of the keys to sustainability. Dense vertical cities like Manhattan may be the urban design for a truly green city. Paolo Soleri, the visionary architect, designed and argued for what he called arcologies. These were to be massive cities that were one giant building with nearly all the systems closed loops like a space station. Only some food and sunlight would come from the outside. He died while trying to build the first one from scratch in the desert of Arizona. They might actually develop as urban buildings gradually merge into vast superstructures that resemble Soleri's vision, in impact if not in his elegant and aesthetic designs. These hyper-dense city buildings would be a half-mile high, like today's Burj Khalifa in Dubai. They would have populations like today's nations with over a hundred million people. If population reaches thirty billion by 2300, that implies about 300 cities of about 100 million each. The goal would be to concentrate the human population and minimize their footprint on the land, all of which would be needed for food production and ecological services.

While arcologies may become technologically possible, their continued functioning will also depend on high social discipline. People will have to behave properly or decay and disorder will set in fast. One of the only institutions that has been successful in compelling and constraining human behavior over very long time frames is religion. The Pope, the Imam, and the Guru can all counsel their people that their God wants them to take care of her Earth and to manage human consumption sustainably. Political leaders always promise the alternative—more stuff, not more discipline. Only the priests can argue for the virtue of abstemiousness. So religion is at the core of this scenario and indeed finally provides a motive for starflight.

But before we get to starflight, let's take a quick look at the history of the future in the world of religion and high population. Over the course of the century ahead, population growth comes back with a vengeance as faith and fertility go together. By 2100 the population of Earth is at twelve billion, by 2200 twenty billion, and by 2300 thirty billion—all living in the mega-arcologies.

Carving up the world into contained zones of belief began with the Great Expulsion in 2050. Decades of religious strife, rising anger, and deep racism led the European Christians to expel the Muslims. This time there were no concentration camps and ovens for human beings. All the rights of Muslims were taken away—owning property, having assets, education, citizenship, employment—were all legally denied to Muslims. They had no way to survive. The only thing they were granted were one-way tickets to Muslim nations, most of which were not very welcoming. By 2060, Europe had been cleansed of Muslims and was once again a unified Christian nation. Much violence accompanied the Expulsion because not everyone went gracefully. And of course that set in motion similar dynamics and much more violence around the world so that by the end of the century the world was divided into three great religious zones, the Christian, the Muslim, and the Hindu. After the hot religious wars of this century, a cold war and intense competition shaped their relationships over the next century.

Suppose in this scenario we invent the technology of long-duration fairly high-speed starflight and can build a generation ship. Religious orders such as monasteries have a long history of maintaining discipline, and might just be what is needed to provide order over the centuries of interstellar travel. Among the motivations for the great voyages of exploration in the last millennium was finding new adherents for whichever god was behind the voyage.

So imagine what might happen in this scenario if intelligent life is discovered within reach of a generation ship from Earth. The missionaries would be outbound as fast as they could get organized. Their goal would be to bring the word of their god to the heathens of this unenlightened star system. It is not hard to imagine a race of three interstellar missions of the faithful, with the prize being the souls of the aliens and further proof of the superiority of their faith over all others. Their faith would sustain them and provide the necessary discipline and continuing motivations to each successive generation for the mission, to survive the centuries of starflight.

Interstellar Trillionaires

This final scenario is based on a slowly shrinking population on a world of great wealth, powered by huge advances in science and technology. And it takes us on two very different routes to the stars. A set of major scientific advances lead to a very different future than most people expect.

By mid-century, biological and medical advances lead first to more people living healthy lives much longer. Life expectancies rise to well over a hundred. But then come breakthroughs in life extension, leading to some people living to over 150 with no obvious end in sight. As with every species with long life comes falling birth rates until the planet is below replacement. The Chinese experience of one-child families becomes the norm, not by policy but by choice. Children become rare and precious, capturing a great deal of investment. Education and dependence stretch into the forties as the norm.

Several technologies of sustainability begin to transform the planet by the end of century. Nuclear fusion, very efficient solar power, and synthetic biology combine to end the fossil fuel era. Similar advances in synthetic biology lead to a revolution in industrial processes and materials. Robotic technologies lead to a radical shift in the structure of industry, as almost anything can be assembled locally in small-scale manufacturing cells. Quantum computers of huge power will enable extremely complex process control and perhaps even meaningful machine intelligence. The energy and industrial revolutions mean that environment and resources are no longer barriers to the continued increase of wealth. The huge leaps in technology make a very high growth scenario over the long run quite plausible Over the next two centuries average global growth of 5 percent means a doubling of wealth every fifteen years or a multiple of thirty each century, so that by the year 2200 the world is nearly a thousand times wealthier than today with half as many people. A less-optimistic growth rate of say 3 percent would still yield a vast increase in wealth, a multiple of hundreds, over three centuries.

One of the remarkable economic developments of the last twenty years was the huge growth in the number of billionaires. Silicon Valley, investment banking, and explosive growth in China, India, Russia, and Brazil have all combined to create literally hundreds of billionaires. And some of these new super-wealthy have turned their attention to space. Elon Musk of Space X, Richard Branson of Virgin Galactic, and Jeff Bezos of Blue Origin are the most visible. But other spaceships are on the way from today's billionaires. So imagine the kind of wealth that the super wealthy have in a world that has experienced growth of a factor of 1,000. This low-population-growth world, where the super-wealthy marry and combine their fortunes before handing off to their one child, will almost certainly produce at a least a few trillionaires with an interest in the pursuit of the stars.

It may only be long-lived trillionaires who can fund a sustainable effort to build a starship. And they don't need the motive of conversion of aliens or human survival to drive them. At least some have the energy, vision, and fascination with the human frontiers that they are willing to drive the new era of exploration. And it is here where the scenario bifurcates into two very different routes to the stars and fates of humankind.

While a scientific and technological revolution was transforming life on Earth, it was also opening up new possibilities for space travel. As we develop ever more powerful energy sources, I can imagine precisely the kind of stepwise course to the edges of our solar system and just beyond that Freeman Dyson described in his essay here. It is not hard to imagine that the fusion reactors of the middle of the next century might power a starship to a high fraction of the speed of light. We can call such a relativistic starship a Tau Ship because for the crew the most important variable shaping their lives is Tau. As the velocity of their ship gets closer to light, Tau shrinks. Indeed, Tau only reaches zero at the speed of light. But if Tau is very small, the passage of time on board relative to Earth will slow dramatically. Voyagers to the stars will age only very slowly. So an exploratory journey to nearby stars might take a decade or two to get there and back even as the crew has only experienced a year or two. This was the premise of *Tau Zero*, Poul Anderson's wonderful tale of a hydrogen ramjet unable to shut down, that goes faster and faster, but not faster-than-light, until they zip through galaxies in what appears to them only seconds and feels like a cosmic speed bump.

So one variation of this scenario is the interstellar trillionaires themselves heading out on their Tau starships to an ever-expanding sphere of stars while many decades and even centuries pass back on Earth. As today's boy billionaires compete on who is doing the coolest thing with their wealth, the competition of the trillionaires for the fastest Tau ships and the best star systems could drive the new wave of exploration.

But there might be a different route to the stars for the trillionaires, based on two radical leaps in science and technology. First let us suppose that over the next century or two The 100 Year Starship Project learns how to copy a human consciousness into some machine-like device—call them iPeople, if you like. An iPerson will be able to control some avatar-like device to function in the real world, if it so chooses. Indeed, they could have the ability to function in environments inhospitable to human beings, like the deep ocean or the moons of Saturn. As interstellar explorers, iPeople would not be limited to Earthlike planets. Much more of the universe would be of interest to them. It is of course possible that an iPerson of a trillionaire will elect to go voyaging on Tau ships. But there is a still more radical scenario possible.

Suppose Saul Perlmutter is right, that today's weirdness of strings, multiple universes, and dark energy are all signs that we are going to reinvent our model of the universe. And suppose further (which Saul does not) that in this new model we find a way to exceed the speed of light and build an FTL starship. Now, this dream verges on a hallucination. Nearly every recipe for a starship begins with something like, "build yourself a black hole and then punch a hole through the fabric of space and time." But if I had argued in the middle of the nineteenth century that we would one day be able to send moving pictures through the walls of buildings on invisible waves, it would have seemed equally implausible and fantastic.

So perhaps in the century after the reinvention of physics and a great deal of engineering, the 100 Year Starship project funded by a few trillionaires is able to build an FTL ship. It is likely that whatever the technology needed to break the light barrier, it is probably not conducive to human health. Imagine accelerations of thousands of Gs, for example. So it is the iPeople of trillionaires who become star travelers, able to go wherever their curiosity or interests take them. This scenario always leads in science fiction to the story of the FTL starship that arrives in time to welcome the slower explorers on generation or sleepships at their destination, Gregory Benford's story in this volume for example.

At the destination worlds of FTL starships, they are likely to be able to grow new human flesh avatars for their iPeople. So if they wish to copy themselves into bio-people who colonize distant worlds, that becomes an option. And of course, with the ability to modify human beings we will almost certainly create radical variations on the species, capable of surviving in a great variety of environments. This was in effect the scenario in the film *Avatar*. So finally this is a scenario of humankind spreading itself in many forms outward into our galaxy, carried among the stars by iPeople, and inhabiting the many worlds within our reach in human forms suited to their unique conditions.

Conclusions: Many Paths

What do we learn from this exploration of the possible scenarios for starships? Most of all, we can see many pathways to the stars.

The scenarios I described are clearly not the only possible futures. They were interesting and somewhat surprising, so there was something to learn from them. Some aspects of the scenarios were quite surprising. The likelihood of voyages to the stars is a function of what happens here on Earth. The conventional wisdom is that nation states will drive the exploration of space. But in our set of scenarios it is religion and vast private wealth, not only nationalism, that provide both the motive and the means for getting to the stars. Clearly, context matters. Are the people of Earth struggling or prospering? Are we converging or diverging? The outward drive can exist in many scenarios, but not all, and it may be that the religious drive along with visions of the super wealthy may be the most likely route to the stars.

So *who* matters. Is it simply about aspiration, or is it about competition to save souls, find new homes, or prove your superiority to your competitor? So *why* matters.

It is also clear that the scientific and technical challenges are huge. Which capabilities actually develop will shape the options we have. Are we stuck in our solar system and its vicinity? At the core of interstellar travel as well as the fate of Earthly human civilizations is the issue of energy. Can we develop a technology for powerful, clean, cheap energy to drive the economies of our planet? And can we develop the power sources needed to achieve interstellar speeds, including FTL? If the answer to both questions is *yes* then we are looking at fairly optimistic scenarios. If the answer to both is *no* then we have some big problems. If we can get to the stars but not save Earth, then at least we may have a survival option. And if we really do have sustainable energy, then the need for starships may not be so great.

The exploration of these scenarios for starships and the fates of humankind leave me very optimistic about the future of our species, if not Earth. Clearly there are at least several plausible fates of humankind and starships do matter. There are at least several plausible pathways out into the universe for our species. We can't be very sure of the timing. It depends upon context and discoveries in science and technology, which are unpredictable.

But that we can see so many different combinations as plausible suggests that eventually it is *nearly inevitable.* One way or the other, on generation ships, sleepships, Tau ships, or FTL starships we will get there. The *we* that gets there may be from the great faiths, a few enduring nations, or grand dreamers—but the drive for the stars will almost certainly exist. So the bottom line is that sooner or later, for good or ill, we will build starships and we are headed out into the galaxy. It is the fate of humankind to play on the stage of a wider universe.

Few of us pause amid the hubbub of our days and think, This will be history. Yet we make it every day. Historians often convey the impression that the past, since it is now fixed, was a neat, cut-and-dry time. This mistake makes the present seem messy in comparison. The past is a far country, but the distance should not confuse us about its turbulent nature. We begin this book's fiction with a story stemming directly from the 2011 100 Year Starship Symposium—which is now history.

Cathedrals
Allen M. Steele

Florida Ballroom 5 was a ballroom in name only; Frank doubted that it was large enough for even a half-decent foxtrot. He found himself envying Orange Ballroom D just down the hall, where the keynote speeches had been held; it had a couple of thousand chairs, with screens big enough to be seen from the back of the room. But the conference's breakout sessions—as many as five occurring at any one time—had been scheduled for these smaller rooms where there were only a few dozen seats, often forcing many of the attendees to stand along the walls or sit on the floor.

At one of yesterday's sessions, Frank had heard John Cramer call the conference "Woodstock for nerds." Frank had missed Woodstock—in 1969 he'd been at Marshall Space Center, helping NASA plan those last two Apollo missions that the Nixon administration killed—but he supposed that the analogy was as good as any. Sure, the conference didn't have tents, mud, brown acid, or Jimi Hendrix playing the "Star Spangled Banner" at sunrise, but nonetheless there was a sense of freewheeling, wide-eyed possibility that a hippie would have grooved on. He smiled at the thought. Did anyone still say groovy? Or was he just showing his age?

At the podium beside him, Jim Benford was wrapping up his introduction for this afternoon's session on breakthrough propulsion systems. Frank shuffled his notes as he pretended not to notice that the room was half-empty. Only about thirty or so people had shown up to hear him speak. Which wasn't bad, really—before he'd retired from Stanford, his lectures typically had only a dozen or so grad students—but this morning's talk on faster-than-light travel had been standing-room only. Once again, he scolded himself for being a bit reserved in his choice of title for his presentation. "Warp Field Mechanics" sounded a little more intriguing than "Quantised Inertia and FTL," but wasn't anywhere near as interesting as "Terraforming Planets,

Geoengineering Earth" or "Hostile Journey and Destination—Yes, But Weapons?" And none of them rocked—oh, yeah, that was the word that replaced groovy—like "Did Jesus Die for Klingons, Too?" Frank had no idea what that was about, other than it belonged to the program track for philosophical and ethical considerations, but he had little doubt that it had jammed Florida Ballroom 7.

His thoughts were interrupted by scattered applause from the audience, and he looked up to see Jim stepping away from the podium. Frank picked up his notes, pushed back his chair, and sauntered toward the podium, exchanging a brief smile and a nod with Jim as they walked past each other. He laid his notes upon the podium and picked up the wireless remote used to operate the projector in the back of the room. Half of the conference speakers had run into trouble with the damn thing—apparently it used a version of PowerPoint few of them were familiar with —and he prayed that he wouldn't embarrass himself, at least not that way.

Never mind that now. The next twenty minutes were his. Frank glanced at the longhand notes he'd written on a yellow legal pad—an explanation of the Alcubierre metric; a discourse on the problems inherent in generating a warp bubble within spacetime; speculation on the creation of boost shells within pseudo horizons; the possibility of constructing a toroidal capacitor ring as a means of harnessing negative energy—and realized that it wouldn't be wise to jump straight into his presentation's most technical material. This wasn't Stanford; if he wanted to keep his audience from wondering what Jesus might say to a Klingon, he'd better open with something a little more compelling than equations.

A remark someone said to him over breakfast that morning came back to him. Frank wasn't above cribbing from his colleagues, so he stole it.

"You know," he began, "I've been thinking about how we should go about interstellar travel, and it occurs to me that we do this much the same way the great cathedrals of Europe were built...."

As he spoke, his voice was amplified by the podium mike, while his image was caught by the Sony digital camera mounted on a platform in the back of the room. Both fed straight into one of the half dozen or so DVD recorders saving the conference's sessions for posterity.

The camera softly purred as it focused upon him, capturing the projection screen in the background....

When Karen Cho's grandfather died, his passing was newsworthy enough to be mentioned on the front page of the *Liberty Post*: DAVID CHO, ORIGINAL ALABAMA COLONIST, DIES. The story went on to explain that Dr. Cho was one of the fabled "d.i.'s" who hijacked the URSS *Alabama* in 2070 and was therefore among the 109 people who, after a 230-year journey, reached the 47 Ursae Majoris system to establish the first colony on Coyote.

The reporter who'd written the story apparently thought his readers needed a history refresher, because he went on to reiterate—rather unnecessarily, in Karen's opinion—what every child who had been born and raised on Coyote had been told since their first days of school: how a small group of conspirators led by Captain R.E. Lee had managed to substitute half of the ship's military crew with d.i.'s who had been arrested and were on their way to a government re-education camp. The article went on to explain that one of them had been Karen's grandfather, an engineer who'd worked for the Federal Space Agency until his questionable loyalty to the United Republic of America had caused him to be branded as a dissident intellectual—the term for those who hadn't toed the Liberty Party line—and dismissed from his job. Nonetheless, Dr. Cho had been among those responsible for the design of humankind's first starship, and so when Captain Lee and his closest associates hatched the conspiracy, he and her grandmother (who'd died before Karen was born) had belonged to the civilians who were smuggled aboard the *Alabama*.

Karen knew all this, of course. She'd grown up hearing the stories, from both her parents and from her grandfather himself. However, what she'd been told as a child had exaggerated her grandfather's role in the hijacking, making it seem as if he'd charged aboard the ship, fighting Federal Service soldiers shoulder-to-shoulder with heroes like Capt. Lee, Tom Shapiro, and Jorge Montero. It wasn't until years later that she understood how this was an embroidered version of how things actually happened. The hijackers hadn't taken *Alabama* by storm, but instead had come aboard rather quietly, using subterfuge rather than force. There had been no fight in the main airlock, as legend has it, and her grandparents had done nothing more than go to their berths, strap in, and wait for others in the command center to usurp control of the ship's computers and engage the launch sequence. Indeed, it was this discovery that led Karen to become a historian; she decided that she'd rather know banal reality than glorious fantasy.

All the same, she'd honored and respected her grandfather, and after her own parents were killed in a ferry accident on the Great Equatorial River, she'd cared for the old man during his last years. By then, Gran'pa barely remembered Earth at all, and even had trouble recalling the details of how he'd come to Coyote. Unfortunately, this sort of senescence wasn't uncommon among original colonists; doctors at the University of New Florida believed that degeneration of neural tissue was a long-term aftereffect of the centuries-long biostasis they had endured aboard *Alabama*. Even Marie Montero, the younger sister of former Coyote Federation president Jorge Montero, had been afflicted by this.

So Karen had brought Gran'pa into her home while she finished her graduate studies at the university, and it was shortly before she received her master's that the old man quietly passed away while taking an afternoon nap. Karen had been expecting this for quite some time, and although she grieved at the death of the last member of her immediate family, there was comfort to be taken from the fact that Gran'pa's death had been peaceful and without pain.

His memorial service was remarkably well-attended, even if she didn't recognize many of those who showed up. Gran'pa had few friends left, and although Karen had her own supporters, she suspected that most of the people there simply wanted to pay respects to one of the *Alabama* colonists. Karen spoke for a few minutes, and then David Cho's shroud-wrapped body was placed upon a wooden bier and set afire. The service was followed a wake, but Karen showed up just long enough to accept a few condolences before going home.

As her grandfather's sole survivor, it fell to her to settle his final affairs. Gran'pa had left everything to her, so it became mainly a matter of going through his belongings and deciding what to discard, what to give away, and what to keep. This was a sad but rather mundane chore until she came upon a small nylon bag among the things he kept in his desk.

The bag was threadbare, dusty, and obviously quite old. When Karen opened it, she discovered that it contained a thin plastic cartridge with a plug and a serial port at one end. It took a few minutes for her to recognize what it was: an antique backup drive, the sort once used to store digital information.

Karen realized that this was an artifact from Earth that Gran'pa had carried with him aboard the *Alabama*. For a historian such as herself, such a find was priceless; there was no telling what might be stored on it. The technology was obsolete, of course, but the University's history department possessed a twenty-first century computer, itself another *Alabama* relic, which was carefully maintained for the express purpose of reading such data caches when they were found.

A few days later, once she had the time, she took the old drive to the University, where she showed it to the history curator and asked if he could find out what it contained. After warning her that it was entirely possible that its memory may have decayed to the point of uselessness, the curator carried the backup drive to a dust-free room where an impossibly old comp stood upon a table. He ran a cable from the comp to the drive, then carefully typed instructions into the brittle keyboard that opened the drive's file directory.

As it turned out, the curator's warning had been correct. Of the dozens of files contained on Gran'pa's drive, all but a few had eroded over time, leaving behind only a list of filenames—*notes. doc, SPsched, phonelist,* and so forth—which offered nothing but tantalizing clues as to what they'd been about. Yet among those still readable was one with an intriguing name: *100YrSS— Grandfather.*

Karen asked the curator to open this one. Instead of text, they discovered that it contained an old-style video wavefile. Was this something Gran'pa had once recorded? This was Karen's first thought, until she realized that her grandfather would have never referred to himself as such. And what did "100YrSS" mean? "100 Year" was the most likely interpretation for "100Yr," but the letters "SS" were mysterious.

The chronometer indicated that the wavefile had once been a little more than twenty minutes long, but what had survived was much shorter than that: less than two minutes. When the curator played it, they saw:

An elderly gentleman, wearing a dark sport coat and necktie of late twentieth-century style, stood before a podium. The word *Hilton* was visible on a plaque attached to the front of the podium, but this was less obvious than what projected on the large video screen beside him: *100 Year Starship,* the words superimposed upon an stylized celestial compass, obviously a logo of some sort. The video was apparently shot from a camera located in the back of a meeting room. At the bottom of the screen were the backs of a dozen or more heads; at one point a young woman briefly walked in front of the camera, obscuring the imagine for a second.

"You know," the man at the podium said, *"I've been thinking about how we should go about interstellar travel, and it occurs to me that we should do this much the same way the great cathedrals of Europe were built. Not as a short-term project, with goals that can be achieved only within a few years and everything else pushed back to a hazy and not well-conceived timeline, but rather as a*

long-term initiative that may not be completed until our children's or even grandchildren's time. When we currently think about the logistics of space exploration, such as returning to the moon or going to Mars, we tend to fall into a pattern that we used during Project Apollo. That was good for putting men on the moon within a decade, but it's not so appropriate when confronting the challenges of sending a vessel to another star system a hundred years from now, if not sooner...."

He raised his left hand toward the screen, fumbled with something he was holding. The logo disappeared, replaced an instant later by a diagram that looked like a pair of parabolic curves placed against each other on either side of a horizontal line. *"Indeed, the technology for building a practical warp-drive engine may not come into existence for another couple of generations. However, because we can project with some degree of confidence just what such a drive might entail, we don't necessarily have to push this into our grandchildren's laps. We can begin thinking about it now. The Alcubierre Metric, for example, postulates using negative energy to generate a field, or warp bubble, around a spacecraft which would it allow it to...."*

The video came to an abrupt end. The curator tried to salvage the remaining data, but was unable to retrieve anything more than harsh static and a few grainy images. The rest was lost to time.

What little they saw, though, was enough to cause Karen to stop breathing for nearly two full minutes. For in those few precious moments, she realized that the man at the podium wasn't her grandfather, but rather *his* grandfather, and that he was speaking to her from across a gulf of nearly 400 years.

❧

"There's not much family resemblance, is there?" Sitting up in bed, Arturo studied the image of Karen's great-grandfather, which her data pad had projected on a wall screen. He smiled as he glanced at her. "You're sure he's related to you?"

"I'm sure." Karen sat on the edge of the bed, wrapped in a hemp bathrobe and drying her damp hair with a towel. "You told me yourself...Dr. Frank O'Connell, Ph.D., professor of physics, Stanford University, and NASA." She frowned. "That's the National Aeronautics and Space Agency, isn't it?"

"National Aeronautics and Space Administration. Otherwise, yes, you got it...that's what I found when I double-checked the historical records." He gazed at the image again. "I'm just saying, he doesn't...he didn't...look a lot like you."

"The Asian bloodline hadn't entered my family's gene pool quite yet." Karen laid down the damp towel, then picked up a brush and began combing out her long, black hair. "From what I remember of my family history, my grandmother married a guy from Seoul. That's how Koreans came to belong to an otherwise Irish-American family." She nodded toward the holo. "I see the resemblance. You just have to look hard, that's all."

"Hmm...I suppose." The smile reappeared, a little more coy this time, as Arturo's hand slid across the bed. "Perhaps I should make a more thorough investigation," he said, taking hold of the sash that held closed the front of her robe.

"Quit." Karen gave his hand a gentle slap. The two of them had been sleeping together for the past several months, having met at the University, but this was the first time she'd felt comfortable having Arturo share her bed. Because she didn't want to have her grandfather hear them from the next room, they'd always made love at his place instead. "I'm serious. I know so little about him. That's why I asked you to look into it for me. So you found out his name and that he'd worked at Stanford and NASA…."

"Sure. That was easy enough." Arturo let go of her sash, picked up her pad and used it to expand the image until the 100 Year Starship logo filled the wallscreen. "Once I knew where the vid was recorded, I just had to dig into the history archives until I found the conference proceedings. And there he was, listed right there on the agenda. 'Warp Drive Mechanics,' F. O'Connell…12:50 pm, Florida Ballroom 5."

"And the date was…?"

"Saturday, October 1, 2011." He ran a hand through his dark brown hair. "Can I use the shower next? I'd like to freshen up a bit, too."

"Let's talk first." Karen laid down her comb. She wanted to get dressed, but she was afraid that if she removed her robe, Arturo would get distracted and…well, they'd never get through this. "So you got his name, and from his biograph in the published proceedings you got the rest."

"Uh-huh." He lowered an eyebrow and stared at her intently, an expression that she'd learned to recognize as inquisitive puzzlement. "I don't understand. How can you realize that he was your ancestor, but know so little about him? His name, for instance?"

"My grandparents left Earth with almost nothing but the clothes they were wearing. Everything they owned was left behind, and the government of United Republic of America confiscated all that once they were gone. So what little I know about my family has been pretty much what I've been told."

"I understand," he said. "Same sort of thing happened to my family."

Arturo's parents had come to Coyote only about twenty years earlier, during the second great immigration wave following the collapse of Earth's major governments which occurred in the wake of the global environmental catastrophes that made humankind's homeworld all but uninhabitable. It wasn't quite the same thing—those immigrants had been able to bring a few belongings, at least—but Karen wasn't about to argue the point with him.

"Sure," she said. "Anyway, Gran'pa used to tell me that *his* grandfather had been one of the first scientists to work on interstellar travel, and that his work had led to Project Starflight, which in turn led to the construction of the *Alabama*."

"But since that was all hearsay, you had no way of knowing for certain." Arturo nodded. "All right, I understand. So now you do…."

"No, I don't…not completely, I mean." Karen let out her breath in a frustrated sigh. "Look, I know there's not much here, but from what I can tell, my great-grandfather didn't have anything to do with the *Alabama* or Project Starlight. He was involved in something else entirely, the theoretical development of a warp drive…and that's not what was used by the *Alabama*."

"No, it wasn't." Stretching luxuriously beneath the sheets, Arturo folded his arms behind his head. "*Alabama* used a RAIR engine…a ram-augmented interstellar rocket. Sort of a variation on a Bussard ramscoop. The designers decided on that because no one yet knew how to build a warp engine…those were still many years away from being built or perfected…and starbridges were only a vague possibility." An ironic chuckle. "You know what's funny? Of all the possible propulsion systems discussed during that conference, Bussard rams were not among them. I don't know why, but…." He shrugged.

"Then my great-grandfather didn't contribute anything." Karen's voice echoed her disappointment. Another story about her family debunked. All of a sudden, she wondered if Gran'pa had been little more than a freeloader who'd managed to get aboard the *Alabama* simply out of blind luck.

"I wouldn't say that." Arturo shook his head. "Warp drive was eventually developed, wasn't it? Maybe it wasn't used by the *Alabama*, but the subsequent ships sent by the Western Hemisphere Union used that technology."

"The Millis-Clement Drive, sure. But my great-grandfather…."

"Might have had something to do with it. Who knows? I'd be willing to bet that, if you were to investigate the history of its research and development, you'd find citations to theoretical work by Dr. Frank O'Connell. And even if not…well, consider him one of the cathedral's anonymous bricklayers."

She looked at him askance. "Come again?"

"Think about what he said, when he compared building a starship to building a cathedral." Pushing aside the sheets, Arturo swung his legs over the side of the bed. "The cathedrals of Earth…the ones built in Europe during the Renaissance, I mean…were the result of generations of labor. Those who built them had to work with the available technology of their time, which was not much more advanced than bricks and mortar, ropes and pulleys. So the ones who started work on cathedrals like Notre Dame knew that their work probably wouldn't be finished in their own lifetimes, but probably by their children or grandchildren. And they were content with that."

"You think so?" Karen couldn't help but watch Arturo as he rose from bed. She enjoyed seeing him nude. "I mean, to spend your life working on something, knowing that you won't benefit from it yourself…."

"I don't know about you, but I think it would give my life purpose. I think that's what people like your ancestor were doing when they went to that conference…starting something which they intended to be finished by their children or grandchildren, if not themselves. Like a cathedral."

"You make it sound almost religious."

"Faith doesn't need religion. Just the ability to believe in something greater than yourself." Arturo stood up, started to head for the bathroom. "Does that make sense to you?"

"Umm-hmm." She couldn't take her gaze off his buttocks. "It does, I suppose."

"Good." He glanced back at her, then paused in mid-step. "But I think we were talking about something else before then. Family resemblances? Expanding the gene pool? That sort of thing?"

She felt a smile coming to her face. "Yes, we could have been."

Arturo turned around to walk back to her. She stood up from the bed, and he began untying the sash of her robe. "Maybe we ought conduct a little experiment of our own…."

After he finished his talk, Frank had about five minutes to take questions from the audience. There were only a few, but nonetheless they were worth answering, giving him a chance mostly to explain a few of the more technical aspects of his presentation. Then Jim reminded everyone that they were scheduled for a fifteen-minute break before the next sessions began, and there was a brief smattering of applause before everyone stood up to leave the room.

All the sessions let out at the same time, and the convention center mezzanine was filled with attendees, each with laminated name badges dangling around their necks. Frank took a Coke from one of the beverage carts the Hilton staff positioned in the mezzanine during breaks and stood off to one side, decompressing from the efforts of the last hour.

A knot of people stood in front of the British Interplanetary Society table, examining their exhibit on Project Icarus. Over here, Jill Tartar chatted with people who'd been at her talk. Over there, Douglas Trumbull was deep in conversation with another person. A group of science-fiction writers—instantly recognizable as such because nearly all of them had beards and were losing their hair—were clustered together, perhaps discussing what they were managing to learn from these sessions that they would later use in their stories. Frank had just spotted Stewart Brand walking by when the cell phone in his coat pocket purred.

He dug out his Android and glanced at the screen: L. CHO 415-555-0994. Smiling, he ran his finger down the screen, then held the phone to his ear. "Hello, Lisa," he said, raising his voice a little so he could be heard above the crowd around him.

"Hi, Dad. How's the conference going?"

"Pretty well. I just finished my presentation a few minutes ago."

"Great! How did it go?"

"Umm…not bad. Could have used a few more people in the room, but there's a lot going on here, so I had competition."

"I'm sure they were all there for you."

Frank grinned. That was his daughter: always trying to be optimistic. "Yeah, well…how's things with you this weekend?"

"Good. Kim went off to the gym, so I'm taking care of David this morning. We're going to the zoo in a little while."

"Sounds like fun." It had taken Frank a few years to get used to the fact that his daughter had married an artist from South Korea, but Kim Cho seemed like a nice enough fellow. He spent a lot of time working out at the Y, though. "I'm sure Davy will like the zoo."

"Yeah." A sigh of motherly exasperation. *"I know he will…he insists on going at least once a month."*

"Maybe he needs to expand his interests."

"He's four, Dad…give him time. Anyway, when are you flying back tomorrow? Kim and I were thinking about picking you up at the airport, then going out for dinner after that."

"That would be great. I'd like that." Frank rubbed his forehead, trying to remember his flight schedule. "Umm…I think I get in around 4:30 pm your time. Is that too early for you?"

"Not at all. Once we get through traffic, it'll be about time to eat."

"All right, then. If you'll pick me up outside the United baggage claim area, we can…."

A child babbled from somewhere in the background. *"Oh, all right,"* Lisa said, a voice a little distant, then she returned again. *"David wants to talk to you."*

Frank grinned. "Put him on, by all means."

He heard hands fumbling at a distant cell phone, then a small boy's high-pitched voice. *"Gran'pa!"*

"Hello, David," he said, turning away and placing a hand over his left ear so that he could hear better. "How are you?"

An uncertain pause. *"Fine."*

"Good, that's good. I hear you're going to the zoo today."

Another pause. *"Yeah."*

"Oh, that's good. I'm sure you'll enjoy that. Do you know where I am?"

A couple of seconds passed. *"No."*

"I'm in Florida, talking about starships."

For a second or two, he was unsure whether David had heard him. Then his grandson piped up again. *"What's a starship?"*

Frank smiled. "Well, then…I suppose I'll have a lot to tell you when I see you tomorrow."

The EASS Galileo, capable of near-light speed travel, as described in Allen Steele's novel Spindrift. The ring contains an Alcubierre-type warp drive (see chapter by John Cramer), which here is folded and stowed in the ship's midsection until its nuclear main engine accelerates the vessel to cruise velocity. At that time, the drive torus would be inflated to generate a warp bubble around the ship. (Thomas Peters)

PATHWAYS
TOWARD
STARSHIPS

Recently Neal Stephenson began Hieroglyph, a project to envision truly large-scale, new ideas for our future. A major goal is to bring constructive criticism to our ideas of the future themselves, not just pointing out problems. As his collaborator Ed Finn put it, "they seek a more optimistic, realistic approach—fewer zombies and man's folly-style catastrophes, more creative inventions and solutions." Stephenson's story here will later appear in a Hieroglyph volume. Here, "thinking big" is quite literal. Gregory Benford adds a short coda to it as well, advancing the concept into the starship era.

Atmosphaera Incognita
Neal Stephenson

It's called soil," I told him, for the third time.

Carl didn't even like to be told anything *twice*. He drew up short. "To me," he said, "it's all dirt."

"Whatever you call it," I said, "it's got a certain ability to hold things up."

I could tell he was about to interrupt, so I held up a hand to stifle him. Everyone else in the room drew in a sharp breath. But none of them had known Carl since the age of five. "All I'm saying," I said, "is that civil engineers happen to be really, really good at building things on top of *dirt*—" (this was me throwing him a bone) "—and so rather than begin this project—whatever the hell it is—by issuing a *fatwa* against dirt, maybe you should just trust the engineers to find some clever way to support *whatever the hell it is you want to build* on top of *whatever kind of soil* happens to cover *whatever the hell site* you want me to buy."

Carl said, "I don't trust dirt to support a tower twenty kilometers high."

That silenced the room. With any other client, someone might have been bold enough to raise their hand and ask if he'd really meant what he said.

Or, assuming he had, whether he was out of his mind.

No hands went up.

"Okay," I said finally, "we'll look for a site where bedrock is near the surface."

"Preferably *is* the surface," Carl said.

"I'm just saying that might be tricky," I pointed out, "combined with your other requirements. What were those, again?"

He sighed and fixed me with a baleful look. *I know that was just a rhetorical question, Emma. But for you—*

"Direct access to a Great Lake," he said, as if reciting a nursery rhyme to a roomful of Montessori kids. "Extra points if it has a steel mill on it."

"What if the steel mill isn't for sale?" someone was dumb enough to ask.

"It will be," I said, before Carl could.

With me and Carl it was one of those relationships where we went for a quarter of a century without having any contact at all and then picked up right where we'd left off at the age of twelve. We'd gone to the same schools and scuffled together on the same playgrounds and even advanced as far as some exploratory kissing, which, for reasons that will shortly become self-evident, hadn't actually gone very well. Then the coach of the middle school football team had refused to let me participate in that sport, save as manager or cheerleader, and my parents had yanked me out of the place and home-schooled me for a year before sending me to a private academy. This had led to college and grad school and a long dispiriting run of un- and under-employment, since the economy didn't seem to be terribly interested in comparative religion majors. I'd moved to California with a girlfriend during a window when gay marriage was legal, but broken up with her before we could tie the knot—for something about knowing you *could,* really focused one's attentions on what life would be like if you *did*—then met Tess and married her instead. Tess was making decent money as a programmer for a series of tech firms, which left me as one of those stay-at-home spouses with nothing to pass the time except yoga. Eventually, as an alternative to simply going crazy, I had gotten into the real estate business. I had learned that I was good at all parts of it except dealing with silly homeowners-to-be who couldn't make up their minds which house they wanted to buy.

Commercial real estate had turned out to be my ticket. Those buyers knew what they wanted, and I liked such people.

People like Carl.

I'd followed his career like everyone else: the cover stories on the business magazines, the photos of him opening the New York Stock Exchange. I hadn't realized that he was Carl, the kid from the playground, until he'd become a billionaire, lost most of it, and become a billionaire a second time: exhibiting a tolerance for risk that fit in perfectly with his behavioral profile during recess.

One year I'd gone back home for Christmas. My mom, busy in the kitchen, had dispatched me to the grocery store to buy a can of cranberry relish. I'd found myself standing next to Carl in the checkout line. He was holding a tub of sour cream for, I guessed, his mother, and a six-pack of beer for, I guessed, himself. Just me and the eleventh-richest man in America standing there waiting for Old Lady Jones (as we had known her three decades earlier) to finish her interminable coupon-sorting. Carl and I had strolled across the parking lot to the Applebees and spent a while catching up. I told him about my marriage. Carl just nodded as if to say yeah, that would be you. This created an immediate and probably stupid feeling of gratitude and loyalty in my breast that saw me through a lot of the crazy stuff that happened afterwards.

Then some internal timer seemed to go off in his head. Maybe he sensed that the sour cream and the beer were both getting warm, or maybe that's just how guys like Carl are hooked up. He turned into a grownup again. Asked me what I did for a living. Asked me *a lot* of questions about it, then interrupted my answers when they reached the point of diminishing returns. Requested my business card.

A week later I was back in the Bay Area, finding Carl a hangar to store his collection of restored World War I biplanes. After that it was helping one of his companies move to a new facility in Redwood Shores. Then finding an office building for his microfinancing venture.

And it was always easy between us. Even when he was impatient or downright pissed off about something, it was always Emma and Carl, twelve years old again. Even—no, *especially*—when he came to me with a very twelve-year-old look on his face and said "I've got a weird one for you."

"You weren't kidding about the weird part," I told him, after the engineers and bankers and lawyers and a single lonely astrophysicist had all filed out of the room, looking a bit shellshocked.

"I was going to keep it secreter, longer," he admitted, "but people can't make good decisions if I don't tell them the plan."

"Is it a plan?" I asked. "I mean, how much of this have you figured out?"

"I've had civil and mechanical engineers on it for a few months," he admitted, "a small team. What I haven't figured out yet is—" and here, uncharacteristically, he was at a loss for words.

"Why it makes sense?" I prompted him.

"Ah, I knew there was a reason I hired you."

HOW STEEL IS MADE sounds like the title of one of those earnest educational films that Carl and I had respectively slept through and watched in fourth grade. If you're of a certain age, you can see that film in your mind's eye: the grainy black-and-white footage, the block-letter title cards, the triumphant soundtrack trying to blow out the tiny speakers of your classroom's AV cart. Here I'm using it as a kind of placeholder for the first six months of my tenure in Carl's organization. There was no point in even starting to think about building a twenty-kilometer-high steel tower until we had figured out where the steel was going to come from.

Making no pretenses to narrative coherence, here's that six months broken down into six bullet points:

- There's a reason most of the steel mills were around the Great Lakes. These seemed to have been designed by God to support the production of steel on a massive scale. Iron ore from northern Minnesota came together with coal from Appalachia (or, later, from Wyoming) and poured into mills dotted around the shores of those enormous bodies of water. To you and me, "lake" might mean fishing and water skiing, but to industrialists it meant "infinitely wide superhighway for moving heavy things around."

- Unfortunately most of those mills were obsolete.
- The steel industry was, in Carl's unkind phrasing, "the Jurassic Park of the business world." Cutting-edge nerds and ambitious entrepreneurs gravitated toward other industries. It took a long time to pay off the massive capital investment needed to build a new mill, so owners were resistant to change. Innovation tended to be forced on them by early adopters, elsewhere in the world, who had nothing to lose.
- Speaking of which, China was kicking the crap out of us. Most of their mills were new, which meant that they produced better product: more consistent, higher quality, easier to work with. They were getting their ore from Australia and their coal domestically and they weren't encumbered by regulations.
- None of the existing U.S. mills were making the stuff we were going to need.
- As a little side project en route to building his tower, Carl was going to have to reboot the American steel industry.

Our initial idea, which we quite fell in love with, was to plant the tower along the shore of a Great Lake and basically extrude it out the top of a brand-new steel mill. Needless to say, we got a lot of love from Chambers of Commerce in that part of the country until our structural engineers finally achieved mind-meld with some climate scientists, and called us in for a little meeting.

The engineers had been getting more and more nervous about wind. It had been clear from early on that the big challenge, from a structural engineering point of view, wasn't supporting the self-weight of the tower. The amount of steel needed to do that was trivial compared to what was needed to prevent its being knocked flat by the upper-altitude winds. Kavanaugh Hughes, our head structures guy, had an effective demo that came to be known as "I am the wind." He would have you stand up in a normal, relaxed attitude, feet shoulder-width apart, and then he would get to one side of you and start pushing. First he would get down on his hands and knees and push on your ankle as hard as he could. "Low level winds," he explained. No one had trouble resisting a force applied that close to the floor. Then he'd rise up to a kneeling position, place his hands on your hip bone, and push. "Note the transfer of weight," he'd say, and he'd keep urging you to articulate what you were feeling until you got the right answer: Your "downwind" leg and foot were bearing more weight, your "upwind" leg and foot were more lightly loaded. Because your only way to resist the force of Kavanaugh was that differential push-pull between one leg and the other—the "couple," as he called it. "The downwind leg has to be stronger to take that extra force. But since we don't know which direction the wind might blow from, we have to make all of the legs stronger by the same amount. That means more weight, and more steel." Finally, Kavanaugh would stand up, put his hand on your shoulder, and push. It didn't take much force to knock the average person off balance. Short of that, other things were going on: not just the intensifying "couple" between the upwind and downwind feet, but some internal strains in the torso. "My physical trainer is always nagging me to activate my core," Kavanaugh said, "and what that means to me is a system of internal cross-bracing that makes it possible for me to transfer stresses from one part of my body to another—and eventually down into one of my feet." Then he would push you until you were forced to hop away from him. "The problems are two," he explained. "First, all of that cross bracing requires more steel—and more steel catches more wind, and increases the force!"

"Shit, it's an exponential," Carl said.

"Yes it is," said Kavanaugh. "Second, the most powerful winds aren't down at ankle height, where it's easy to resist them. They're up near the top—the worst possible place."

"The jet stream," Carl said.

"You got it. Now, I'm not saying we can't build a tower capable of resisting the jet stream. We can do anything we want. But common sense tells us to avoid places where the jet stream is powerful and frequent." He nodded to one of his new climate scientist buddies, who flashed up a map of the world showing where the jet stream wandered most frequently. And it was immediately obvious that the upper Midwest and the industrial Northeast were the worst places in the whole world to construct our tower.

Near the equator and near the poles tended to be better. For reasons I only found out later, relating to rocket launch, Carl nixed the poles. So we were left staring at a band of latitude that, roughly speaking, corresponded to the tropics.

"I know what some of you are thinking," Carl said, after studying it for a minute, "but no, I'm not going to build this tower in some Third World hellhole only to have it end up being the property of the first junta that comes along."

A few of the people in the room had actually been born and raised in what Carl considered to be Third World hellholes, and so, not for the first time, I felt my butt sliding forward on the chair as I tried to make my face sink below the horizon of the table.

Carl was oblivious. "Political stability and property rights are non-negotiable site selection criteria."

"The northernmost capes of Australia look ideal, then," someone pointed out. And for a minute we were all ready to purchase stylish hats and join the Qantas frequent-flier program, until someone in the climate science group mentioned that those areas tended to get hit by cyclones.

"Okay," Carl said, "we need a place with boring weather at all altitudes, and political stability."

The answer was the southwestern United States, with California's Central Valley being ground zero. Of course, there was quibbling. Left-leaning people denied that the U.S. was a politically stable entity. Right-leaners took issue with the premise that Americans really had property rights. And Californians seemed offended by the assertion that their climate was boring. People in every part of the world, it seemed, like to complain about their local weather. We shouted them down by pointing out their lack of tornadoes, hurricanes, and blizzards, and then began to search outward from the Central Valley. Could we find a location with better seismic stability? Better access to heavy freight transport? A nice high-altitude plateau, perhaps, so that we could get an extra height boost?

In due time we found promising locations in central California. Southern Nevada. Central Arizona. Southwest Texas. Every time we found a place that would work, my acquisitive instincts kicked in, and I started pestering Carl with text messages and emails, wanting to go in for the kill. But it seemed that all he wanted was to string these people along for as long as possible. Hoping to play them off against each other and drive the price down, I reckoned.

Then one day the following text message showed up on my phone

BUY ALL 4

To which I replied

LOL REALLY?

And he answered

AS LONG AS YOU THINK THEY CAN BE RESOLD WITHOUT SERIOUS LOSS.

And, moments later,

DON'T SPEND IT ALL IN ONE PLACE

Referring, I guess, to the fact that I was about to collect four commissions on four separate purchases—and perhaps as many as three more when he decided to resell the ones he didn't want to use ("losers" in Carl-speak).

At this point I was beginning to suspect that the tower was just a ruse and that he was actually making some kind of incredibly complicated play in desert real estate.

The reality became clearer to me when Carl actually bought all of those properties and then began to visit those towns and show the locals the dog-and-pony show his engineers had been preparing on the subject of why it was such a great thing to have a twenty-kilometer-high tower in one's community. Lots of PowerPoint slides explaining, in the most soothing possible way, why it was impossible for the thing to fall over and crush the town. Yes, even if it got hit by a 747.

I ended up going on many of these dog-and-ponies. Which made very little sense given my official job description. I had already done the part I was qualified to do. But my job title kept morphing as the project developed. For Carl was no respecter of titles and credentials. Whomever he trusted, was in his field of vision, and hadn't said anything colossally stupid recently, tended to end up being assigned responsibilities, whether or not they had the background. So I ended up becoming one of the advocates for this thing, completely trashing my regular business (it was okay, we worked it out in the aftermath) and had to buy a pocketbook to contain all of my loyalty program cards for Hertz, United, Marriott, et al. Then a purse to contain the pocketbook. Then skirts to go with the purse. Which I mention because I'd always been a wallet-in-the-pocket-of-my-jeans kind of girl. Tess watched my sartorial transformations with a mixture of amusement and alarm, accusing me of traveling to the Intermountain West in drag. It actually became a little tense between us until one day the light bulb came on and I explained something important: "They don't give a shit that I'm gay."

"Really?"

"Really. They actually think it's kind of cool. Most of them."

"I just thought—"

"No. The clothes are about being taken seriously."

Tess was mollified but not fully convinced.

"They're afraid it's going to fall over on them. The explanation of why this is never going to happen needs to come from someone who is not wearing any black leather."

I could do the PowerPoint in my sleep. As a matter of fact I often *did* do it in my sleep, tossing and turning in my hotel bed. We'd hired a graphics firm to make a nice animated film showing the transformation of the site. Leveling the ground. Planting trees to make it purty. A new railway line, lollipop-shaped, terminated with a perfectly circular loop nine miles in diameter. Extending inward from that, the spoke lines. Half a dozen of them, one for each of the Primaries—the primary supports that would hold the tower up. A little homily here on the subject of "why six?" In theory you could build a stable tower with only three. But if something happened to one of them—I didn't have to mention a jumbo jet strike, since everyone was clearly picturing it their heads—the other two wouldn't be able to hold it up. You might be able to make it survivable with four, but it would take some structural legerdemain. Five was a safer bet. Six gave you even more of a safety margin as well as some benefits resulting from symmetry. The greater the number of Primaries, the closer each was to its neighbors, and that simplified, somewhat, the problem of webbing them together structurally. So six it was.

The next step was to construct the foundation strips: six reinforced-concrete tracks, each straddling one of the spoke lines. This part of the presentation went pretty fast; there wasn't much interesting you could say about pavement.

The concept of a rolling factory was a little harder to explain. Factories they got, of course, but no one had ever seen one crossed with a main battle tank the size of a shopping mall. This was where the computer-graphics renderings really came in handy, showing how the thing was built from the ground up on huge steel treads, how it accepted its inputs (steel! steel! And more steel!) from the railway line that ran right through the middle of it, assembled them into trusses, connected them to the bottom of the Primary and then pushed them straight up through a hole in the roof. It was all reasonably easy to follow, once you got the gist of it. The one part that was a little hard to convey was that each of these six rolling factories—one for each Primary—was also a structural foundation supporting its share of the tower's whole weight. The factory didn't just have to roll (slowly!) along the runway. It didn't just have to assemble trusses and feed them out its ceiling. It also had to contain hydraulic rams for pushing the whole tower up, transmitting its share of the weight down through its structure into the big steel tank treads and from there into the foundation strip and finally into Carl's precious bedrock.

Having gotten those preliminaries out of the way, I was able to proceed to the big all-singing all-dancing animation (complete with moving symphonic music) showing the six Struders (as we had come to call the truss-extruding factories) poised at their starting positions at the innermost extremes of the spokes, nearly touching each other. Six trains came chugging up the lollipop handle and went their separate ways around the rim line. A seventh went straight into the center, headed for a central, non-moving Struder designed to extrude the tower's core. Once each of the seven had been supplied by its own train, steel trusses—kinda like radio towers—began to emerge from the holes in their roofs, growing upward like stalks from magic beans. There was a pause as cranes went to work framing in a platform that joined the six Primaries with the core. This was my opportunity to wax poetic as I marveled over the fact that this platform would one day be twenty thousand meters above the ground, for all practical purposes in outer space, where the sky was black and the curvature of the Earth visible. Honeymooners would luxuriate in pressurized suites, astronomers gaze at the universe through glass eyes undimmed by atmospheric pollution. Rockets would launch from it and extreme skydivers would jump off.

And yet 99 percent of the workers who built it would never have to leave the ground.

The reaction to *that* was mixed. Oh, everyone understood why it made sense—you couldn't have a large workforce commuting straight up into the sky every day, breathing from oxygen tanks and swaddled up in space suits. But it did take some of the romance out of it. At some level I think that every blue-collar worker who ever attended one of these presentations was telling himself that he would be one of the tiny minority of employees who would actually get to go up high on the tower, inspecting and troubleshooting.

The rest of the movie was, I guess, predictable enough. Everyone who'd watched it to this point knew how it was going to end. The trains kept rolling in, the Struders kept extruding, pausing from time to time so that the freshly extruded Primaries and Core could be webbed together with stiffening trusses—Kavanaugh's "core muscles." We speeded up the movie, of course, once people got the gist of it. Push, pause, web. Push, pause, web. With each push the factories rolled outward imperceptibly on their tracks, moving about one meter for every fifteen meters of stuff they extruded, keeping the tower's height-to-width ratio pretty much fixed. Though, toward the end, they started moving a bit faster, making the base splay out, giving it a bit of Eiffel Tower feel. Even the people who walked into the room claiming to be worried that it would fall over were somehow convinced by this; it had a wide enough stance that it just *looked* stable. Up and up went the steel as I recited lore that I had picked up from Wikipedia and from meteorology textbooks and long conversations with Ph.D. metallurgists about the different layers of the atmosphere and the varying challenges that the tower would have to contend with: Down below, rain and rust. Up higher, icing. Higher yet, wind loading, the possibility of contact with a wandering jet stream (or a wayward jet). Profound cold that would render the metal brittle if we had been dumb enough to use the wrong alloys. Thermal expansion and contraction as the unfiltered sun shone on its higher reaches in the day and then disappeared at night. Each of these challenges an opportunity to generate energy with photovoltaics (up high) or convection ducts (down low) or wind turbines (in the middle).

So much for the pitch.

And so much (almost) for my marriage, which barely survived all of the absences, all of those nights in chain hotels far from home, all of those alarming changes in wardrobe and hairstyle.

Looking back on it later, if I were to write a book about building the tower, I'd here interpolate a three-year-long chapter titled "Politics and Lawyers." Halfway through it, I got a text message from Carl:

CALI = LOSER

Meaning, "sell the property in California."

My response, of course, was:

OK

but my reaction was a little more complicated. I had known all along, of course, that we'd end up selling at least three of the four properties. But I'd spent time in each of those places and had made friends with the locals and I didn't look forward to breaking them the news that their bid—for by this point, each of these things had mushroomed into a complicated bid package binding together state and local governments, unions, banks, and other worthies—had been rejected.

The answer, simply, was that the tower was going to be visible from the exurbs out to the east of Oakland and Berkeley: a spendy part of the world where lots of rich people were accustomed to looking out the windows of their nice houses and seeing the landscape. And *only* the landscape. They didn't want their views marred by a twenty-kilometer-high "monstrosity" whose "stark, ugly, industrial profile" was going to be "cluttered" with "ungainly industrial encrustations" and "gaudy" with a "Las Vegas-style light show" that would "sully the purity of the skies night and day."

In due course, all of them had to face the very nearly out-of-control wrath of the Central Valley Rotarians, but that's another story.

Southwestern Texas got killed six months later by environmentalists being used as sock puppets by an unholy alliance of—well, never mind. Demonstrating in court that their claims were bogus would have been expensive. Bankrolled as well as they were, they could have stretched the process out forever by filing legal challenges. Arizona was the next domino to fall. It had always been a long shot, but we'd held onto it mostly to give us greater bargaining power over the Nevada site, where local politicians had smelled money and begun to let us know, in various ways, that we were going to have to play ball.

So that was how I came to earn four commissions by purchasing four "losers" for Carl, and another four by selling each and every one of them. We made money on two, lost money on another two, and pretty much broke even on the whole deal.

That was how the project ended up where it did: between an Indian reservation and a decommissioned military bombing range, out in the southwestern desert, in an area that had already demonstrated its openness to radical transformations of the landscape, first by having the crap bombed out of it, then by building a casino complex, and most recently by its wholehearted acceptance of wind farms. At about the same time we closed a deal on an aging complex near the Illinois/Indiana border and got to work building a new kind of steel mill. The Great Lakes were still the best place in the world to make steel. This was a far cry from our original scheme to have the mill on site. But in the intervening years it had become clear that *lots* of people wanted the kind of steel that a new mill could produce. Hard as it was to believe, the Tower had become a minor customer.

Transportation wasn't that big of a deal. Smaller pieces could be shipped southwest on freight trains. Big stuff was barged down the river system to the Gulf, dragged through the Panama Canal, landed at the head of the Gulf of California and then transported overland using land trains.

The site was twenty minutes' drive from a college town, which gave the employees a place to educate their kids and entertain themselves, and gave us a ready supply of fresh young engineering talent.

As well as a cowgirl-themed gay bar. Which became pretty important when Carl told me—as if this should have been obvious from the beginning—that I was moving there to run the whole thing.

"I'm not qualified to construct a twenty-kilometer-high tower," I pointed out.

"Since it has never been done before," he said, a little annoyed at having to make such an obvious point, "no one is."

"The engineering is totally beyond my—"

"We have engineers."

"All the legal ins and outs—"

"Lawyers too."

I was dreading the conversation with Tess but she'd seen it coming long ago. Hell, maybe Carl had even prepped her.

"Let's go," was all she said.

Bless her beautiful heart, I thought. But what I said was, "Huh?"

"I've been looking into it. Pre-cleared it with my boss. I'll telecommute."

Neither of us really believed that, of course. Her job lasted for all of twelve weeks after we moved.

She cashed in some stock options and bought the cowgirl bar for less than what she had spent on her last car. With what was left over she bought a pickup truck from a rancher who had sold his land to Carl.

Five years later, the bar had morphed into *the* hangout for all of the engineers, gay and otherwise, who had moved to the area.

Another five years after that, Tess was operating the First Bar in Space.

Oh, people argued about it. Space tourism had been gathering steam. Queasy/giddy tourists drifted around the tiny envelopes of their suborbital capsules and sucked premixed cocktails from nippled sacs and this got billed as the first bar in space. It became like the debate on who had built the first computer: well, depends on what you consider a computer.

What do you consider a bar? For Tess it had to have a jukebox, a dartboard, and gravity. You can't get a head on your Guinness in zero-G.

At first it was just a shipping container with portholes plasma-torched by Tess's eternally grateful clientele of elite ultra-high-altitude steelworkers. This was back in the early days when the Square Kilometer—as we called the (actually round) platform at the top of the tower—was still only a couple of thousand meters off the ground. Once we broke through 4,000 meters we had to start running oxygen concentrators full-time, even for the altitude-adjusted regulars. On the day the Top Click (as we called the Square Kilometer by that point) pushed up past the altitude of Mt. Everest, we moved the whole operation into a pressurized Quonset hut and filled it with sea level atmosphere. Beyond 10,000 meters we just started calling it the First Bar in Space. There was carping on the Internet—always is—but the journalists and businesspeople who rode the helirail up to the top and sat at the bar taking in the black sky and the curvature of the Earth—well, none of them doubted.

I'm leaving a lot out: five years of starting the project, ten years of riding it up. Tess and I had two kids, raised them to teenagerhood, and went through a spell of personal-life hell when she had an affair with a Mohawk ironworker who drifted in from upstate New York and stormed out a year later when Tess thought better of it. I ran the show for a few years until Carl suddenly announced during a meeting that (a) I had done a fantastic job but (b) I was being replaced effective immediately and (c) he was commencing radiation therapy for prostate cancer forty-five minutes from now. He gave me the world's most unusual commercial real estate gig: selling off the Top Click. Obvious conflict-of-interest issues were raised by my wife's bar; Carl resolved them by giving us a lease in perpetuity, hand-scrawled on the back of a boarding pass.

Shipping materials to the top of the Tower only became more expensive as it went up, so we had framed in the big structures while the Top Click had been on the ground, then stockpiled steel and other goods that could be used to finish it later. All of it got a free—but very slow—ride to 20,000 meters. Additional structural work proceeded at a leisurely pace during the years that the Top Click was rising up through the Dead Zone—the altitudes from about seven to twelve kilometers. Below seven, humans could breathe (though most needed oxygen bottles), move around without pressure suits, and enjoy a decent enough view. It was cold as hell, but you could wear warm clothes; it was like being at a base camp on a manmade Himalaya. Much above seven, there wasn't enough atmosphere to breathe, but there was enough to supply foul weather in abundance. The view down was often blocked by clouds, the view up not yet enlivened by starlight. Past Everest height—nine kilometers or so—we got up into screaming sub-sub-sub-zero winds that, at their worse, were close to jet stream intensity. There wasn't much point in trying to keep glass in window frames. Even heavy-looking stuff like shipping containers had to be welded down or it would blow off, fall a few miles, and break something on the ground. There were ways to deal with it; but it basically led to Top Click operations being put in suspended animation until the Struders on the ground pushed it up out the top of the Dead Zone. During that time, we had other things to think about anyway: sheathing the horizontal braces in giant wings, and getting them to work right. About which more later.

Above the Dead Zone, things got nice in a hurry. The buildings, which had been empty shells for several years, got shelled in by space-suited workers and then pressurized with proper atmosphere so that shirt-sleeved workers could get in, lay carpet, and put on doorknobs.

It was during that phase, when the Top Click was about 17 kilometers above the ground, that we threw a party in the First Bar in Space for the purpose of scattering Carl's ashes. The basic idea being that they would fall for a couple of kilometers, get snatched by the wind, and disperse.

I was the last to arrive, carrying the guest of honor—a Ziploc bag full of Carl—in a messenger bag that looked way too hip for a middle-aged mom in the real estate business. My flight from SFO had been delayed by a monster storm front: the kind of thing that had been sweeping west-to-east across the Great Plains since time immemorial but was rarely seen in our part of the Southwest. But like the proverbial frog in a pan of water, we'd all been getting accustomed to shifts in the climate, and weather events unheard of during the previous century. The airline had found a way to route me around the storm, but as I drove in from the regional airport in Tess's pickup truck, I could see clearly enough that it was determined to catch up with me: an arc of stratocumulus anvil clouds stretching, it seemed, from Baja to Utah, blotting out the late afternoon sun and flashing here and there with buried lightning.

It was the only thing that could make the Tower look small.

One of the engineers, way back at the beginning, had described the Tower as "a gas of metal" which was pretty poetic for an engineer but did convey its gist: the minimum of steel needed to do the job, distributed over the largest volume it could feasibly occupy, but in a specific way meant to solve a host of structural problems. At night, when the lights came on, it looked far more substantial than during the day, when it was a glinting cloud that rose up out of the desert like an inverted tornado. If you let your gaze be drawn up high enough—astonishingly high, far above most clouds—you could see its ladder of wings cruising in the jet stream, like a set of venetian blinds hanging inexplicably in space. Above that, frequently obscured by haze and clouds, was the flare at the very top where it broadened to support the Top Click.

Even though I'd been living with—and on—this thing for going on twenty years, I was still impressed with its scale when I approached it as I was doing now. But having Carl's earthly remains on the passenger seat somehow drew my attention to his ground-level legacy, which now spread out from the base of the Tower to a radius of ten or twenty miles. Fanning away to the east-southeast was an expanse of open rangeland, inhabited only by bison, groundhogs, and a few back-to-the land types: *vaqueros* and the Indians who had always lived in those parts. Part of it was bombing range. The rest I had acquired, one ranch at a time, using shell companies so that the landowners wouldn't gang up on me. Because one of the questions people had asked was "what if it gets rusty and falls over?" and Carl's answer had been "then we'll use demolition charges to fell it like a tree down the middle of the Swath" as this territory had come to be known. Which to me had seemed like bending over backwards for the NIMBY types, until I'd understood that Carl had always intended to use the Tower as a catapult for launching space vehicles, whose trajectories, for the first twenty miles, would pass right down the middle of the Swath, which he therefore needed to keep clear anyway so that failed rockets would have a place to crash.

The new highway from the airport ran along the Swath's northern border for the last few miles and so as I drove in I enjoyed, to the left, a vista of grazing bison and the occasional horse-riding Indian, and to the right, a generic exurban sprawl of strip malls and big-box stores that had sprung up to fill the needs of all the people who had moved here. Behind that line of development I could hear the long blasts of a locomotive whistle: another huge train rolling in from Chicago carrying prefab steel trusses to feed into the Struders.

The ring line encircling the base was discernible as a crescent of five- to ten-story commercial buildings adorned with the logos of the tech firms and contractors that had set up shop here. Mixed in were hotels and apartment buildings housing temporary residents as well as the younger, more urban crowd who wanted to be close to what had developed into a passable nightlife and entertainment district. From their windows they could look out over residential developments spreading away along the state highway connecting to the college town. All of this had a temporary feel, since it was understood that when the Tower topped out and the Struders ground to a halt, the bottom kilometer would develop into a vertical city, a much cooler place to live—climatically as well as culturally.

For now, though, the Tower's lower reaches were a web of bare trusses with steelworkers, and their robots, crawling about. Welding arcs hung in it like bottled fireflies, and cranes pivoted and picked like hollow mantises. In most building sites, a crane had to be capable of hoisting itself higher as the building grew beneath it, but here the cranes had to keep working their way *down* the structure as it pushed up from the Struders. It wasn't rocket science but it did make for some crowd-pleasing erector-set gymnastics, watched by vacationing families and know-it-all retirees from covered viewing platforms spotted around the ring road. We made money from those by putting in vista-viewing bars and restaurants.

Rocket science was the domain of the innermost core, a ten-meter-diameter chimney running all the way up the Tower's central axis. During the first couple of years I had pestered Carl with questions about what specifically was going to go into that empty space—that perfectly round hole at the center of every floor plan.

"You're assuming I have a secret plan," he had said.

"You usually do."

"My secret plan is that I have no secret plan."

"Wow!"

"I am going to sell—*you* are going to sell—that right-of-way to the highest bidder. On Ebay if necessary."

"And what is the highest bidder going to put in it?"

"I have no idea. Since it is twenty kilometers long and pointed straight up, I'm going to make a wild guess that it will be something connected with hurling shit into space."

"But you really don't know what exactly?"

He had thrown up his hands. "Maybe a giant peashooter, maybe a railgun, maybe something that hasn't been invented yet."

"Then why did you pick ten meters for its diameter?"

"It was easier to remember than eleven point three nine zero two four…."

"Okay, okay!"

The secret plan had worked. The people who had won the bidding war—a coalition of commercial space companies and

Tower (Keith Hjelmstad)

defense contractors—had given the Tower a useful shot of cash and credibility at a time when both had been a little tight.

Now cutting across the ring road, working against an outflow of traffic—workers coming off the day shift, headed home for the weekend—I passed a security checkpoint and rolled across a flyover that had been thrown across the circular railway line. This ramped down to ground level and became a road paralleling the Northwest Spoke.

Instead of paving the Spokes all at once—which would have been a huge up-front expense—we had been building them just in time, a few meters and a few weeks in advance of where and when they would be needed. This made it possible to keep the paving crews employed on a steady full-time basis for years. So, directly in front of the astonishing bulk of the Northwest Struder was this fringe of preparatory activity: orange flags marking the locations of soil samples, graded and tamped earth, a grey haze of webbed rebar, plywood forms, freshly poured concrete. The giant linked treads and looming hulk of the Struder rising just behind.

On the top of the Struder, evening-shift workers in safety harnesses were ascending from dressing rooms below to busy themselves on the most recently extruded truss section, inspecting, x-raying, installing sensors and lights and wires. A lot of that work had been done hundreds of miles away when the trusses were being prefabbed, but there didn't seem to be any computer-driven process that couldn't be improved upon by humans crawling around on the actual structure and writing on it with grease pencils. As the Tower had risen up from the desert, data pouring in from its millions of strain gages, thermocouples, cameras, and other sensors had given up oceans of information about how the models had, and hadn't, gotten the predictions right, creating a demand for "tweaking crews" to make adjustments to newly extruded work before it got pushed so high into the sky that it became hard to reach.

There was more than one way to the top. Climbing hand-over-hand had become a new extreme sport. Helipads were available at various altitudes, and work was underway to build a new regional airport at what would eventually be the 2500 meter level—airplanes would land and take off by flying into and out of apertures in the Tower's side, saving them huge amounts of fuel as they avoided the usual ascent and descent.

Surrounding that ten-meter right-of-way in the middle were vertical elevator shafts. But the primary transport scheme was the helirail: a cross between a train, an elevator, and an amusement park ride that corkscrewed up the periphery of the structure, ascending at a steady twenty-degree angle. It was really just a simple ramp, about sixty kilometers long, that had been wrapped around the tower as it was built. Cut a triangle out of paper and roll it around a pencil if you want the general idea. Special trains ran on it with tilted floors so that you always felt you were on the level. Train stations were built around it every two thousand meters' altitude. One of those—the second to last—had just been roughed in, and was dangling there a few meters off the ground. I was able to clamber up into it via some scaffolding and catch the next up-bound train.

Actually "train" was too grand a word for this conveyance, which was just a single car with none of the luxury appurtenances that would be built into these things later when they were carrying droves of tourists and business moguls. All of the regulars knew to empty their bladders first and to bring warm clothes even if it was a warm day at ground level. I shared my car with

Joe, an aeronautical engineer who was headed up to fourteen-thousand meters to inspect the servomechanisms on a wing; Nicky, an astronomer going to the Top Click to work on the mirror stabilization system for the big telescope a-building there; and Frog, a video producer readying a shoot about the BASE jumping industry, which was already serving a thousand clients a year. After peppering us with recorded warnings, the car began to hum up the helirail, banking slightly on its gimbals as it picked up speed. A recorded message told us where to look for motion sickness pills and barf bags, then moved on to the more serious matter of what to do if we lost pressurization.

This was a pretty white-collar passenger manifest, but this was Friday afternoon and most of the workers were headed down for the weekend—as we could see when we looked into the windows of the crowded down-going coaches spiraling the opposite way. Their descending mass provided most of the energy to raise us, through the electromagnetic equivalent of a counterweight system, like on old San Francisco cable cars.

For the hundredth time since leaving the funeral home, I reached out and patted the bundle of ashes in my bag. Carl had a lot to be proud of and had not been shy about taking credit when earned. But I knew, just from watching his reactions, that he took special pride in having created countless blue-collar jobs. His family back home had been steelworkers, electricians, farmers. Carl had always been more comfortable with them than with the crowd at sun Valley or TED, and when he had passed, the outpouring of grief from those people had been raw and unaffected.

As we spiraled up, we revolved through all points of the compass every four minutes or so. Views down the brown expanse of the Swath alternated with the panoramic storm front now blotting out the evening sun. The top fringes of the anvils were still afire with bent sunlight but their bases were hidden in indistinct blue-grey murk, cracked open here and there with ice-white lightning.

The car began to hum and keen as it pushed its way up into an eighty-mile-an-hour river of air. Bars of shadow began to flash down over us as we passed upwards through a structure that resembled a six-sided ladder, with each rung a giant wing. For this part of the tower was not so much a structure in the conventional sense as a stationary glider. Or perhaps "kite" was the correct word for it.

The idea dated back to the very first months of the design process, when the engineers would work late into the evening tweaking their models, and wake up in the morning to find long emails from Carl, time-stamped at three in the morning. The weight of the Tower—what Carl called the Steel Bill—kept growing. Sometimes it would creep up stealthily, others made a sickening upward lurch. The problem was wind. The only way to win that fight—or so the engineers thought at first—was to make the Tower beefier, so that the downwind legs could push back. Beefy steel catches more wind, increasing the force that's causing the problem in the first placc. Not only that but it demands more steel below to support its weight. This feedback loop produced exponential jumps in the Steel Bill whenever anything got adjusted.

It wasn't long before someone pointed out that, from an aerodynamics standpoint, the Tower was a horror show. Basically every strut and every cable was a cylinder—one of the draggiest shapes you can have. If we snapped an airfoil-shaped fairing around each of those cylinders, however, leaving it free to pivot into the wind, the drag went down by an order of magnitude and the Steel Bill dropped like—well, like a wrench dropped from the roof of a Top Click casino. And those fairings would have other benefits too; filled with lightweight insulation, they would reduce the thermal ups and downs caused by sunlight and direct exposure to space. The steel would live at a nice in-between temperature, not expanding and contracting so much, and brittleness would be less of a problem.

Everyone was feeling pretty satisfied with that solution when Carl raised an idea that, I suspect, some of the engineers had been hiding in their subconscious and been afraid to voice: why not fly the tower? If we were going to all the trouble to airfoilize everything, why not use the kind of airfoil that not only minimizes drag, but also produces lift?

Wings, in other words. The Tower's lateral braces—the horizontal struts that joined its verticals together at regular intervals—would be enclosed in burnished-aluminum wings, actuated by motors that could change their angle of attack, trimming the airfoils to generate greater or lesser amounts of lift. When the jet stream played on the Tower's upper reaches like a fire hose slamming into a kid's Tinkertoy contraption—when, in other words, the maximum possible crush was being imposed on the downwind legs—the wings on that side would be trimmed so as to lift the whole thing upwards and relieve the strain. Performing a kind of aerodynamic jujitsu, redirecting the very energy that would destroy the Tower to actively hold it up. The Tower would become half building, half kite.

People's understandable skepticism about that scheme had accounted for the need to maintain a huge empty swath downwind of it. Many took a dim view of a building that wouldn't stand up without continuous control system feedback.

When we had boarded the helirail I'd exchanged a bit of small talk with Joe, the engineer sitting across from me. Then he had unrolled a big display, apologizing for hogging so much table space, and spent most of the journey poring over a big three-dimensional technical drawing—the servomechanism he was going to take a look at. My eyes wandered to it, and I noticed he was studying me. When I caught him looking, he glanced away sheepishly. "Penny for your thoughts," he mumbled.

"Oh, I spent years talking to concerned citizens in school gyms and senators in congressional hearings, selling them on this idea."

"Which idea?"

"Exactly," I said, "you've grown so used to it you don't even see it. I'm talking about the idea of flying the Tower."

He shrugged. "It was going to require active damping anyway, to control oscillations," he said.

"'Otherwise, every slot machine on the Top Click will have to come equipped with a barf bag dispenser.' Yeah, I used to make a living telling people that. 'And from there it's a small step to using the same capability to help support the tower on those rare occasions when the jet stream is hitting it.'"

Joe was nodding. "There's no going back," he said, "it snuck up on us."

"What did?"

He was stumped for an answer, and smiled helplessly for a moment. Then threw up his hands. "All things cyber. Anything with code in it. Anything connected to the Internet. This stuff creeped into our lives and we got dependent on it. Take it away and the economy crashes—just like the Tower. You gotta embrace it."

"Exactly," I said. "my most vociferous opponent was a senator who was being kept alive by a pacemaker with a hundred thousand lines of embedded code."

Joe nodded. "When I was younger I was frustrated that we weren't building big ambitious stuff anymore. Just writing dumb little apps. When Carl came along with the Tower idea, and I understood it was going to have to fly—that it couldn't even stand up without embedded networks—the light went on. We had to stop building things for a generation, just to absorb—to get saturated with—the mentality that everything's networked, smart, active. Which enables us to build things that would have been impossible before, like you couldn't build skyscrapers before steel."

I nodded at the drawing in front of him, which had been looping through a little animation as we talked. "What's new in your world?"

"Oh, doing some performance tweaks. Under certain conditions we get a rumble in the Tower at about one tenth of a Hertz—you might have felt it. The servos can't quite respond fast enough to defeat it. We're developing a workaround. More for comfort than for safety. Might force us to replace some of the control units—it's not something you can do by sitting on the ground typing." He nodded toward the luggage rack by the door where he had deposited a bright yellow plastic case, obviously heavy.

"That's okay," I said, "sitting on the ground typing wasn't Carl's style."

As if on cue the car dinged to warn us of impending deceleration. Joe began to collect his things. A minute later the car stopped in the middle of a sort of pod caught in the fretwork of the Tower like a spider's egg case in a web. Lights came on, for it was deep dusk at this point. A tubular gangway osculated with the car's hatch, its pneumatic lips inflating to make an airtight seal. Air whooshed as a mild pressure difference was equalized, and my ears popped. The door dilated. Joe nodded goodbye and lugged his bag case out into the station, which was a windowless bare metal tub. A minute later we were on our way again.

"I just wanted to introduce myself. Nicky Chu." This was the astronomer en route to the Top Click. "Sorry, but I didn't realize who you were until I heard your story."

"Have you spent much time up there, Nicky?"

"Just once, for orientation and safety briefings."

"Well, you're always welcome in the bar. We're having a little private observance tonight, but even so, feel free to stop in."

"I heard," Nicky said, and, perhaps in spite of herself, glanced toward the messenger bag. "I only wish I could have shared one of these rides with the man before…."

The pause was awkward. I said what Carl would have said: "Before he was incinerated in a giant kiln? Indeed."

A senator had once described the Internet at a series of tubes, which didn't describe the Internet very well but was a pretty apt characterization of the Top Click. "Shirtsleeves environment" had been the magic buzzword. I knew as much because Carl had once banned the phrase from PowerPoint slides—shortly before he had banned PowerPoint altogether, and then attempted to ban all meetings. "The Cape of Good Hope is not a shirtsleeves environment. Neither was the American West. The moon. The people who go to such places have an intrepid spirit that we ought to respect. I hate patronizing them by reassuring them it's all going to be in a shirtsleeves environment!"

This sort of rant had terminated some awkward conversations with casino executives and hoteliers. I had donated a small but significant chunk of my life to getting him to admit that my job would be easier and the Top Click would be more valuable if it had a breathable atmosphere that wouldn't cause simultaneous frostbite and sunburn. Building One Big Dome on the top was too inflexible and so we had ended up with a mess of cylindrical and hemispherical structures (because those shapes withstood pressurization) joined by tubular sky bridges. The Top Click helirail terminal was a hemispherical dome, already awe-inspiring in a Roman Pantheon way even though it was just a shell. Radiating from it were tubes leading to unfinished casinos, hotels, office buildings, and the institute that Carl and some of his billionaire friends had endowed. The observatory was there, and so that was where Nicky went after saying goodbye and exchanging contact data with me. I shouldered the bag with Carl in it and hiked down a tube to a lobby where I changed to another tube that took me to the First Bar in Space. Frog, the video producer, walked with me; having slept most of the way through the ride, he was in the mood for a drink. I helped him tow his luggage: a hard-shell case full of video gear and a Day-Glo pink backpack containing the parafoil he intended to use for the return trip.

It was a pretty small party. Carl didn't have a lot of friends. Alexandra, his daughter from a long-dead marriage, had flown in from London with her boyfriend, Roger, who was some sort of whiz-bang financial geek from a posh family. Tess was there to greet me with a glass of wine and a kiss. Our kids were off at college and at camp. Carl's younger brother Dave, a college volleyball coach, had come in from Ohio. He was already a little tipsy. Maxine, the CEO of Carl's charitable foundation, and her husband Tom, a filmmaker. We took over one corner of the bar, which was pretty quiet anyway. Marla the bartender and Hiram, one of the regulars, were watching a Canadian hockey game on the big screen. Hiram, a teetotalling Mohawk ironworker, was knocking back an organic smoothie. Frog grabbed a stool at the other end of the bar, ordered a Guinness, whipped out his phone, and launched into a series of "You'll never guess where I'm calling from!" calls.

It was lovely to be home in my bar with my wife. I was just a few sips into that glass of wine, starting to wish that all of these nice guests would go away and leave us alone, when heads began to turn and I noticed that we had been joined by a woman in a space suit.

Not totally. Nicky Chu had had the good manners to remove the helmet and tuck it under her arm. She said, "Sorry but I think we all need to get under cover. Or *over* cover is more like it."

"Over cover?" I asked.

Roger broke the long silence that followed. I don't know, maybe it was that British penchant for word play. "What's coming up from beneath?" he asked. "And how's it going to get through the floor?"

"High-energy gamma ray bursts," Nicky said, "and some antimatter."

"Antimatter?" several people said at once.

"I'll explain while you're donning," she said, and started backing toward the exit. "I'm afraid you're going to have to put down your drinks."

Donning, as most of us knew, meant putting on space suits. It was to living on the Top Click what the life vest drill was to an ocean cruise. Thanks to the space tourism industry, it had become pretty idiot proof. Even so, it did take a few minutes. They were stored in a vestibule, which for very sound reasons was a sealable windowless capsule. Nicky insisted we drag them out into an adjacent sky lobby—a future restaurant—with big west-facing windows. She wanted us to get a load of the storm.

And this bore no resemblance to watching a storm approach on the ground. We've all done that. From a distance you can look up into the structure of the high clouds, but as it gets closer all is swallowed in murk. Lightning bolts, hail, torrents of rain, and wind gusts jump out at you from nowhere.

Here, we were miles above the uppermost peaks of the anvil clouds, enjoying an unobstructed view of outer space. The Milky way shot up like an angled fountain above the storm front, which from this height looked like a layer of ground-hugging dry ice fog in a disco. Sometimes it would glow briefly.

Nicky's warning had put me in mind of nuclear war and so I had to wait for my logical mind to catch up and tell me that those flashes were nothing more than lightning bolts, seen from above.

Nicky had turned toward me—but she wasn't looking at me. She was staring unfocused. "They're easiest to see in your peripheral vision," she remarked. "They'll be very high up—in space."

"What are you talking about?" Dave demanded. He wasn't handling this especially well. Fortunately Hiram interceded. "Sprites," he explained. "We see 'em all the time."

From anyone else this would have provoked a sarcastic rejoinder from Dave. Coming from a crag-faced, 250-pound Mohawk, it took on more gravity.

"Oh!" Tess exclaimed.

I heard the smile in Nicky's voice. "Big one there!"

"Where?" people were asking.

"It's already faded," Nicky said. But then I saw a disk of red light high up, which expanded while darkening in its center, becoming a scarlet halo before it winked out.

I turned back to Nicky to ask a question, which never made it out of my mouth as something huge registered in my peripheral vision: a cloud of red light, jellyfish-like, trailing hundreds of streaming filaments. By the time I had snapped my head around to focus on it, this had shrunk to a tiny blob that went dark.

Within a minute, everyone had witnessed at least one of these sprites and so all questions as to Nicky's credibility had gone away. For the most part they all faded in the blink of an eye.

But sometimes, ghostly orbs of blue light would scamper up the red tendrils for a few moments afterward, prompting gasps of delight. These I heard over the wireless voice com system built into my suit—for by this point I had my helmet on.

Nicky was watching their reactions uneasily, clearly wishing they would take this a little more seriously. "A couple of decades ago," she said, "some of our orbiting gamma ray observatories began picking up incredibly powerful bursts. Long story short, it became obvious that these were coming not down from deep space but up from *below*—from Earth. So powerful that they maxed out the sensors, so we couldn't even tell how massive they actually were. Turned out they were coming from thunderclouds. The conditions in those storm towers down there are impossibly strange. Free electrons get accelerated upward and get kicked up into a hyper-energetic state, massively relativistic, and at some point they bang into atoms in the tops of the storm towers with such energy that they produce gamma rays which in turn produce positrons—antimatter. The positrons have opposite charge, so they get accelerated downward. The cycle repeats, up and down, and at some point you get a burst of gamma rays that is seriously dangerous—you could get a lifetime's worth of hard radiation exposure in a flash." She paused for a moment, then stared directly at me with a crazy half-smile. "The Earth," she said, "is an alien world."

The story was jogging some memories. This was one of those "gotchas" that had come along halfway through the project and precipitated a crisis for a few weeks. The hard part, actually, had been getting Carl and the other top decision-makers to believe that it was for real. The engineering solution hadn't been that complicated—shield the floors of the buildings with radiation-stopping materials, and, during thunderstorms, evacuate any parts of the structure that hadn't been shielded yet—

"Thank you," I said, "you're a sharp one."

Nicky nodded.

"What is your idea? Why are we donning?" I asked her.

"There's a place over in the depot where some plate steel has been stacked up—I reckon if we huddle on top of that, it should stop most of the gamma rays coming up from the storm."

Roger had been listening intently to all of this. In a weird burst of insight, I understood why: Alexandra was pregnant. She wasn't showing yet. So there was no real evidence to support my intuition. But she had declined the offer of a drink, which was unusual for her, and there'd been something in the way she and Roger looked at each other…Carl's grandchild was up here, taking shape in her womb. We had to get her over cover.

The depot that Nicky had spoken of was the old construction materials dump, near the middle of the Top Click. It was all out in the open, which was why we'd had to don in order to get there. In a few minutes' time we were all able to put on our helmets and let our suits run through their self-check routines. We stepped out into an airlock and experienced the weird sensation of feeling the suits stiffen around us as the outside pressure dropped. They were awkward to move in, and none of us was really trained in their use—they were for survival purposes only. For that reason a number of electrically powered scooters were parked in their charging docks right outside the exit. We merely had to waddle over to them, climb aboard, and then steer them away. In an ungainly queue we followed Nicky in a circuitous path among buildings-in-progress.

Shortly we arrived at the depot and followed her to a place where corrugated steel floor panels had been stacked up in neat rectangular blocks almost as tall as our heads. Hiram, who of all of us was most adept at moving around in a suit, clambered up onto the top of a stack and then reached down to pull the rest of us after him.

The view here wasn't as good, but none of us doubted Nicky's judgment as to the gamma rays, so we didn't mind. Soon the most intense part of the storm was passing beneath us. We could tell as much from the fact that the red sprites were directly overhead. Most of us ended up lying on our backs so that we could gaze straight up and watch the light show.

We had taken all of these precautions for one reason: to avoid exposure to gamma radiation. The storm was nowhere near us. Far below, winds were buffeting the lower structure, but we didn't even feel it. Our view down blocked by tons of steel, our only clue that a storm was in progress was the sprites blooming tens of thousands of meters overhead.

All of which made the superbolt just that much more surprising.

Of course, we didn't know it was something called an "upward superbolt" until much later. At the time, I just assumed that I was dead for some reason, and that the transition to heaven, or hell, was a much more jarring event than what tended to be described by survivors of near-death experiences. My next hypothesis was that I was still alive, but not for long—I remember reaching up to touch my helmet, fearing that it had popped off. Nope, it was still there. Then, for a minute or two, I was simply convinced that terrorists had set off a small nuclear device somewhere on the Top Click. Buildings were damaged, debris—glowing hot—was cascading to the deck. Finally my ears recovered to the point where I could hear Nicky saying "lightning" and despite all of this chaos some part of my brain was registering the schoolgirlish objection that lightning was a cloud-ground interaction, and we were not between the clouds and the ground, so how could that be? Now, of course, I know more than I want to about upward superbolts: another fascinating middle-atmospheric phenomenon that Nicky hadn't gotten around to lecturing us about.

Anyway, shock and awe only last for so long and then you begin to take stock of reality.

The First Bar in Space was mostly gone. The superbolt had melted a hole through its floor and roof and it had explosively decompressed, vomiting its contents—barstools, dartboards, Carl's ashes—into space. So Nicky's precaution about the gamma rays had not only saved us from a stiff dose of radiation but, by dumb luck, kept us from being blown to bits and sprayed across the desert.

Several other buildings were no longer shirtsleeves environments. Some looked undamaged. Safety doors ought to have slammed down in the tube network, preventing depressurization of the whole Top Click. When the storm had passed and the danger of gamma rays was over, we'd move into one of the undamaged structures and wait for rescue.

In the meantime, we were in for a wild ride, because the Tower had begun swaying and shuddering beneath us.

Panic would have been too obvious. I confess I was headed in that general direction, though, until Hiram's voice came through: "I don't care how big the lightning was," he said, "I rode that steel up and there is just no way it could have taken that much damage." He rolled off the stack of plates and let himself down to the deck. "I'm going to go have a look."

What the heck. My kid-having days were over. I had responsibilities. First and foremost to Carl's grandkid-in-the-making.

"Emma, what the *heck*!?" Tess called.

"The storm is over, baby," I said, and followed Hiram, who let me down easy to the deck.

I guess some touristy impulse led us straight to the wreckage of the First Bar in Space. As we got closer to the hole in the deck, we dropped to our knees and approached it on all fours.

Staring down through the hole, we could see, a couple of kilometers below us, something like a venetian blind that had been attacked with a blowtorch. Several of those burnished aluminum airfoils had been blasted by the superbolt. A huge strip of aluminum had peeled away from one of them and gotten wrapped around the one below it.

"Well, there's your problem," Hiram pronounced. "No wonder the Tower ain't flying right." As if on cue, the structure lurched beneath us, eliciting squeals of horror from our friends back in the depot.

"Is it dangerous?" I asked him. Because one of the advantages of being a middle-aged chick in this world was the freedom to ask questions that a young male would be too insecure to voice.

"To people below? If some of that crap falls off? Sure!" he said. "To us? Nah. Structure's fine. Just ain't flying right. Only real risk is barfing in our helmets."

I wasn't about to second-guess Hiram, who had ridden the Tower up from day one and had a strong intuitive feel for what made it stand up. But uneasy memories were stirring of briefings, years ago, about top-down failure cascades. The classic example being the Twin Towers, which had collapsed in toto despite the fact that all of the initial damage had been confined to their upper floors. Debris from a high-level event could damage structural elements far below, with incalculable results. The thin, almost nonexistent atmosphere up here would allow debris to fall at supersonic velocity. Energy, and damage, would increase as the velocity *squared*....

Where was the problem, I wondered, that was preventing the tower from "flying right?" Had the superbolt fried the electronics? Jammed a mechanism? Bent a control surface?

Which caused me to remember one other detail from earlier in the evening....

"Crap," I said, "There's a guy down there. Working on the system. Joe. An engineer. I hope he's okay."

"I wonder if there's any way to reach him?" Tess asked. Reminding me that all of us up here were linked in a single conversation by a wireless mesh network.

"First things first," I said. "The absolute top priority is patrolling the edge of the Top Click to make sure no debris falls off the edge. Spread out and do that. If you see anything, report it to Hiram. Hiram, you'll have to jury-rig something...."

"I'll get a welding cart. Should be great for tacking things down."

"Everyone should be spreading out now," I insisted. "Storm's over, folks. I see clear air below, headed our way. Look for anything that might go over the edge, or get jostled loose by all this shaking. And while you're at it, look for buildings that are still pressurized, where we can take shelter before our suits run out of air."

"Good news, bad news," Dave announced a minute later. "I found a little dome—a construction shack—that still has pressure. But this thing next to it got zapped real hard—it's about to shed a roof truss over the edge."

"I know the place. On my way," Hiram announced.

It didn't rain up here, so "roof" meant a piece of steel for reducing the flux of cosmic rays. The point being, a roof was a heavy object, and not something we wanted to drop off the top of the tower.

Toddling around in my suit, I caught sight of Hiram trundling a welding cart as fast as he could manage it. These weren't like terrestrial welders. They had built-in engines to generate power, fueled by liquid propane—but since there was little air up here, they also carried their own supply of liquid oxygen. It made an unwieldy package even for a big man like Hiram, but presently Dave waddled over to lend a hand.

Inside, our little refuge was like any other construction site office, decorated with freebie catalogs from tool companies and beer posters featuring sexy babes. Once we got through the airlock and pulled off our helmets, we were able to sit down and take stock of things. Outside, Hiram was tack-welding the wayward truss into place, assisted by Dave and a contractor who had joined up with our little band. He'd been working late on the top floor of a casino when the lights had gone out and a six-foot long arc had jumped out of a nearby faucet and connected with an electrical outlet, passing close enough to his head to singe his hair. Once we'd heard his story, we understood why we couldn't get Internet. But we could make cell phone calls, connecting, albeit patchily, with towers on the ground.

Tess and Marla the bartender, who knew their way around the Top Click as well as anyone, swapped in new oxygen supplies and went out to reconnoiter for injured stragglers. I was able to get through on the phone to Joe's boss and hear the troubling, but hardly surprising, news that he hadn't checked in.

Which was how Roger, Frog, and I ended up abseiling three kilometers straight down the Tower's central shaft on a rescue mission.

Why us three? Well, Frog was the veteran BASE jumper, a little past his prime and above his station, but—bottom line—the only way to prevent him from coming would have been to hit him on the head with a wrench and zip-tie him to a girder. Roger, true to his pedigree, had mountaineering experience.

And I knew as much about the Tower's weird little ins and outs as anyone.

The central shaft, which would one day be filled with some as-yet-unbuilt space launch tech, was empty. Ever-pragmatic construction workers had strung long cables down it, like mountaineers' ropes. They used mechanical descenders to glide down when they wanted to get lower but didn't have time to mess about with helirails. There were also elevators, but these had all shut themselves off when the superbolt had fritzed their electronics, and there seemed no way to reboot them without getting through to a customer service rep in Pakistan. Roger, the biggest member of our group, slung an extra spacesuit on his back (these folded up, sort of, into backpackable units, bulky but manageable). This was just in case Joe was in need of one. Hiram showed us how to harness the descenders to the outside of our suits. Each of us connected to a different cable—there were plenty of them—and then we backed off the rim of the shaft and let ourselves drop.

So this was a tubular vacancy, ten meters across and (currently) seventeen thousand meters long. Walking to the edge of it and looking straight down had become a popular tourist activity on the Top Click. To put this in perspective, it had the same relative dimensions as a forty-foot-long soda straw. As a rule, it was perfectly straight, as if it had been laser-drilled through a cube of granite. Quite a trick given that it was held in place by a wind-buffeted gas of metal.

Tonight, though, it was sashaying. You couldn't see the bottom from the top. It was like staring down the gullet of an undulating snake. Because, in Hiram's phrasing, the Tower wasn't flying right. Its accustomed straightness was a *process*, not a *state*; it was made straight from one moment to the next by a feedback loop that had been severed.

As alarming as it looked, the undulation wasn't as huge as it appeared from above, and once we had adjusted the tension in our descenders, we were able to plunge more or less straight down. In a forty-foot soda straw, even a little bend looks enormous.

The self-weight of the cables became a problem after a while and so they terminated every hundred meters, forcing us to stop and transfer to new ones. It took us thirty transfers, and as many minutes, to get down to the altitude where Joe had detrained some hours ago.

This brought us into the Neck, the skinniest part of the Tower, but in some ways the most complicated. The Top Click was destined to be the domain of gamblers and scientists. The bottom kilometers would be a city with an airport on its roof. The central core, a somewhat mysterious ballistics project. But the Neck was the domain of engineers: mechanical, control-system, and aeronautical. That's because it was here that the wind stress was at its peak, and here that it had to be addressed with what were called "active measures." The most conspicuous of these were the airfoils, large enough for people to walk around inside of them. At one level there was also an array of turbofan engines, the same as you see on airliners, which had been put there as a last-ditch measure in the event of a full-on jet stream hit. If that ever happened we would just fuel them up, turn them on, and run them full blast, thrusting back against the force of the wind, until the jet stream wandered away, a few hours or days later.

All of this gear for playing games with enormous forces had made the Neck beefier by far than the rest of the Tower, and so as we descended silently into it, our view of the stars and of the curving horizon was interrupted, then cluttered, and finally all but blotted out by a mare's nest of engineering works, most of it wrapped in streamlined airfoils to make it less draggy.

At our target altitude, six horizontal braces radiated from the core to the six primary legs of the Tower. These were trusses, meaning webs of smaller members triangulated into rigid systems, looking a bit like radio towers laid on their sides. Plastic tubes had been built around them, forming airtight corridors. Those in turn had been encased in aerodynamic sheaths. Six of those converged like spokes on the place where we stopped our descent and unhooked ourselves from the descenders. Moving deliberately, clipped to safety rails, leapfrogging from one handhold to the next—for the wind was fearsome—we made our way to the airlock that afforded entry to the southeastern strut/truss/tube/airfoil. Based on information from Joe's boss, I believed we would find him at the end of it. So I was dismayed when the airlock's control panel gave us the news that the tube was depressurized. This thing was supposed to be full of a proper atmosphere so that engineers could move along it without having to leave that all-important shirtsleeves environment. But apparently the superbolt had caused it to spring a leak. This was okay for me, Roger and Frog, but I didn't know what it might portend for Joe.

In any case, opening the door was easy since we didn't have to cycle the airlock. We were confronted by a view down a straight tube a thousand meters long, illuminated dimly by blue LEDs. The steel truss had been equipped with plastic catwalk grating. We started walking. This would have been a lot easier in an atmosphere. As it was, I wished we'd had some of those electric scooters, like on the Top Click. The designers of those suits had made the best of a tough design challenge, but at the end of the day they were made for passive survivors awaiting rescue by people in *real* space suits. Hiking down a catwalk wasn't in the design spec. It was like wading through wet cement and feeling it start to harden whenever you planted a foot. I wanted to break the mood with a joke about what great cardio this was, but I was too out of breath, and judging from the sound effects in my headphones, Roger and Frog weren't doing much better.

In fact I was just about ready to start whining about how hard it all was when we got to the end of the tube—meaning we had reached the southeastern Primary—and walked through another dead airlock into the pod where Joe had been working.

The pod was spherical. A floor and a ceiling had been stretched across it to turn it into a round room about the size of a two car garage. The dome-shaped spaces above and below were packed respectively with electronics and with survival gear. The first thing I noticed when I walked in was an open floor hatch, which gave me hope that Joe had had time to yank it open and grab a suit.

But Joe wasn't in here.

My eye was drawn to a scarlet flash on the other side of the darkened room. I realized I was looking straight out through a hole that had been blasted in the spherical shell. The red flash had been one of those sprites, off in the distance, high above the top of the thunderstorm as it migrated eastward.

Frog bent down and picked up an overturned swivel chair. Its plastic upholstery was patchy where it had melted and congealed.

On the workspace where Joe had been seated, and on the jagged twists of metal around the rupture, was a mess that I couldn't identify at first—because I didn't *want* to. And when I did, I almost threw up in my suit. Joe hadn't opened the floor hatch, I realized. It had been blown open when this whole pod had explosively decompressed. The atmosphere had blasted out the hole, taking Joe with it. Later forensic analysis suggested he'd been killed instantly by the superbolt, so at least he'd been spared the experience of being spat out, fully conscious, into free fall. But none of that changed the fact that, through no fault of his own, he'd been sitting in the wrong place at the wrong time.

He had become the third accidental fatality on the Tower construction project. Number one had been early—a forklift mishap, moving some steel around. Number two had been only a couple of months ago: a taut cable had been snapped by a wayward crane, the broken end recoiled under tension and struck a worker hard enough to break his neck. Joe was number three, killed instantly by an Upward Superbolt: a species of upper-atmosphere monster of which we had known only traces and rumors when the Tower had been designed.

What we did next got described all wrong in the news reports. Oh, they weren't factually incorrect, but they got the emotional tenor wrong. Yes, seeing that the southeastern control node had been blitzed off the network, we concluded that its responsibilities would have to be shunted to other nodes on the same level that still had luxuries such as power and atmosphere. Lacking communication with the ground, we had to make do with a few erratic cell phone conversations. Roger, Frog and I spread out to the south, northeast, and north control nodes on the same level—lots more cardio—and finally took those cursed suits off and, following instructions from the ground, repatched cables and typed in arcane computer commands until control had been transferred. The Tower stopped swaying, and as the control loops recalibrated to its new aerodynamics, stopped vibrating as well. All of that was true. But the news feeds described it as an Apollo 13 type of crisis, which it never was. They made it sound like we were doing really cool, difficult work under pressure, when in reality most of it was sitting in shirtsleeves (sorry, Carl!) and typing. And they totally failed to understand the context and the tone that had been set by the death of Joe.

The one thing they got right was what happened in the wee hours that followed: Hiram and Frog going out on the damaged airfoil to corral loose pieces of metal that were banging around in the wind and that could have inflicted catastrophic damage had they come loose. That was really dangerous work, performed at great personal risk without proper safety lines and, because it took longer than expected, with dwindling air supplies and cold-numbed digits. Frog, true to BASE jumper tradition, went out the farthest, and took the biggest risks—maybe because he had a parasail strapped to his back. And, though he later denied it, I think he had a plan. Only after all of the loose debris had been securely lashed and tack-welded down did he "fall off" in an "unexpected wind-gust" and free-dive for a few thousand meters before deploying his parasail and enjoying a long ride down to terra firma. You've seen the YouTube of him touching down in the desert at dawn, popping off his helmet, gathering up his chute, and striding toward the camera to make the grim announcement that a man had died up on the Tower last night. Standing there in his space suit, unshaven, exhilarated by his "fall" but sobered by the grisly scene he'd witnessed in the pod, he looked like nothing other than an astronaut. And an astronaut he was, on that morning. One without a rocket. Exploring, and embracing the dangers, not of outer space but of the atmosphæra incognita that, hidden from earthlings' view by thunderheads, stretches like an electrified shoal between us and the deep ocean of the cosmos.

Coda: Atmosphaera Incognita
Gregory Benford

…and so it came to this.

I squinted against harsh sunlight as we floated in blissful zero-g, a consummation devoutly to be wished by those of sore joints and flagging arteries. We had come storming up through the magcat tunnel and popped out into the stratosphere in the latest model slimship rocket, and I had felt only a mild weight from the thrust. Those new water immersion beds worked wonders for weary, antique bones.

I swam over to the big oval port, feather-light. Through it I could see the scoop sail, blossoming like a colossal ivory vacuum flower. It turned slowly, basking in the constant sun. It was so far away I could see flight teams like ants flexing it. Last minute checks. Trying out the tensile muscles in its smart fiber, running code through its myriad carbon-tube circuits, tuning it for the interstellar deeps.

They had it in full parabolic reflector mode, a beautiful curved funnel shape it would have as it swam away from the sun. A mutter in my ear said this was a last "rigor test" and by chance I was looking down the throat of it. I could see the intense sunlight focused on a large solar panel down at the focus. As I watched, the majestic flower slowly, slowly flattened itself into a vast disk, the solar panels standing atop its payload. Ready for the stars.

But first, it had to fall deep and long and loop around the sun. To sear away its smear of blow-off paint, like a rocket, to gain huge delta-Vs…and soar outward. Beyond Mars it would focus down to the parabolic reflector again, using focused sunlight to power its ion thruster. But would be a kind of rocket, too. The scoop-sail had magnetic fields embedded in it. Those would grab ions from the stellar wind and focus that incoming plasma. Its drive, powered by sunlight, would accelerate the whole big shebang, driving it forward with a tight, sea-blue exhaust forking behind.

I let myself float, savoring the view, ignoring the buzz of talk around me.

Tess would've loved this, for sure. The somber star-sprinkled dark, cut by the sharp hard white of a "smart sail" as they called it, kilometers across. I *so* wish she could've made it through that last long hard bout with cancer. Just to see this. The thought of her made me close my eyes, recalling the bright full days we had together, when the Tower was rising and our hopes with it. Ripe days….

"Getting close to power-on," an impossibly fresh-faced kid said to me.

I blinked. "So we have to back off?"

"'Fraid so."

"Then do it." I groped for my couch.

A slight tap of gravs and we slid away from the vast sheet of brilliant sail. Time to get out of the way of the big coherent microwave antennas that would power *Starsail One* out of low Earth orbit. That invisible beam would hammer at the craft and loft it beyond the crescent moon.

A monotonous voice counted down to Blast Time. There were display panels making diagrams and running simulations on the walls and I watched them, not trying to follow the details.

A mere hour or two of microwave drivers would push *Starsail One* hard. Those big orbiting emitters had been around for a decade or two, to power fast sails into the outer solar system. Everything here was built up a step at a time, each stage useful. Those beamers would give *Starsail One* enough momentum to arc out on a long sloping hyperbola, free of Earth—and into interplanetary space. Then it would plunge down the grav well for a tight, hot parabolic loop around the sun. A half year to fall, then the serious business would begin.

"Ma'am, need any meds?" my personal assistant asked.

"I'm fine. How about champagne?"

"The autodoc says—"

"Screw that software. I know we have it aboard."

"But—"

"I'm on holiday, see?" I gave her the old lady glare.

"Um. Uh. Yes."

Then she went shooting away with the required worried frown. Of *course* I was going off the reservation. Sometimes it's gotta be fun to be an icon.

This was a celebration, one Carl and Tess should've seen. Carl had missed the big repair-genes revolution by barely a few years, too.

The party behind me murmured on but they left me in my peace. When you're 135 years old, everybody treats you like a national monument. Getting frisky with a tad of alcohol was what monuments need, but personal assistants don't get the memo. She was a kid, maybe forty, and could not recall when the Tower was a shiny new idea. Now it flung slimships into orbit in dozens per day.

I could hear some news jockey sounding off in the distance about the first starship leaving Earth, burbling like one of those earnest educational films Carl and I had slept through, back when we were just out of diapers. I clicked free of the couch belt and let myself drift light and easy over to the big bay port again. One last look.

Starsail One basked in the sun like a porcelain saucer as we drifted over Africa. The Blast Time count droned on and then there were shouts behind me. I kept silent, remembering. Then came a bong.

A glossy fog rose from *Starsail One* as the microwave beam struck her. The vac-chemists had prepared a thick skin that now blew off at high speeds, driving *Starsail One* hard away from us. It left streamers of vapor, trailing the great white saucer like flags flapping in farewell. *Bon voyage!*

Carl and all the others had not built the Tower to deliver huge rolls of smart carbon into orbit, or the robot teams that assembled those into huge sheets, or the dazzling fiber tech. He had left the big cylinder hole running up the Tower as a blank space, waiting for ingenuity and economics to fill it with something marvelous that nobody could plan. The magnetic catapult won out because it could throw a slimship out the pressure membrane at twenty klicks' altitude, with a velocity better than a klick a second. Not even the ground launcher running up the slope of Mount Kilimanjaro could beat that.

Nobody back then thought that assembly in orbit, fed by slimships, could make sprawling solar harvesters, either. Now they were delivering gigawatts to Earthside. Nobody back then thought to make the smart ion thrusters that *Starsail One* carried. Or to build magnetic funnels able to grab plasma from thin interstellar spaces and hurl it out the back. All that was whiz-bang stuff to me, still. Yet I was quite sure that tech could drive forward across the light-years, into the Alpha Centauri system, to map and analyze. For the men and women to follow, yes.

A basic law about the future: Make a big new thing and it will spawn more big new things you never saw coming.

I thought of Frog, the BASE jumper. He became an emblem of the Tower's promise. It always helps to be brave, smart, and good looking.

No human would ride to the stars today, but the payload held an artilect that might as well be human, given its range of abilities and knowledge. It will carry out explorations of the Centauri system, swooping on wings through alien sunlight. I wished they had named it Carl….

Starsail One shot away from us. My assistant put a champagne bulb in my clawlike hand and I lifted it in tribute. "To Carl!" I croaked. "To Tess! To the stars!"

Cheers rolled over me like a roaring wave.

Nobody much noticed me in that moment. I was just an old somebody who had the pull to come up one last time and watch the consequences of my life play out.

My bleary eyes could barely see the white dot disappear among the spray of stars. As if joining them.

Historically, coal and the railroad train enabled much of the industrial revolution. Both came from the underlying innovation of steam engines. Coal was the new wonder fuel, far better than wood though harder to extract, and it made continental scale economies possible. Synergistically, coal drove trains that in turn carried coal, crops, and much else. A similar synergy may operate to open the coming interplanetary economy, this time wedding nuclear rockets and robotics. These could operate together, robot teams carried by nuclear rockets to far places, and usually without humans, who would compromise efficiency. Mining and transport have enormously expanded the raw materials available to humanity, and the rocket/robot synergy could do so again. As such fundamentals develop in space, other businesses can arise on this base, including robotic satellite repair/maintenance in high orbits, mining of helium-3 on the moon, and metal mining of asteroids. Finally, perhaps snagging comets for volatiles in the outer solar system will enable human habitats to emerge within hollowed-out asteroids, and on Mars and beyond. Such ideas have been tried out in the imaginative lab of science fiction, exploring how new technologies could work out in a future human context. Here Robert Zubrin describes the engineering and economics of such an era, starting with an appreciation of how far we have come already in modern times.

On the Way to Starflight: The Economics of Interstellar Breakout
Robert Zubrin

According to the best current archeological evidence, until around 100,000 to 50,000 years ago, the human race, modern *Homo sapiens,* was confined to a small region surrounding the Rift Valley in eastern Africa. The climate was favorable, the game fairly abundant, and non-technological humans were more than adequate for the challenge of the environment. As a result, the Rift Valley dwellers were able to get by with a tool kit limited to little more than the same split-rock hand-axes that had served their *Homo erectus* ancestors for the previous million years. For some unknown reason, however, a few bands of these people decided to leave this relative paradise and travel north to colonize Europe and Asia, eventually going on from there to cross the land bridge into the Americas.

They went *north,* into the teeth of the ice ages, into direct competition with giant carnivores and stocky Neanderthals who had already adapted to life in the cold. They went north, into a world of challenge, where fruit, vegetables, and game were not available all year long, and where efficient weapons, clothing, and housing were necessary. In abandoning Africa, they embraced a wider world that could only be survived through the development of technology. Thus was born *Homo technologicus,* man the inventor, amid ice and fire. Thus humanity transformed itself, from an East-African local curiosity to the dominant species on this planet.

In a sense, the Biblical tale of Genesis tells this story but has it backwards. It was not eating of the Tree of Knowledge that forced mankind to leave Paradise. Rather, it was the abandonment of Paradise that forced humanity to seek the forbidden fruit.

Many years ago the Russian space visionary Nikolai Kardashev outlined a schema for classifying civilizations. Adopting Kardeshev's scheme in somewhat altered form, I define a Type I civilization as one that has achieved full mastery of all of its planet's resources. A Type II civilization is one that has mastered its solar system, while a Type III civilization would be one that has access to the full potential of its galaxy. The trek out of Africa was humanity's key step in setting itself on the path toward achieving the mature Type I status that the human race now approaches.

The challenge today is to move on to Type II. Indeed, the establishment of a true interplanetary space-faring civilization represents a change in human status fully as profound—both as formidable and as pregnant with promise—as humanity's move from the Rift Valley to its current global society.

Space today seems as inhospitable and as worthless a domain as the wintry wastes of the north would have appeared to a denizen of East Africa 50,000 years ago. But yet, like the north, it is the frontier arena whose possibilities and challenges will allow and drive human society to make its next great positive transformation.

However, I am going to focus not on humanity's current challenge—the transition from Type I to Type II—*but on how the further development of the Type II civilization, whose birth is now immanent, can prepare the way for the breakout to Type III,* thereby opening the path to an infinite human future among the stars.

The Economic Basis for Interstellar Flight

Let us consider the requirements of an early interstellar mission. The ship has a velocity of 10 percent the speed of light, allowing it to reach nearby stars on a timescale of decades, and a dry mass of 1,000 tons, permitting it to support a few score people carrying the skills and cultural heritage necessary to establish a small colony. Such a vessel would have a kinetic energy of 125 trillion kilowatt-hours. Assuming a price of 10 cents per kilowatt hour, the cost of this much energy, if transmitted to the ship at 100 percent efficiency, would be $12.5 trillion. But energy cannot be used to accelerate a spacecraft at 100 percent efficiency, and beyond its energy requirement, the expedition has many other costs, notably including its technology development and hardware manufacture. These would likely multiply the expedition cost by an order of magnitude, yielding a rough cost estimate of $125 trillion for the mission. For comparison, the cost of the entire Apollo program from 1961 to 1973, in today's dollars, was about $120 billion, an amount equal to 1 percent of the $12 trillion GDP of the entire human race in 1968. But the cost of the interstellar mission is 1,000 times that of Apollo. Therefore, if the cost of interstellar mission is not to exceed Apollo levels in proportion of its society's wealth, the human race at the time of its launch will need to have a GDP amounting to 1,000 times that of 1968, or about 200 times greater than that of today.

But is such a society possible? According to Malthusian thought, it is not. However, as we shall see, Malthus was entirely wrong.

The Malthusian Error

Writing circa 1800, Thomas Malthus said that production rises only arithmetically as population rises exponentially. If this were true, then GDP per capita would collapse as both population and GDP rise, implying a total impoverishment of humanity long before world GDP rose to the level required for starflight.

But if we examine historic data, we can see that the theory of Malthus is at complete variance with the facts.

We begin with Fig. 1, which shows data for world population, global gross domestic product (GDP), and GDP per capita from the year 1500 to the year 2000.

Examining Fig. 1, we can see that while human population has certainly increased over time, GDP has increased even more, and the key metric of average human well-being, *GDP per capita, has gone up as population has increased,* rather than down, as Malthusian theory would predict.

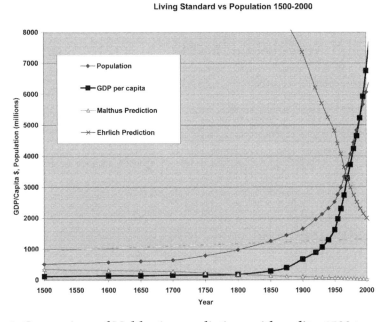

Fig. 1. Comparison of Malthusian predictions with reality, 1500 to present.

In Fig. 1, the actual per capita GDP is shown by the thick black line marked with squares. We also show with the thin line marked with triangles, the predictions provided by Malthusian theory at the time of Malthus's writing circa 1800. According to Malthus, the sixfold increase in population between his time and the year 2000 should have resulted in a disastrous drop in human living standards. Instead, global per capita GDP actually increased almost fortyfold, from $179 annually in 1800 to $6,757 by 2000. In short, Malthus was wrong.

Well, everyone has the right to be wrong about predicting the future. But the reader will note that I have taken the liberty of extending Malthus's prediction into his past. The world population in 1500 comprised 500 million people, just half of that prevailing in Malthus's day. If living standards go down with increased population, they should go up with decreased population. Thus, according to Malthusian theory, the world should have been much richer in 1500 than it was in 1800, with per capita GDP in the range of $360, instead of the $114 in actually was. *Thus Malthus was not just wrong about predicting the future, he was wrong about predicting the past,* and not by a small variance, but by a factor of three.

Even more absurd were the predictions based on Malthusian theory widely published by Paul Ehrlich, both alone and together with his protégé John Holdren (currently President Obama's science advisor) as well as the Club of Rome during the 1968-1972 timeframe.

Writing circa 1970, all of these groups predicted that the "population explosion" would cause a catastrophic collapse of human well-being by the turn of the century, exactly the opposite of what actually happened. If Ehrlich et al had been right, average world GDP per capita would have fallen from $3,200 in 1970 to less than $2,000 today. Instead rose to over $8,500 by 2010.

This fourfold error in predicting the future was really unforgiveable, because Ehrlich et al. had the advantage of hindsight in knowing about the wild inaccuracy of Malthus's original prediction, which by 1970 had already been shown to be off by a factor of fifty (an actual eighteen-fold increase in GDP per capita instead of a predicted threefold drop.) But even worse, it is evident that Ehrlich, Holdren, and the Club of Rome studiously ignored data readily available to them about the economic history of the recent past. That is, if one takes the trouble of extending their predictive theory backward in time (as shown by the thin line marked with "x's" in Fig. 1.), we see that it predicts an average global GDP per capita of over $7,300 for the year 1900 instead of the $670 that is was in reality. Now Ehrlich was born in 1932 (for which his theory predicts an average GDP of $5,700, instead of the $1,060 it really was). If a wealthy world had actually existed in the lesser-populated 1930s, he should have been able to witness it himself. Instead, within his own lifetime (by 1970), he had seen the world population double and the global standard of living more than triple at the same time. He didn't even have to open an almanac to know he was wrong. Instead, the data that disproved his theory was readily available to him by direct observation. The same can be said of Aurelio Peccei and the band at the Club of Rome. They simply ignored the facts.

Having dispensed with such charlatans, let us examine the data ourselves, to see where it can lead us in developing a true theory that accurately predicts the relationship between human well-being and population size.

One of the first things any real scientist would do in trying to discover how one variable within a system changes with respect to another is to graph the first variable against the second and see if a clear relationship emerges. So let's follow this tried-and-true standard technique, and graph world population size and per capita GDP but against each other, using the data of the last 500 years of human history. The results are shown in Fig. 2.

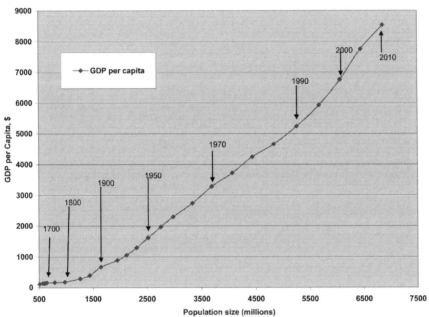

Fig. 2. How per capita GDP has changed as population has grown, 1500-2010.

Well, there certainly seems to be a pattern here, which obviously is *not* the Malthusian claim that living standards decrease as population grows. Rather, what we see is GDP per capita *increasing* with population, with the line curving upward for the past two centuries. Since 1800, the world population increased sevenfold, while the global GDP per capita has increased fiftyfold. Put another way, over the past 200 years, GDP per capita has risen in proportion to population size *squared,* while the total GDP has risen in proportion to the size of the population *cubed.* Some of this profound progress may be due to a qualitative change in global communication and transport that made the growth rate of the effective population of world civilization significantly faster during the earlier part of this period than the gross numbers alone would imply. However, if we take the more limited period since 1950 as our relevant sample, we still find GDP per capita increasing in proportion to population to the 1.6 power.

"But that makes no sense!" the Malthusians cry. It doesn't matter. That's what the data says, and science is about accounting for reality. So how can we explain the *fact* that as the number of human beings on the planet has grown, we've nearly all become much better off? Why should there be more of everything to go around, when there are more of us to feed, clothe, and house?

There are a number of very good reasons why this should be so. As economist Julian Simon has noted in his indispensible book, *The Ultimate Resource,* a larger population can support a larger division of labor, and so it is more economically efficient. Ten people with ten skills, working or trading together, can produce far more than ten times as much as one person with one skill. A larger population also provides a larger market which makes possible mass production and economies of scale. This is extremely important, as we can see by comparing the price of an RL-10 rocket engine with that of a small car. The RL-10, a tried-and-true thruster that has been in production since the 1960s, contains less metal (about 500 pounds) and is significantly less complex than a typical small car. Yet RL-10s sell for around $3 million each, while a new compact car can be obtained for less than $10,000. This is because the market is for only a few RL-10s per year, while cars are sold by the million. Because they represent a larger market, larger populations drive investment in new plant and equipment much more forcefully than small populations. If the market for an item is small, no one is going to build a new factory to produce it, or spend much money on research to find ways to improve it. But if the sales opportunity is big, the necessary investment will occur instantly as a matter of course.

A larger population can much better justify and afford to build transportation infrastructure, such as roads, bridges, canals, railroads, seaports, and airports, all of which serve to make the economy far more efficient and productive. A larger population can also better afford to build other kinds of highly productive economic infrastructure, including electrification and irrigation systems. It also can better afford infrastructure necessary for public health, including hospitals, clean water, and sanitation systems, and act far more effectively in suppressing disease-spreading pests. It takes a large-scale effort to drain a malarial swamp, a reality that puts such projects beyond the capability of small highly dispersed populations such as still persist in many parts of Africa. Furthermore, human boots on the ground are necessary to patrol the regions in which we live to prevent ponds and puddles from being used by mosquitoes and other disease carriers as breeding grounds. A thin population will thus in many cases tend to be a much sicker population than a dense population which enjoys the safety than only numbers can provide against humanity's deadly natural enemies. Moreover, a healthy population will be more productive than a sick population, and reap a much better return on the investment it chooses to make in education (and thus be able to afford more education), since more of its young people will live to employ their education, and be able do so for longer life spans.

That said, it is clear that the actual causative agent for higher living standards is not population size itself, but the overall technological development that it allows. The average living standard is defined by GDP available for consumption per capita, which is equal to the production per capita, which is determined by technological prowess.

If we choose to be mathematical, we could even write this down as an equation. Let L=Living standard, P=Population, G=Gross Domestic Product, and T=Technology. Then we have:

L=G/P Living standard=GDP/Population

G=PT GDP=Population X Technology

Putting these equations together, we find that:

L=T Living standard=Technology

Furthermore, based on our observations of Fig. 2, we note:

T~P^n (with 1.6<n<2)

If population is growing exponentially, that means that the rate of growth of population, which we will signify $\underline{P}$, is proportional to the size of the population itself. Put mathematically:

$\underline{P}$~P

If this is so, we can put equations (4) and (5) together, and then, using some fancy math, we find that the rate of growth of technology, $\underline{T}$, is given by:

$\underline{T}$~PT$^{(n-1)/n}$

If n=2, this simplifies to:

$\underline{T}$~PT$^{1/2}$

Does this make sense? Well, let's ask ourselves, what causes the advance of technology? Clearly, technology does not come from land, or other natural resources, it comes from *people*. It is the product of *human work*. Furthermore, the more skilled and educated people are, the more inventive they are likely to be. Equation (7) puts this in mathematical terms, saying that the rate of growth of technology is directly proportional to the population size multiplied by a weaker function (the square root) of their living standard. The most general way to measure human work is in terms of man-years. So, to check this relationship, let's graph the growth of technology (measured in dollars as GDP per capita) against human man-years expended, counting from the year 1 AD to the present. The results are shown in Figs 3 and 4. (I've used two graphs to show this to avoid the necessity of using logarithmic scales, which are harder to read.)

Let's look first at Fig. 3, (following page) which shows the growth of human technology worldwide from the time of the Roman and Han Empires to the late nineteenth century. There's a lot of very interesting history to be seen here, but basically it breaks down into three periods: before 1500, from 1500 to 1800, and after 1800.

From 1 AD to 1500, technology does grow, but only at a very slow rate of 17.5 percent over 460 billion person-years, or an average of 0.035 percent per billion person-years. Between 1500 and 1800, the pace picks up substantially, with GDP per capita increasing by 58 percent in 200 billion person-years, or 0.23 percent per billion person-years, a more than sixfold increase over the preceding period. Then, around 1800, technology literally takes off, with GDP per capita

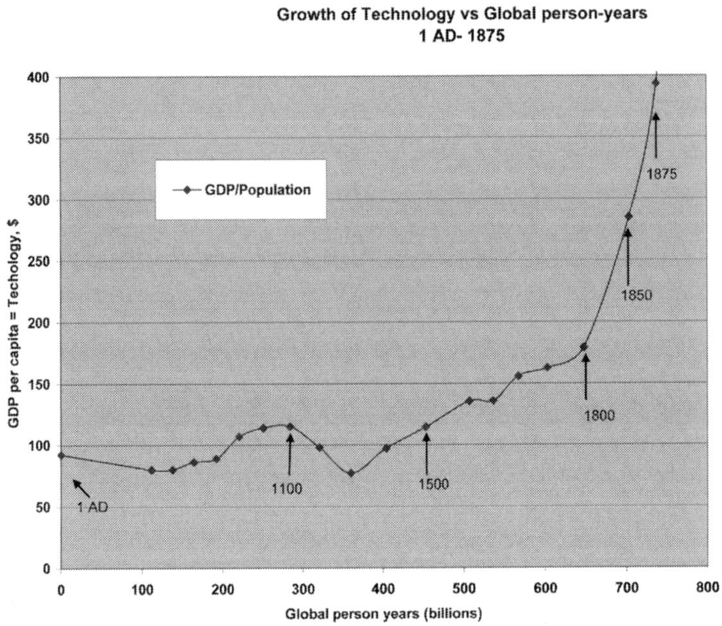

**Growth of Technology vs Global person-years
1 AD- 1875**

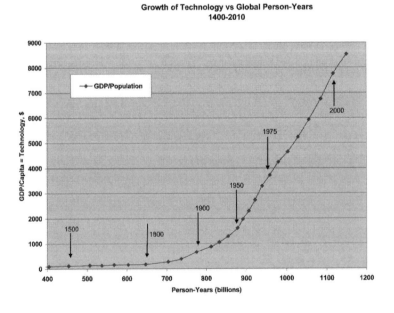

Fig. 3 Growth of technology with respect to person-years, 1 AD to 1875.

growing 116 percent over the next 90 billion person-years. As shown in Fig. 4, this growth continues, showing a 4,700 percent increase over the entire 500 billion person-year span from 1800 to 2010, for an average growth rate of 0.8 percent per billion person-years.

**Growth of Technology vs Global Person-Years
1400-2010**

Fig. 4 Growth of technology with respect to person-years, 1400-2010.

These results make perfect sense. Before 1500 there really wasn't a world economy in any substantial sense because long-distance trade and communication was so limited. Rather than a world economy, what existed was a number of disparate civilizations including European Christendom, the Islamic world, India, China, Mexico, and Peru, each with their own economy. Important innovations made in one civilization could take centuries or even millennia to propagate to the others. Thus, for example, it took hundreds of years for such important Chinese inventions as paper, printing, and gunpowder to reach Europe, and thousands of years for European domesticated horses, wheeled vehicles, and numerous other technologies to reach the Americas. Thus, the relevant inventive population size driving the advance of each civilization was not the whole world population, as small as it was, but the much smaller population of the civilization itself.

But around 1500, following the voyages of Columbus, Vasco de Gama, and Magellan, European long-distance sailing ships unify the world economy, creating vastly expanded markets for commerce, and making it possible for inventions made anywhere to be rapidly implemented everywhere. Thus the effective inventive population for each civilization was radically expanded virtually overnight to encompass that of the entire world, creating a sixfold increase in the rate of progress per person-year compared to that of prior history. With more people engaged, the world advanced faster. Furthermore, it was precisely those countries with the greatest contact with the largest number of people worldwide, i.e. the European seafaring nations, that advanced the fastest.

Then, around 1800 the industrial revolution begins, and the average rate of progress per person-year of human effort quadruples yet again. This occurs not only because the application of steam allowed human beings to wield vastly greater mechanical power than had ever been possible before, but because particular technologies, most notably steamships, railroads, and telegraphs, radically increased the speed and thus the effective range of transportation, commerce, and communication. By the mid 1800s, innovations made by anyone, anywhere, could spread around the world extremely rapidly, defining a new global reality of accelerated progress that continues to the present day. The fact that any technological advance can now have immediate global impact makes human creativity today far more powerful, and thus valuable, than ever before.

The critical thing to understand here, is that *technological advances are cumulative*. We are immeasurably better off today not only because of all the other people who are alive now, but because of all of those who lived and contributed in the past. If the world population had been smaller in the past than it actually was, we'd be much worse off now. Just consider what the world today would be like if the global population had been half as great in the nineteenth century. Thomas Edison and Louis Pasteur were approximate contemporaries. Edison invented the electric light, central power generation, recorded sound, and motion pictures. Pasteur pioneered the germ theory of disease that stands at the core of modern medicine. Which of these two would you prefer not to have existed? Go ahead, choose.

Human beings, on average, are creators, not destroyers. Each human life, on average, contributes towards improving the conditions of human life. This must be so, or our species would long since have disappeared. We live as well as we do today, because so many people lived in the past and made innumerable contributions, big and small, towards building the global civilization that we enjoy. If there had been fewer of them, we today would be poorer. If we accept Malthusian advice, and act to reduce the world's population, we will not only commit a crime against the present, but impoverish the future by denying it the contributions the missing people could have made.

If we are going to make it to the stars, we are going to need more people.

Expanding Human Civilization

An early interstellar civilization will need a GDP about 200 times that of today. Provided that we can continue to expand GDP in proportion to the cube of the population for the next two centuries, as we have done for the past two, then a sixfold increase of humanity, to a total population on the order of 40 billion, should be sufficient to make that possible. If instead we assume the weaker progress of the period since 1950, during which humanity's GDP grew as the 2.6th power of the population size, then an eightfold increase, to about 54 billion people, will be needed. In either case, the increase required is roughly in the same range as the sevenfold multiplication humanity has experienced over the past 200 years since Malthus wrote his spectacularly wrong tome, so it should clearly be possible. The two main limiting factors are space and energy.

Whether or not a world of 40 to 50 billion people would necessarily seem excessively crowded is hard to say. Most of Earth's surface today is largely uninhabited. This is true not only of the oceans and the high latitude regions, but even of the temperate and tropical lands (as can be readily ascertained by anyone who looks out the window during the large majority of long-distance airplane flights.) Earth only seems crowded to many today because most people choose to live in the tiny fraction of the planet comprising cities. If it were an independent country (as distinct from the entire United Kingdom) England, with a population density of 395 people per square kilometer, would be the most densely populated nation on Earth. Yet it is generally considered to be one of the most pleasant. The Earth has a land area of 128 million square kilometers. Populated at the same average density as England, its continents could readily host 50 billion. Yet, as anyone who risks driving around its rustic countryside can see, even England is thinly populated in many places.

That said, should folks begin to feel crowded, there is one resource that space offers in infinite abundance, to wit: space. Once a society becomes truly space-faring, the notion that its growth could be limited by physical overcrowding becomes intrinsically absurd. On the contrary, to the extent that any sense of overcrowding develops, whether due to physical crowding or, far more likely, social-regulatory overcrowding, such stimuli could only result in the explosive expansion of humanity into space. Indeed, it is the quest for freedom that will ultimately be the decisive driver in assuring interplanetary and then interstellar settlement.

Energy is the other most frequently cited limit to the continued growth and development of humanity. This, however, is an entirely incorrect view of the matter. *Energy does not limit human technological development. Human technological development unleashes energy.* This can be seen clearly in the history of humanity's technological advance, which has enabled, and in turn been accelerated by, ever-increasing energy utilization.

If we consider the energy consumed not only in daily life, but in transportation and the production of industrial and agricultural goods, then Americans in the electrified 1990s used approximately three times as much energy per capita as their predecessors of the steam and gaslight 1890s, who in turn had nearly triple the per capita energy consumption than those of the pre-industrial 1790s. Such rising levels of energy consumption have historically correlated directly with rising living standards, and, if we compare living standards and per capita energy consumption of the advanced sector nations with those of the impoverished third world, the correlation continues today. This relationship between energy consumption and the wealth of nations will drive further radical expansion of the set of accessible energy resources. Simply to raise the entire present world population to current American living standards (and in a world of global communications it is doubtful that any other arrangement will be acceptable in the long run) would require increasing global energy consumption at least ten times. However, world population is increasing, and while global industrialization is moderating this trend, it is likely that terrestrial population levels will at least triple over the coming century. Finally, current American living standards and technology utilization are hardly likely to be the ultimate (after all, even in early twenty-first-century America, there is still plenty of poverty) and will be no more acceptable to our descendants a century hence than those of a century ago are to us. All in all, it is clear that the exponential rise in humanity's energy utilization will continue, as they must if a civilization capable of interstellar colonization is ever to develop. In 2000, humanity mustered about 15 terrawatts of power (one terrawatt, TW, equals one million megawatts of power). At the current 2.6 percent rate of growth we will be using nearly 200 TW by the year 2100, and 2,500 TW by the year 2200. The total anticipated power utilization and the cumulative resource used (starting in 2000) is given in Table 1.

Table 1: Projected Human Use of Energy Resources

Year	Power	Energy Used after 2000
2000	15TW	0TW-years
2025	28	519
2050	53	1,491
2075	101	3,350
2100	192	13,700
2125	365	26,400
2150	693	50,600
2175	1,320	96,500
2200	2,500	

By way of comparison, the total known or estimated energy resources are given in Table 2.

Table 2: Solar System Energy Resources	
Resources	**Amount**
Known Terrestrial Fossil Fuels	1,200TW-years
Estimated Unknown Terrestrial Fossil Fuels	7,000
Nuclear Fission, U Fuel, no reprocessing	700
Nuclear Fission, U Fuel, with reprocessing	50,000
Nuclear Fission, Th Fuel, with reprocessing	200,000
Nuclear Fusion Using Lunar He3	10,000
Nuclear Fusion Using Jupiter He3	5,600,000,000
Nuclear Fusion Using Saturn He3	3,040,000,000
Nuclear Fusion Using Uranus He3	3,160,000,000
Nuclear Fusion Using Neptune He3	2,100,000,000

In Table 2, the amount of He-3 given for each of the giant planets is that present in their atmospheres down to a depth where the pressure is ten times that of Earth's at sea level. If one extracted at a depth where the pressure was greater, the total available He-3 would increase in proportion. If we compare the energy needs for a growing human civilization with the availability of resources, it is clear that, even if the environmental problems associated with burning fossils fuels and nuclear fission are completely ignored, within a couple of centuries the energy stockpiles of Earth and its Moon will be nearly exhausted. Large-scale use of solar power can alter this picture somewhat, but sooner or later the enormous reserves of energy available in the atmospheres of the giant planets must and will be brought into play.

Thermonuclear fusion reactors work by using magnetic fields to confine a plasma consisting of ultra-hot charged particles within a vacuum chamber where they can collide and react. Since high-energy particles have the ability to gradually fight their way out of the magnetic trap, the reactor chamber must be of a certain minimum size so as to stall the particles' escape long enough for a reaction to occur. This minimum size requirement tends to make fusion power plants unattractive for low-power applications, but in the world of the future where human energy needs will be on a scale tens or hundreds of times greater than today, fusion will be far and away the cheapest game in town.

A century or so from now, nuclear fusion using the clean-burning (no radioactive waste) deuterium-helium-3 reaction will be one of humanity's primary sources of energy, and the outer planets will be the Persian Gulf of the solar system.

The Persian Gulf of the Solar System

Today, Earth's economy thirsts for oil, which is transported over oceans from the Persian Gulf and Alaska's North Slope by fleets of oil-powered tankers. In the future, the inhabitants of the inner solar system will have the fuel for their fusion reactors delivered from the outer worlds by fleets of spacecraft driven by the same thermonuclear power source. For while the ballistic interplanetary trajectories made possible by chemical or nuclear thermal propulsion are adequate for human exploration of the inner solar system and unmanned probes beyond, something a lot faster is going to be needed to sustain interplanetary commerce encompassing the gas giants.

Fusion reactors powered by D-He-3 are good candidates for very advanced spacecraft propulsion The fuel has the highest energy-to-mass ratio of any substance found in nature, and, further, in space the vacuum the reaction needs to run in can be had for free in any size desired. A rocket engine based upon controlled fusion could work simply by allowing the plasma to leak out of one end of the magnetic trap, adding ordinary hydrogen to the leaked plasma, and then directing the exhaust mixture away from the ship with a magnetic nozzle. The more hydrogen added, the higher the thrust (since you're adding mass to the flow), but the lower the exhaust velocity (because the added hydrogen tends to cool the flow). For travel to the outer solar system, the exhaust would be over 95 percent ordinary hydrogen, and the exhaust velocity would be over 250 km/s (a specific impulse of 25,000 s, which compares quite well with the specific impulses of chemical or nuclear thermal rockets of 450 s or 900 s respectively.) Large nuclear electric propulsion (NEP) systems using fission reactors and ion engines, a more near-term possibility than fusion, could also achieve 25,000 s specific impulse. However, because of the complex electric conversion systems such NEP engines require, the engines would probably weigh an order of magnitude more than fusion systems. As a result, the trips would take about twice as long. If no hydrogen is added, a fusion configuration could theoretically yield exhaust velocities as high as 15,000 km/s, or 5 percent the speed of light! Although the thrust level of such a pure D-He-3 rocket would be too low for in-system travel, the terrific exhaust velocity would make possible voyages to nearby stars with trip times of less than a century.

Extracting the He-3 from the atmospheres of the giant planets will be difficult, but not impossible. What is required is a winged transatmospheric vehicle that can use a planet's atmosphere for propellant, heating it in a nuclear reactor to produce thrust. I call such a craft a NIFT (for Nuclear Indigenous Fueled Transatmospheric vehicle). After sortieing from its base on one of the planet's moons, a NIFT would either cruise the atmosphere of a gas giant, separating out the He-3, or rendezvous in the atmosphere with an aerostat station that had already produced a shipment. In either case, after acquiring its cargo, the NIFT would fuel itself with liquid hydrogen extracted from the planet's air and then rocket out of the atmosphere to deliver the He-3 shipment to an orbiting fusion-powered tanker bound for the inner solar system.

In Table 3 we show the basic facts that will govern commerce in He-3 from the outer solar system. Flight times given are one-way from Earth to the planet, with the ballistic flight times shown being those for minimum energy orbit transfers. These can be shortened somewhat at the expense of propellant (gravity assists can help too, but are available too infrequently to support regular commerce), but in any case are too long for commercial traffic to Saturn and beyond (even if the vessels are fully automated, time is money). The NEP and fusion trip times shown assume that 40 percent of the ship's initial mass in Earth orbit is payload, 36 percent is propellant (for one-way travel; the ships refuel with local hydrogen at the outer planet), and 24 percent is engine. Jupiter is much closer than the other giants, but its gravity is so large that even with the help of its very high equatorial rotational velocity, the velocity required to achieve orbit is an enormous 29.5 km/s. A NIFT is basically a nuclear thermal rocket with an exhaust velocity of about 9 km/s, and so even assuming a "running start" air speed of 1 km/s, the mass ratio it would need to achieve such an ascent is over 20. This essentially means that Jupiter is off limits for He-3 mining, because it's probably not possible to build a hydrogen-fueled rocket with a mass ratio greater than six or seven. On the other hand, with the help of lower gravity and still large equatorial rotational velocities, NIFTs with buildable mass ratios of about four would be able to achieve orbit around Saturn, Uranus, or Neptune.

Table 3: Getting Around the Outer Solar System

Planet	Distance from Sun	Ballistic			Velocity to Orbit	NNIFT Mass Ratio
		Ballistic	NEP	Fusion		
Jupiter	5.2 AU	2.2 yrs	1.1 yrs	1.1 yrs	29.5 km/s	23.7
Saturn	9.5	3.0	1.5	1.5	14.8	4.6
Uranus	19.2	5.0	2.5	2.5	12.6	3.6
Neptune	30.1	6.6	3.3	3.3	14.2	4.3

Titan

As Saturn is the closest of the outer planets whose He-3 supplies are accessible to extraction, it will most likely be the first of the outer planets to be developed. The case for Saturn is further enhanced by the fact that the ringed planet possesses an excellent system of satellites, including Titan, a moon which, with a radius of nearly 2,600 km, is actually larger than the planet Mercury.

It's not just size that makes Titan interesting. Saturn's largest moon possesses an abundance of all the elements necessary to support life. It is believed by many scientists that Titan's chemistry may resemble that of Earth during the period of the origin of life, frozen in time by the slow rate of chemical reactions in a low-temperature environment. These abundant pre-biotic organic compounds comprising Titan's surface, atmosphere, and oceans can provide the basis for extensive human settlement to support the Saturnian He-3 acquisition operations.

Because of its thick cloudy atmosphere, the surface of Titan is not visible from space, and many basic facts about this world remain a mystery. Here's what we know:

Titan's atmosphere is composed of 90 percent nitrogen, 6 percent methane, and 4 percent argon. The atmospheric pressure is 1.5 that of Earth sea level, but because of the surface temperature of 100° K (-173° C), the density is 4.5 Earth sea level. The surface gravity is 1/7th that of Earth, and the wind conditions are believed to be light. The latest evidence from radiotelescopes using Earth-based radar and the Cassini probe indicates the surface consists of a mixed terrain, including at least one solid continent in a methane ocean. The presence of higher hydrocarbons and other organic compounds within the bodies of liquid methane is nearly certain, but the precise chemical nature of the mixture is unknown. Hydrocarbon and ammonia ice may also exist.

The same nuclear thermal rocket engines that power NIFT vehicles mining Saturn's atmosphere could employ the methane abundant in Titan's atmosphere as propellant to enable travel not only all over Titan, but throughout most of the Saturnian system. For example, because of Titan's thick atmosphere and low gravity, an eight-tonne nuclear thermal flight vehicle operating in an air-breathing mode in Titan's atmosphere at a flight speed of 160 km/hr would require a wing area of only four square meters to stay aloft—in other words, almost no wings at all. Employing the methane as rocket propellant in an NTR engine, a specific impulse of about 560 seconds (5.5 km/s exhaust velocity) could be achieved. The ΔV required to takeoff from Titan and go onto an elliptical orbit with a minimum altitude just above Saturn is only 3.2 km/s. Because the specific impulse of the rocket is high and the required mission ΔV is low, the mass ratio of the Titan-Saturn NTR ferry would only have to be about 1.8, which means that it could deliver a great deal of cargo. The cargo shipped downward to Saturn would be released in pods equipped with aeroshields that would allow them to brake from the elliptical transfer orbit down to the low circular orbit of a Saturn He-3 processing station, which supports the operation of the Saturn-diving NIFTs. After releasing the cargo pods, the ferry would continue on its elliptical orbit until it reached its apogee at Titan's distance from Saturn, just six days after its initial departure. Because Titan's orbital period is sixteen days, it would not be there to meet the ferry. So a small rocket burn would be effected that would raise the orbit's periapsis (low point) a bit, thereby adjusting the orbital period of the ferry to ten days, allowing it to rendezvous with Titan and aerobrake and land on the next go-round. Most of the cargo delivered to low Saturn would be supplies or crew for the orbiting NIFT base. However, some could be pods filled with methane propellant. These could be stockpiled at the orbiting station. When enough are accumulated to enable the nine km/s ΔV needed to travel from low Saturn orbit onto a trans-Titan trajectory, a ferry could aerobrake itself and go to the station, and then be used to ship crew or cargo back to Titan.

Alternatively, it might also be found desirable to use some of Saturn's lower moons (several of which are quite sizable and may represent developable worlds in their own right) as intermediate bases. This could make ferry operations a lot easier.

The propulsion requirements to travel from Titan to Saturn's other moons is shown in Table 4. Each excursion involves landing on the destination moon twice, engaging in activity at two locations separated by up to 40 degrees of latitude or longitude, and then returning to aerobrake and refuel at Titan.

Since methane is more than six times as dense as hydrogen, NTR vehicles using methane propellant should be able to achieve mass ratios greater than eight. It can be seen that with such capability, Titan-based NTR vehicles will be able to readily travel to and from all of Saturn's moons, except Mimas.

In certain ways, Titan is the most hospitable extraterrestrial world within our solar system for human colonization. In the almost Earth-normal atmospheric pressure of Titan, you would not

Table 4: Titan-Based Methane Propelled NTR Excursions to Saturn's Other Satellites

Destination	Distance from Saturn	Radius	ΔV	Mass Ratio
Mimas	185,600 km	195 km	13.17 km/s	11.0
Enceladus	238,100	255	11.25	7.77
Tethys	294,700	525	10.05	6.24
Dione	377,500	560	8.60	4.79
Rhea	527,200	765	6.91	3.52
Titan	1,221,600	2575	0.00	1.00
Hyperion	1,483,000	143	3.84	2.01
Iapetus	3,560,100	720	6.90	3.52
Phoebe	12,950,000	100	8.33	4.56

need a pressure suit, just a dry suit to keep out the cold. On your back you could carry a tank of liquid oxygen, which would need no refrigeration in Titan's environment, would weigh almost nothing in the light gravity, and which could supply your breathing needs for a week-long trip outside of the settlement. A small bleed valve off the tank would allow a trickle of oxygen to burn against the methane atmosphere, heating your breathing air and suit to desirable temperatures. With 1/7th Earth gravity and 4.5 times terrestrial sea-level atmospheric density, humans on Titan would be able to strap on wings and fly like birds! (Just as in the story of Daedalus and Icarus—though being more than nine times distant from the sun than Earth, such fliers wouldn't have the worry of their wings melting). Electricity could be produced in great abundance, as the 100 K heat sink available in Titan's atmosphere would allow for easy conversion of thermal energy from nuclear fission or fusion reactors to electricity at efficiencies of better than 80 percent. Most importantly, Titan contains billions of tons of easily accessible carbon, hydrogen, nitrogen, and oxygen. By utilizing these elements together with heat and light from large-scale nuclear fusion reactors, seeds, and some breeding pairs of livestock from Earth, a sizable agricultural base could be created within a protected biosphere on Titan.

Moving Iceteroids

As one moves out through the asteroid belt toward Jupiter, and then beyond, we find bodies increasingly composed of volatile material. This is to be expected. An asteroid made of ice would vaporize if it orbited for long near Earth, and other volatiles, such as ammonia or hydrocarbons, would evaporate from asteroids orbiting near Mars or even in the central Main Belt. It is therefore reasonable to assume that this trend continues beyond Jupiter, and that the outer solar system should be rich in asteroid-sized objects consisting almost entirely of frozen volatiles, such as water, ammonia, and methane ice. As of this writing, only one major ice asteroid or "iceteroid" is known, but that one, Chiron, orbiting between Saturn and Uranus, is rather large (180 km diameter) and it's a rule of thumb in astronomy that a lot of small objects can be found for every big one. In all probability, the outer solar system contains thousands of asteroids that we have yet to discover because they shine so dimly compared to those in the Main Belt. (The brightness of an asteroid as seen from Earth is inversely proportional to the fourth power of its distance from the sun.) It is known conclusively that starting not far beyond Neptune (which orbits at 31 AU) there is an enormous zone known as the Kuiper Belt that contains millions of ice objects. Beyond that lies the astronomically still vaster domain of the Oort Cloud, stretching out more than a light-year (64,000 AU) and home to trillions of frozen objects. It is from these regions that comets originate.

Once nuclear fission or fusion thermal rockets become available, an object made of volatiles is basically an object made of rocket propellant. Such objects, including very large ones, can therefore be moved about the solar system in accord with human designs.

The late twenty-first century will see widespread human activity throughout the inner solar system ranging from mining asteroids and the moon to terraforming Mars. Many of these activities, especially terraforming, will require the importation of large quantities of volatiles for their support. The easiest way to move a lot of stuff around the solar system is in the form of an asteroid. But why go to the outer solar system for it? The reason, strange as it may seem, is that it is easier to move an asteroid from the outer solar system to Mars, for example, than it is to do so from the Main Belt or any other inner solar system orbit. This odd result follows from the laws of orbital mechanics, which cause an object farther away from the sun to orbit it slower than one that is closer in. Because an object in the outer solar system moves slower, it takes a smaller velocity change (or ΔV) to alter its orbit from a circular to an elliptical shape. Furthermore, the orbit does not have to be so elliptical that it stretches from Mars to the outer solar system—it is sufficient to distort the object's orbit so that it intersects the path of a major planet, after which a gravity assist can do the rest. For example, moving an asteroid positioned in a circular orbit at 25 AU, by way of a Uranus gravity assist to Mars, requires a ΔV of only 0.3 km/s, compared to a 3.0 km/s ΔV to move an asteroid directly to Mars from a 2.7 AU position in the Main Belt.

Consider an asteroid made of frozen ammonia with a mass of 10 billion tonnes (2.2 *trillion* pounds) orbiting the sun at a distance of 12 AU. Such an object, if spherical, would have a diameter of about 2.6 km, and changing its orbit to intersect Saturn's (where it could get a trans-Mars gravity assist) would require a ΔV of 0.3 km/s. If a quartet of 5,000 MW nuclear thermal rocket engines powered by either fission or fusion were used to heat some of its ammonia up to 1,900° Centigrade (5,000 MW fission NTRs operating at 2,200° C were tested in the 1960s), they would produce an exhaust velocity of four km/s, which would allow them to move the asteroid onto its required course using only 8 percent of its material as propellant. Ten years of steady thrusting would be required, followed by a about a 20 year coast to arrival. If the object were suitably fragmented in advance, it could be allowed to enter and vaporize in Mars' atmosphere. In the course of doing so, it would release about ten TW-years of energy, enough to melt one trillion tonnes of water (a lake 140 km on a side and fifty meters deep). In addition, the ammonia released by a single such object would raise the planet's temperature by about 3° C and form a shield that would effectively mask the planet's surface from ultraviolet radiation. Forty such missions would double the nitrogen content of Mars' atmosphere by direct importation, and could produce much more if parts of the iceteroid were kept big enough to hit the ground and targeted to hit beds of nitrates, which they would volatilize into nitrogen and oxygen upon impact. If one such mission were launched per year, within half a century or so most of Mars would have a temperate climate, and enough water would have been melted to cover a quarter of the planet with a layer one meter deep. Of course, Mars colonists might be a bit leery of using big iceteroid chunks as surface impactors, and even mere wholesale mass atmospheric entry of small fragments might make some squeamish. Even so, they could be accommodated. By using Jupiter, Venus, and then Mars itself in a succession of gravity assists, the object could eventually be brought into a sedate inner solar system orbit where its contents could be chipped off and shipped off at will in suitably sized packets to meet a host of human purposes.

The farther out one goes, the easier it becomes to ship larger and larger masses inward. So inevitably, the vast reaches of the Kuiper Belt and the Oort Cloud will become arenas of human economic activity, and then settlement.

For of all human possessions, the most precious is freedom. The same wanderlust and reach for new possibilities that drove the old folks to settle Mars and the asteroid Main Belt in the twenty-first century will move their descendants a century or more later to try their luck among the million untamed worlds of the Kuiper frontier. Why go? Why stay? Why live on a planet whose social laws and possibilities were defined by generations long dead, when you can be a pioneer and help to shape a new world according to reason as you see it? The need to create is fundamental. Once started, the outward movement will not stop.

The Road to the Stars

The two main obstacles to settling the outer solar system are power and transportation. As mentioned earlier, solar energy in the realm of the gas giants and beyond is negligible. However, in the era we are discussing, we can expect that fusion, powered by He-3 will be the dominant energy source. Indeed, the need to acquire He-3 to fuel such systems will be one of the prime motivations for the colonization of the far worlds of the outer solar system.

As to the issue of transportation, current space transportation systems may be considered First Generation. These are sufficient for launch into Earth orbit, for manned missions to the moon, Mars, and near-Earth asteroids, and for limited-capability unmanned probes to other planets. For colonization of the inner solar system, out to the Main Belt, we need to move on to Second Generation systems, typified by nuclear thermal rocket propulsion, nuclear electric propulsion, and advanced aerobraking technology. Such Second Generation systems also open up capabilities for vastly expanded unmanned exploration of the outer solar system. They are, however, marginal for manned colonization of Titan, as the three- to four-year one-way flight times they impose upon this mission are excessive. However, as the fusion economy initiated by the moon's supply of He-3 grows, demands will develop that can only be satisfied by the vastly larger stocks of this substance available in the outer solar system. By improving the in-space life support systems associated with Second Generation technologies (and by moving from the Second Generation's simple air and water recycle in the direction of closed cycle ecology as the basis for very long-term life support) a few pioneers will make their way to Titan using Second Generation transportation technologies. Once even a small base is established on Titan, there will be a tremendous incentive to develop Third Generation systems, such as fusion propulsion (especially since we will then have the abundant He-3 supplies needed to fuel them). This will allow for quick trips and rapid development of Titan and the rest of the outer solar system. Such Third Generation propulsion systems, however, together with fully Third Generation closed cycle ecological life support, will enable travel beyond the nine known planets to the Oort cloud, and, when advanced to their limits, create a basis for interstellar missions, with flight times to nearby stars on the order of fifty to 100 years.

Humans will go to the outer solar system not merely to work, but to live, to love, to build, and to stay. But the irony of the life of pioneers is that if they are successful, they conquer the frontier which is their only true home, and a frontier conquered is a frontier destroyed. For the best of humanity then, the move must be ever outward. The farther we go, the farther we will become able to go, and the farther we shall need to go. Ultimately the outer solar system will simply be a way station towards the vaster universe beyond. Just as Columbus's discovery of the New World called into being the full-rigged sailing ships, steamers, and Boeing 707s that allowed the rest of humanity to follow in his wake, so those brave souls who dare the great void to our neighbor stars with ships of the Third Generation will draw after them a set of Fourth Generation space transportation systems, whose capabilities will open up the galaxy for humankind.

Such is the adventure we now begin, as we midwife the birth of Type II civilization. Today, that newborn infant is still weak and vulnerable, its true potential barely imaginable. But given time, it will mature into something truly grand. Occupying vast reaches of space, wielding matter and energy in enormous quantities, and advancing its knowledge and abilities at an ever accelerating pace through the enabled talents of tens of billions of free people, interplanetary humanity will seek out ever greater challenges. Type III will beckon.

For while the stars may be distant, human creativity is infinite.

Habitats can have hidden snares, and some we may have no intuition about until they occur: a cautionary tale.

Knotweed And Gardenias
Nancy Kress

*T*he Western Alliance Space Agency ship *Javelin* moved silently through the void, five AUs out from the sun. The first to attempt the long journey to Titan, she was a thrust into the darkness, a hope of glory and riches, mankind's first bid to leave the inner planets. Her mission was clear: set up a mining colony so that robotic craft could reap D-He-3 from Saturn to feed clean fusion reactors on a power-hungry Earth.

Aboard her, a nuclear reactor powered her variable specific impulse magnetoplasma drive. Aboard her, three-quarters of the crew lay in rotating cold sleep. Aboard her, there flourished the most advanced closed ecology ever built, a moveable Eden.

Aboard her, Master Chief Petty Officer Arvid Larsson wielded a laser cutter, doing his best to destroy the hull.

"This is not supposed to happen," Erik said, and glared at Marianne. That was the moment she knew. Certainty was reinforced when he added, "Dr. Gioserro, tell us whatever you can."

Dr. Gioserro, not *Marianne.* Informality among the officers had become the rule aboard ship, from the first cycle on. (*"Informality, is that what you call it in space?"* her sister Jeanine's voice jeered inside her head. Which was, of course, another data point in itself.)

"Well?" Erik snapped, and she disliked his tone, disliked his German accent, disliked him. *More data points. Pay attention.*

Marianne said, "I don't have any information yet. Not until I talk to Larsson." What she meant was, *You are not going to make me the scapegoat for this.*

Except that Captain Colonel Erik Lorenz did not create scapegoats. He was a fair, calm, respectful leader, a stable extrovert strong on cooperation, empathy, and libido. Marianne, ranking ship psychologist for the *Javelin,* knew the first eight characteristics from batteries of tests, observations, brain scans, genetic analyses, neurotransmitter measures—every tool known to the twenty-first-century human desire to understand itself. The last quality she had determined privately. Although not lately.

Another data point.

Erik said, "Obviously you have not talked to Larsson, since he's still unconscious. But surely you at least have some up-to-date observations on how this could happen? That is your function here, after all."

Erik, usually formal, seldom sounded *that* stiff. Yet another data point.

To Marianne's left, David Chung, the other American present, gave a small start of surprise. His long-fingered hands, perpetually black with soil or green with algae, twisted tightly together. David tested very high on sensitivity to conflict. He was also a genius with closed environment agriculture. Marianne needed him now, just as she needed Juana, who was attending to Larsson. Karen sat across the table with her head lowered and her shoulders hunched as if she were not in the wardroom at all.

It was a pleasant place, although not much like the wardroom Marianne remembered from the last cycle. During her three months in cold sleep, Crew 3 had taken down two algaplast walls and opened up the conference and dining area to the farm. Across Juana's hunched shoulders Marianne could see the recreation garden, and beyond that, the rows and rows of David's domain, soaring racks and tanks of algae curving completely around the hull and filling all the space in between, crossed by catwalks and pipes. The *Javelin* rotated to create one G, and in the living quarters and garden only half the curving hull had been built upon. The other half was empty, as if it were the sky overhead. But for the precious algae, they needed every inch of usable space.

The wardroom furniture, too, had been replaced. The new table was lower, inlaid with a pattern of bright mosaic-work that somebody had adopted as a hobby, and surrounded by elaborately embroidered, very firm cushions at an optimum ergonomic height. Everything on the central living pod of the *Javelin* was recyclable—walls, furniture, human waste, air, food—and the crew had been selected in part for its ability to keep itself occupied and interested during the one-and-a-half years it would take to reach Titan.

Arvid Larsson, expert machinist, had occupied and interested himself in trying to cut a hole in the hull of the ship.

He had not succeeded, of course. Alarms, security bots, forceful restraint—but not before Larsson had turned the laser cutter, which should not even have been unpacked until Titan, on himself and severed his left arm at the elbow. Dr. Juana Pinero had reattached it and now sat beside Larsson in sick bay, which was where Marianne wished she were instead of here.

"Captain," Marianne said, the word strange on her lips after the things her lips had done to Erik, "Larsson's psychotic episode is only part of a larger issue here. I have tried to bring this up before. I want to run full physicals on everyone awake, immediately. Juana agrees with me. The—"

Erik said, "We had full physicals three weeks ago, at crew shift change. You and Dr. Pinero told me everything looked fine."

"It did, three weeks ago."

"And in three weeks CPO Larsson developed a 'psychosis' and all the rest of us are at risk for also destroying the ship? I am not believing this, Doctor."

Marianne held onto her temper—which should not be flaring at all. She tested just as stable as Erik did, although more extroverted. Juana was the dramatic one. A crew needed a mix of personality types to balance each other. It did not need an arrogant, self-righteous prick like Erik to—

Whoa. Stop.

(*"You're not as smart as you think after all, are you?" Jeanine sneered.*)

"Not at all. But all of us, or anyway most of us, are showing signs of mental destabilization, to various degrees." She looked over at Executive Officer Karen Nelson-Jones, still slumped and silent. "Classic clinical depression: irritability, hopelessness about the future, disturbed eating and sleeping patterns, difficulty concentrating, fatigue, loss of interest in once pleasurable activities"—a pointed pause—"including sex."

Erik's face had darkened as she talked, making Marianne talk faster (*dumb, dumb*)—until she got to the word "sex." He paused, looked at her hard, opened his mouth to say something. Before he could get it out, she said, "And David wants to say something, too."

David didn't look as if he wanted to say anything. His plant-stained hands now twisted together so hard that the knuckles were bloodless, white as lightless roots. But he raised his chin, which trembled, and got out, "The algae."

"What about the algae?" Erik snapped.

"Their division rate has slowed."

Silence. The ship depended on algae, was built around algae. They needed algae the way Earth needed the sun.

Erik said, "Slowed how much?"

"Not lethally. Not yet, anyway. By 2 percent."

"What is causing this?"

"I don't know."

"How in *scheiss* does a slowdown in algae division connect to Marianne's so-called 'clinical depression' in *us*?"

She was "Marianne" again, but now it felt like a demotion, as if he were classing her with children who didn't know what they were talking about. His reversion to German was also a bad sign; Erik only did that under extreme stress (or orgasm, yes). This time Marianne didn't restrain her temper. "I don't know the answer to that, *Erik*. I only know there is a problem—two problems—and that we had better do something about it, starting with gathering hard data. And it's not a 'so-called' depression and destabilization—just look around you, for Chrissake!"

David had slid so far down on his cushion that only his head and bloodless hands were visible above the table edge. Karen Nelson-Jones bit her lip and looked confused, although her IQ was higher than anyone else's. Marianne felt her own face warm with fury. Before Erik could answer, Dr. Juana Pinero strode in through the garden and plopped down on a cushion. "Larsson just died," she said flatly. Then she pounded the table with one fist, so hard that one of the mosaic inlays cracked.

Erik looked slowly at each member of his crew, finishing with Marianne. She saw that he had his irritability under control, submerged under his natural leadership ability. The Erik she knew was still in there. She saw, too, that he finally recognized how unmilitary all their behavior was, how far they had strayed from their usual selves.

"Ja," he said. "Marianne, David—run your tests. When you have data, bring me some hypotheses. Soon."

Juanita's and Marianne's data were conclusive, and troubling. David's were even worse.

Marianne found him forty feet above the deck, working deep amid the dense, soaring arcs of the algae farm. Algaplast, a construction material produced by the very algae it then supported, made the narrow walkways and catwalks. The algaplast fabricator, dubbed TinkerToy, spun algae fibers and glued them together with an alga-based adhesive, each layer rotated 90 degrees to the one below, to make sheets that were strong, flexible, and biodegradable.

The larger plants in the garden provided beauty and some crops, but David's beloved algae were the heart of the ship. They took in CO_2, human waste, and the intense light provided by their own auxiliary nuclear reactor, and gave back clean air and water. They made high-nutrient food for humans, the larger plants, and the shrimp and tilapia in the hydroponics tanks. Genetically modified algae built to overexpress certain genes provided materials for algaplast, for algatex for fabric, and for medicines. Everything on the *Javelin* except fuel came from algae, and with a net nutrient loss of less than 1 percent a year. The whole ingenious, glorious, critical enterprise was continuously monitored by a computer program affectionately known as the Jolly Green Giant. It identified genetic drift and the constant reproductive errors caused by cosmic radiation that got through the water tanks that wrapped around the habitat cluster..

"Ready?" Marianne said to David.

He nodded, but as he followed her, his footsteps lagged. At the doorway to the wardroom he whispered, "Holy shit."

Even if Marianne hadn't seen the data, she would have known from the wardroom what was happening to Erik. In two days, he had had the room utterly changed. Crewmen had replaced the two missing algaplast walls and made the space smaller. Gone were the colorful cushions and inlaid table, undoubtedly recycled. The new table was gray, as were the chairs and the walls. The room felt utilitarian, business-like, sterile.

Depressed.

Erik, Karen, Marianne, David, and Juana took hard and uncomfortable seats, and Marianne forced herself to speak. "Every crew member awake shows all the physiological markers for depression. Radically low levels of serotonin and norepinephrine. Elevated levels of cortisol. Abnormal activity in the amygdala, which controls primitive emotions and—"

"We all have read the data," Erik interrupted. A narrow-eyed glance, tightened lips. "Move on, Dr. Gioserro."

"Very well. My observations have verified various forms of advanced depressive behavior in two-dozen people, including all of us." She looked around: Erik scowling and diamond-filament stiff, trying to compensate for how lousy he felt. Karen slumping, hopeless and foggy-eyed. Juana drumming her fingers on the ugly table—*I wish she'd stop that!* David looking as if he expected to be stomped on.

Erik sat even straighter, something Marianne would not have thought possible. And his lips were white. "Dr. Pinero?"

Juana said, "I want to emphasize that everyone is healthy otherwise. No infections, degeneration, or organ malfunction, including the usual suspects in biological depression. And I agree with everything Marianne just said."

"Should we test the other three crews?"

Juana said, "It can't be done in cold sleep. We could wake them, of course, but Marianne and I have gone through data from when they were last awake, a month ago. Nobody showed any markers for depression. Whatever is causing this problem, it's occurred since then."

"Dr. Chung, I did not receive your report."

"No," David said, barely audible. "I was—"

"Speak up, damn it!"

"I was still checking it. I couldn't believe…it isn't….the algae division rate has slowed down by 6 percent."

Stunned, Marianne swiveled her head to look at him. *Six percent.* They all knew that a nutrient loss of 10 percent was fatal. The ship did not, could not, carry sufficient replacement nutrients. And the air—

"Then we're all going to die," Karen said—a statement so premature and appalling from an executive officer that Marianne knew they were in even deeper trouble than indicated by David's data. Giving up could indicate the most extreme form of depression: suicidal ideation. Karen would need to be watched closely.

Juana, whose reaction to depression was irritability even worse than Erik's, snapped, "We don't need that shit, Karen, we need *ideas.* If you don't have any, shut up."

Erik said coldly, "Dr. Pinero, control yourself. Dr. Chung, what are you doing about the algae?"

"What I can," David quavered. "I adjusted the nutrients and the light levels, trying to increase the reproduction rate. I'll have results in a few hours."

"And Dr. Gioserro, what do *you* recommend?"

"Antidepressants, to start. Juana and I agree on that. But you should know that about 30 percent of people have no response at all to even the best antidepressants, and furthermore that it can take a few weeks to figure out the correct medication and dosage for each person. The brain is still largely a mystery."

"So ordered. But we also need a cause for all this. Dr. Chung?"

"I don't know."

"Damn it, that is not helpful!"

Marianne came to David's rescue. "It must be something ship-wide, Captain. External radiation, air composition, gravity, light spectra or intensity. Those are the only things I can think of that would affect both humans and algae. The last four have remained constant over the nine months we've been in space, so that leaves radiation."

Erik said, "The monitors show no new radiation."

"But it may be that the accumulation over time has reached critical levels in our bodies."

Juana said, "I'm seeing no evidence of radiation sickness, Marianne. None."

"But some new kind of—"

"The human body isn't any 'new kind of,' even if the radiation was! And you're really not equipped to diagnose that."

Marianne held onto her temper. "I know that. I'm just suggesting…all right then, I thought of one more possibility. If no factor aboard ship has changed, then the only thing left is that our response to them has changed over time. Human and plant response. Time is the critical factor. After all, we've been in space much longer than any previous mission."

She watched them consider this. With the advanced high-burn fusion drive, Mars trips were only forty days each way, and until now the dry Martian plains had marked the limit of human colonization.

Juana said, "Circadian rhythms—"

"That what I'm thinking," Marianne said. All the research on human reaction to light deprivation, such as above the Arctic Circle for people not indigenous to the locale, show massive incidences of clinical depression. And experimental volunteers who have stayed below ground for—"

"We do not have light deprivation!" Erik said. "We have night and day in the living quarters and the right light for the garden and farm—do we not, Dr. Chung?"

"Yes," David quavered. "It's optimized for spectral balance, timing, heat generation, and—"

Erik said, "Dr. Gioserro, you are not making sense!"

"I think it's the light," Marianne said stubbornly. "Unless someone else has another idea?"

No one did. Erik told them to design, implement, and report on light changes affecting circadian rhythms. Juana would do functional scans of the crew's suprachiasmatic nuclei, the small bundles of neurons near the optic nerves that largely generated circadian rhythms. She would also scan the two awake pets, Katze the ship cat and Karen's toy poodle, M'sieu Byte. Crew would be asked to keep charts of their own subjective mental states.

"Dismissed," Erik said, and strode back to the bridge. Juana rushed out of the reconfigured wardroom; David crept out. Karen remained until Marianne put a hand on her shoulder.

"Karen, you should return to duty."

The executive officer of the WASA ship *Javelin* looked up, face twisted, and said, "We're all going to die."

Marianne got Karen put on the sick list and confined to quarters under close watch. For a week Marianne sat at her screen reading research reports and conferring with colleagues and mission control on Earth. The hundred-plus-minutes lag made these conferences clumsy; Terran bafflement made them useless. No one had any insights to add to what Marianne already knew about both circadian rhythms and clinical depression.

She assisted Juana in matching antidepressants to officers and crew. And she observed her own behavior.

Atypical depression.

She was sleeping too much, unlike nearly everyone else, who was having trouble sleeping at all. She was eating too much. Most of all, she was feeling too much. Her emotions were not damped down into near-catatonic despair, as Karen's were, nor translated into David's fear, Erik's and Juana's irritability, everybody's leaden grayness. Instead, she was elated by the flight of a butterfly in the garden, devastated by a malevolent look from a crewman wrestling with his own demons. Helplessly she watched her mind convert all this over-reactivity into an intense, consuming, visceral homesickness, a longing for her simpler and protected childhood.

Seattle. The glint of sunlight on Elliott Bay. The clean salt wind from Puget Sound. The smells of rain in summer, apples in autumn, pine trees always. She and Jeanine running free and laughing through the forests on the Quinalt Peninsula, their own eternal Eden….

Stop. Reality check. Jeanine, three years older, had tormented and sneered at Marianne. If they had run free and laughing together, it had happened no more than two or three times. And that was the flip side of this atypical depression: Jeanine back in her head, with all the insecurities and self-doubt that had led Marianne to a career in psychology in the first place.

None of her rational thought helped. The homesickness remained, the longing for Earth. She should never have left it. She was a child of Terra, she belonged there….

Stop.

Around and around, and meanwhile the desperate work went on. Light aboard ship outside of the algae farm had been set to the spectrum of Sol, to the intensity of a summer day at 45 degree latitude, to sixteen hours of "day" and eight of "night." They tried different spectra, as in the special lighting used for sufferers of Seasonal Affect Disorder. They tried different intensities and durations of light. They tried artificial dawns and dusks, gradually brightening or dimming the illumination. They tried antidepressants of various types, in various combinations, in various dosages. They tried acupuncture and meditation and temperature variation.

Nothing helped.

Crew members began to neglect their duties. Some sat slumped in their quarters; some got into fights. Someone began synthesizing alcohol and more than one person, self-medicated, tried to carry out duties drunk. Two women got into a brawl over a chess game and one was struck badly enough to give her a concussion. Marianne found herself crying when she heard about it, which was beyond stupid. She continued to long for Terra in dreams, in reverie, in her heart. We are children of Earth. *We do not belong here.*

Katze lay curled in a corner and refused to eat.

M'sieu Byte bit a crewman, who then kicked the tiny dog.

The algae reproduction rate slowed by 7 percent.

And Lieutenant Commander Karen Nelson-Jones killed herself in her quarters.

The human body did not do well with more than three months at a stretch in cold sleep. Too much loss of muscle tone, too much gum recession, too much risk of nerve-coordination damage. However, cold sleep was by far the best way to preserve resources. More to the point, there was just not enough to do aboard ship for the full crew to be kept awake, no matter how many hobbies, chess matches, or structured studying. So only one-quarter of the crew was awake at any given time, plus the core officers, of which they now needed another one to replace Karen.

Juana, Erik, and Marianne all stood beside Lieutenant Commander Neil Baransky's cold-sleep pod as the new exec was awakened and briefed. Erik informed him of his promotion and told him to report at 0300 to the captain's quarters. Juana, just as curt as the captain, checked Baransky's vitals, ran him through various physical movements, took fluid and tissue samples. Then both rushed off—to where? Marianne wondered irritably—and left Baransky to her.

She was shocked at his appearance, and shocked at her own shock. This, then, was what they had all looked like just six or seven weeks ago: smiling, calm, alert, ready to take on whatever came next. How long would all that last? He was her canary in the coal mine, even as the miners were already breathing poison gas.

"Neil, the Captain told you that Karen Nelson-Jones killed herself, but not why. We have a situation aboard ship." Quickly she filled him in, watching his eyes widen. With surprise, yes, but also with skepticism?

"It's important that I monitor your mental state regularly, since it may provide us with clues about what is happening here. Captain Lorenz wants me to give you tests each week, starting now."

"I'd prefer to hear that order from the captain," he said pleasantly.

"Of course." But instantly her mood plunged; he didn't trust her. She wasn't respected, wasn't accepted…. *Steady, Marianne.*

He comlinked Erik, was barked at, and submitted to the psychological evaluation with good grace. She already had his file, and nothing deviated from it now. Neil Baransky was a stable extrovert with a good grasp of reality, strong will, middle to low empathy, good social skills over an underlying mild narcissism, and a lamentable fondness for puns. He had, however, the good sense to not make any now.

Marianne realized she didn't like him. But, then, she didn't like much of anything these days.

"Oh," jeered Jeanine in her head, *"Little Miss Perfect can't make any more sunshine? Poor poor baby."*

"We're done," she snapped at Baransky, startling him. But she didn't explain, even though she knew she would feel guilty about that later. Guilty, foggy, hopeless.

Stop it. This isn't really you.

But if not her, then who?

The thing about depression, Marianne thought with great unoriginality, is that when it was not suicidal, it was monotonous. For most everyone on board, each small action felt like digging through a brick wall with a teaspoon in gray drizzle: difficult, endless, exhausting. And it hurt.

How many depressive patients had she treated back on Earth, without really understanding what it felt like? She hadn't understood the active anguish, like an icy wind in the mind. The draining anxiety. The simple fact that everything, mind and body, *hurt*.

She caught herself doing less and less. Even reading was an enormous effort, each paragraph a tangle of black thorns not worth the pain of pushing through. "The black dog," a British statesman had called his clinical depression two hundred years ago. He had feared doing what Karen Nelson-Jones had done: "*I don't like standing near the edge of a platform when an express train is passing through. I like to stand right back and if possible get a pillar between me and the train. I don't like to stand by the side of a ship and look down into the water. A second's action would end everything. A few drops of desperation.*"

And yet Winston Churchill had beaten off the fangs of his black dog, again and again. With work, with too many whiskey-and-sodas, with grandiose ambition. Marianne, in her deepest trough yet, should be helping everyone else on board to recover. She should be doing more, thinking more clearly, she should….

She was so tired.

Churchill had not had an entire nation of depressives to work with. There had been at least some whole people around him, some people out of the miserable, drizzling gray fog that filled the *Javelin*. Marianne had no one. No, not true—she had David, Juana….

Juana was increasingly eccentric, David increasingly withdrawn. His last comment to her had been, "Some people are knotweed, which can grow anywhere, and some are gardenias." True but not helpful.

She should get up and back to work. She should….she must…

"*Well, well, stuck at last?*" jeered Jeanine, her personal tangle of thorns, her black dog.

Marianne stood. It physically hurt, as she hurt everywhere. No, she had to get up, had to find an answer—

She lay back down on her bunk, praying for the oblivion of sleep.

In the middle of the afternoon, a knock on her cabin bulkhead woke Marianne, who shouldn't have been asleep in the first place. Groggily, she slid back the door panel. Erik stood there.

"May I come in?"

Wordlessly, sleepily, she stood aside to let him pass. To save shipboard space, cabins were small, with a bunk, desk with data screen, and storage lockers. They were meant for sleeping, studying, reading, and sex, not socializing. Erik, massive and masculine, filled the space and his scent wafted into her nostrils. Marianne woke fully.

"I want to apologize for my behavior lately," Erik said stiffly. "It is not how a captain should treat valued officers."

Marianne felt her spirits soar. Hormones kicking in. *Over-reaction. Atypical depression.* But all she said was, "It's not your fault."

"Yes, it is. I cannot accept anything else, Marianne. Blaming some sort of deficiency in the brain just…do you not see where that leads? Any sort of act—dereliction of duty, cowardice, murder—becomes no more than the product of brain chemistry. It completely leaves out personal responsibility. I am more than my biology!"

She had never seen him like this. Not the depression—she had seen that. The vulnerability.

"I do see what you mean, Erik. I do."

"Good. I merely wanted to apologize." He turned to go.

"No—wait. I want to say something, too."

"What?"

What did she want to say? *Please stay, I miss you, can't we help each other through this, hold me through the long night.* But everything in his body language said that no such declarations would be welcome. So instead she said, in as calm a voice as she could manage, "We are the children of Earth."

"What?"

Seattle, beautiful and lost, the glint of sunlight on Elliott Bay. The clean salt wind from Puget Sound. The smells of rain in summer, apples in autumn, pine trees always. She and Jeanine running free and laughing through the forests on the Quinalt Peninsula, their own eternal Eden….

Then, all at once, she saw it.

The over-reactive atypical depression kicked in again—who cares? She *saw* it—and Marianne laughed and seized Erik's arm. He pulled away. She laughed again, sounding even to her own ears shrill and demented.

"We *are* the children of Earth! Erik—I know what the problem is!"

He blinked.

"Not the solution, not yet—but at least the problem. I know what it is."

She was elated, upbeat—too elated, too upbeat. The others in the small, ugly wardroom that Erik had created out of his depression plus his military attempt to cope with depression, felt claustrophobic. The Marianne that was temporarily manic snapped her fingers at the ugly gray walls. The Marianne that was a rational psychologist, a WASA staff officer, fought for control and lowered her voice.

"We understand quite a bit about circadian rhythms in human beings and in plants," she told Erik, Juana, David, and Neil. "Physiologically, behaviorally, genetically. But chronobiology knows hardly anything about circannual rhythms, except that they exist. Animals mate in spring, leaves change color in autumn, geese migrate south and then north again—all that happens in response to seasonal changes in light. Human beings show some reaction to seasons, too, but not much. We originated in Kenya, where seasonal changes are minimal—we *colonized* the rest of Earth. We don't even know where in the body circannual rhythms are controlled, the way the suprachiasmatic nuclei controls circadian rhythms. The best guess is that control is diffused throughout the body."

Neil Baransky nodded, waiting for the pay-off. Juana was drumming her fingers again. David slumped in his chair as if his bones had all dissolved. Erik scowled, as distrustful of her theory as when Marianne had given it to him in her cabin. Suddenly her mood deflated. The idea now seemed dumb, a desperate grasping after pathetic hope…. She fought that feeling too, and went on.

"The crew awake on the last cycle was not affected. That argues that time was needed to reveal the problem—such as the ten-and-a-half months we've been aboard. Nearly a year. Shipboard environment lacks the clues to circannual rhythm that both animal and plant bodies expected and the reaction is violent enough to—"

"What clues?" Juana interrupted impatiently. "David said he tried duplicating all the yearly light rhythms and nothing improved. Anyway, this environment, light and temperature, was tested for *five years* in closed domes on Earth and nothing like this happened!"

"Because the domes were on Earth, and they—Juana for Chrissake stop that finger drumming!"

Juana snapped, "If you think you can—"

"Stop, both of you," Erik said, and Juana subsided. Marianne could see the effort it cost her. But Juana, no matter how irritably disordered, was still a military officer.

"Sorry, Captain."

"Sorry," Marianne echoed. "As I was saying, we tried to duplicate all the triggers for circannual rhythms that the body needs, in terms of light. But a whole raft of experiments show that animal circannual rhythms operate even in constant light and temperature situations. Regular as clockwork, there are yearly or near-yearly fluctuations in milk production, in molting, in sperm volume. No one knows yet how the brain cooperates with the endocrine and nervous systems to control those rhythms, and they seem to be endogenous. But they might be influenced by factors on Earth that we don't have aboard ship. Tides, caused by the moon. The magnetic field around the planet. Maybe more things that I haven't thought of yet. It's possible that the absence of those is causing—don't look at me like that, all of you! Don't you understand how little we know about the brain, even now? About how it functions away from Earth for this long?"

Neil said softly, "Take an even strain, captain."

Juana said, "How in hell are we supposed to duplicate the tides of the *moon*?"

But David had sat up straighter. He had bones again. "Marianne might have something here. Algae have circannual rhythms, too—seven- or ten- or twelve-month variations in chloroplast movement, mitosis, growth rate, cell aging, all independent of light and temperature. Why, P. californica—"

Erik cut him off. "Why wasn't this taken into account in mission planning?"

Marianne restrained herself from saying, *Because there's never been this sort of mission, you moron!* Erik knew that.

Neil said, "Deer. And cows."

"What about deer and cows?" Marianne said.

"When they're relaxing in a field, they all tend to align their bodies north-south."

She was surprised; she would not have taken him for a country boy. She turned to David. "Can any of this be tested on algae?"

"I recall cell cultures don't grow as well in zero magnetic field. Maybe plants feel some similar effect? Well…it might not be too hard to test the magnetic-field theory. Set up—"

Juana said, "Isn't Earth's magnetic field caused by electric currents in the molten iron core? You think you can duplicate *that*—what are you, God?"

David—surprisingly, everyone seemed full of surprises today—didn't wilt under her scorn. "You could set up a small solenoid generating the same strength of field as Earth, move algae inside it—somebody has to do the math to determine the exact placement—see if the reproduction rate rises…." He looked timidly at Erik.

Erik said, "What personnel do you need?"

"I don't know. A…a physicist and electrician and…. I don't know."

Marianne said, "Lieutenant Commander Addison, the astrophysicist, is in cold sleep."

"Wake him," Erik said.

Juana muttered, "This better work. A *moon,* for Chrissake."

Marianne pressed her lips together and hoped the magnetic field was the answer. Also, that she did not need to go to Juana—scowling, unhappy, distracted—for so much as a hangnail, not until this was all over.

The crew rallied—unevenly, but they got it done. The electrician's mate and a startled John Addison built the solenoid from cargo-stowed materials. Marianne felt proud of them, and the feeling sent her over-reactive depression into manic overdrive, an elation unappreciated by anyone except Neil.

"You're hopeful that this magnetized algae will thrive, aren't you."

"Yes." At this moment, anyway. "Are you?"

"I'm deeply attracted to the idea."

She laughed. "Oh, lame."

"But true. Nothing could kelp me away from it."

"Stop."

"Especially considering the oceans of information locked up in that seaweed."

"Neil—"

"Which I plan on examining with my friend Brine."

This could get tiresome very fast. Marianne snorted and turned away from him.

The solenoid, a long coil of superconducting wire, was the size of a ship's cabin. Inside the coil hung curved racks of plants: algae, radishes, kale genetically modified for yield and enhanced nutrition, even a bonsai that, like the larger trees in the garden, looked sad with wilted and discolored leaves. Sensors relayed critical data to the Jolly Green Giant.

At least it was a clean experiment. The fusion drive behind the hab cylinder had high magnetic fields, screened out by "flux suckers" that also caught stray radioactivity. Ordinarily the crew had no magnetic fields in their hab; now the algae would.

"Here goes, sir," the electrician's mate said, and turned on the solenoid. Marianne—no physicist—had half-expected at least a faint hum, but nothing happened.

"Now what?" Neil said.

"Now we wait on the algae, sir."

"Then I guess weed sea."

No one even smiled.

But two days later, Neil was no longer making puns. He complained first of feeling jet-lagged. "That's because all your organs are trying to re-set, as if to a new time zone," Juana told him grimly, "and they're all doing it at different rates."

"I'm—"

"No better or worse than the rest of us." But five days after awakening, Neil was much worse. Hopelessness settled on him like malodorous fog. He was unable to sleep, unable to eat, unable to concentrate. Marianne had never seen clinical depression come over anyone so quickly. A *gardenia,* she thought. It gave her renewed respect for Erik's sturdiness, and her own. For all of them still functioning, however badly. Beating back the black dog.

And the algae within the solenoid had rebounded.

"Reproduction rate, chloroplast movement, cell aging—all back to normal," David said. "The larger plants, too. It worked."

Juana scowled, in a deep trough. "So what? The solenoid is too small to rescue much algae. And if we walked into that cabin, we'd be smack on the centerline where the magnetic field is most concentrated. Who knows what that would do to our brains!"

"We're not going to walk into that cabin," Erik said.

Marianne looked at him. "What are we going to do?"

"We're going to wrap the whole ship in a solenoid."

Silence. Then she whispered, "Is that possible?"

"I don't know. Dr. Addison?"

John Addison looked stunned. "Well, you'd need a lot of wire.... We have the fabricator to manufacture that, although it's packed in the hold. Nobody thought we'd need it until we arrived at—Captain, you're not suggesting we go EVA and wrap the ship *outside*? Because the fusion drive generates its own field that shielding—"

"No. Can't the coil be constructed inside, against the bulkhead?"

"Possibly."

"All right," Erik said. "Is there anything else that—"

Juana did the unthinkable: she interrupted the captain. "Anything else! There's nothing but 'anything elses!' How will the ship's motion affect this artificial field? By now we're moving at—what percentage of c? And we're spinning to create gravity, for Chrissake! And what if someone drops something metallic toward the centerline of the field—does the whole thing flash off and ignite? It's an insane idea, Erik!"

Silence. No one wanted to say how unhinged this remark was. A small field had no dramatic effect. Then Erik—when had he risen to his feet? Marianne hadn't even registered that—said, "You are dismissed, doctor."

"You can't—"

"Dismissed." His voice was calm, firm, without irritability or despair, and Marianne knew how much it must have cost him for his training and beliefs and sheer will to override his state of mind. Erik had been right—he *was* more than his biology. And so was she.

After Juana had left, Marianne said softly, "Captain?"

"We will do it. Dr. Addison, give Dr. Gioserro a list of all needed personnel. Draw them from those in cold sleep, who will have—" he glanced at Neil, slouching listlessly "—the best chance of completing the project before they begin to succumb to this…this absence of Terran field. Keep a night-and-day watch on everyone on the building project. Awaken whomever you need and trust, including the head nurse to replace Dr. Pinero. And Dr. Chung—"

David looked up.

"Can you do this? With the algae? Whom do you need?"

"Just crewmen who will follow my orders."

"Oh, they will all do that," Erik said grimly. "This is a Western Alliance Space Agency ship. They will all do that, or answer to me."

They all did "that." Marianne marveled at how much everyone did. Not all the time, not with the speed or smoothness once expected, but they did it, and the doing helped. All the older, non-pharmaceutical aids to lifting depression helped the crew more than had their twenty-first-century medicine. Work. Exercise. Music. Perhaps even contact with growing things. Everyone's fingers began to look like David's: black with soil, green with algae.

Which comes first, chicken or egg? Does brain chemistry create emotion and acts, or do emotion and acts create brain chemistry? It was psychology's oldest riddle, and now there was a third player: Terra herself, punishing her children for leaving home.

Well, all kids leave home, Marianne thought, and held a section of algaplast as a crewman bolted it into place. The best way to observe people was to work alongside them.

One day awake, two days, three. The depression usually hit around the fourth day, although there were wide individual differences: sturdy daisies, fragile gardenias. Marianne didn't know what would be the effects of putting someone with a malfunctioning brain back into cold sleep, so she didn't. Those who were gardenias but could keep working under close supervision, did. The others were assigned to basically do the carrying that robots could have done, just to keep the crew active, unless they were incapable of even that. A few people went catatonic. Marianne kept them awake and revived more crew as needed. The ship's resources were strained to the limit. Rations were cut, despite the labor, but most crew had lost their appetite.

And they built.

Algaplast, amazingly flexible, they fabricated into whatever construction materials were needed, none of which were reactive with a magnetic field. The interior walls of the *Javelin* were shifted to create a long closed rectangle running the entire length of the living area straight down its centerline. This enclosed the strongest part of the field. The soaring racks of algae, plus the hydroponic tanks and the pipes that fed them and carried away reclaimed material, had previously filled two-thirds of the ship. The racks had grown "up" from the curving hull in all directions and met in the middle. Now they were crowded around the large enclosure, stacked on top of it until they reached the deck. The human quarters, which had used only half of the hull's curve, still did so, but shared it with more plants placed wherever they could be. The cabins and cold sleep areas along the hull did not, fortunately, have to be moved, except to cut into them to place the superconducting wire of the solenoid at the precise intervals to recreate the Terran magnetic field. The ugly wardroom disappeared. They held meetings in a space cleared between piles of algaplast. Crew slept where they could, as if this were a sixteenth-century sailing vessel. No one bathed enough. M'sieu Byte and Katze lay listless and snapped if anyone came near.

"It's going to be finished on the seventh day," Marianne told Erik. "Do you think that's too much hubris?"

He smiled, the first smile she had seen from him in weeks. Her hyper-reactivity kicked in and her spirits soared. *Steady, Marianne.*

His smile didn't last. "Let's see first if this works."

"If it doesn't—"She stopped, appalled at herself. If it didn't work, they might well all be dead. Nobody needed that said aloud, and a tacit agreement had grown up that no one would. Her mood plummeted.

"If it doesn't work," Erik said evenly—mouth pinched, eyes exhausted, body held upright by naked will—"we'll try something else."

That was the moment she knew she loved him.

Steady, Marianne.

No.

They had carefully moved everything metallic out of the area enclosed by the solenoid: not an easy task on a spaceship. Dr. Addison, pale as ice, stood by the computer to turn it on. The solenoid would generate a field equal to Terra's at the equator, 0.25 gauss. The physicist had carefully explained to her that it wasn't possible to duplicate many features of the Terran field, such as "the westward drift of the non-dipolar structures of the field at its surface." Marianne had stopped listening. They were doing what they could.

She didn't know what she expected: a flash of light, a glow, a burst of glory. Nothing whatsoever happened.

"Well," she said, to say something.

"Dismissed," Erik said. "People, get some sleep."

John Addison's assistant said, "Let there be magnetism and lo, there was magnetism." But then, he had only been awake two days.

Now we wait.

Marianne stumbled to her cabin, two-thirds of its former tiny size to accommodate a coil of wire behind algaplast barriers, and fell into sleep as if pulled by the weight of Earth itself.

She woke to a headache so severe that her stomach heaved. *Migraine,* she thought—but she didn't get migraines. Well, she had one now. Flashes of light pummeled her eyeballs. What if this was permanent?

Lurching from the cabin, she saw crew slumped against walls algae racks, holding their heads. A few had vomited. Plants seemed to swirl above her, below her, around her with claustrophobic pressure. *Erik,* she had to get to Erik and make sure he was all right…. She stumbled and fell off a catwalk built to connect the hull to the top of the enclosure, also crowded with racks of algae.

It was only a short fall, a foot at most, but her ankle twisted. Marianne sat on a rack of algae, rubbing her ankle, while the scent of seaweed rose around her in a miasma. She blinked back tears. Her head hurt so! *Breathe, breathe regularly….*

She fell asleep again.

When she woke, her head felt even stranger. It took her a moment to identify the sensation. Hands to her skull, she realized what it was.

She felt normal.

"Ma'am! Are you all right?"

A crewman, one of the youngest ones, peering down at her from the narrow walkway. His dirty, rosy face creased in concern—but only concern. She saw no despair, no uncontrollable irritability, no hopeless gray fog.

"I think I twisted my ankle," she said. "Help me up."

He did, handling her carefully. Marianne was grateful for the care, and only grateful. Not elated, not overwhelmed with joy—only grateful, which was in itself reason for more gratitude. M'sieu Byte tore past, in pursuit of Katze.

"You need a doctor, ma'am."

"Yes. I'll sit here and you go get help."

"Yes, ma'am." He was off, already moving expertly amid the many obstructions of the reconfigured terrain.

Terrain. That's what it was, all right—a piece of Terra. They could leave Earth, yes, but they were still her children. More than their biology, yes, and on the whole more gardenia than knotweed. But there were also their biology as it had evolved on their mother world. They had better remember that if they wanted to survive in space.

"*Marianne!*" Juana called, rushing toward her. She was less sure-footed than the young crewman; at one point Marianne thought Juana might fall onto the "top" of the centerline enclosure. But she did not. "Are you all right?"

"Yes," Marianne said. "I'm fine."

At least until the next unknown.

Nothing has slowed space development more than the high price of moving mass around the solar system. Using two stages to get into low Earth orbit may make substantial improvements, and beyond that the right answer may lie in nuclear rockets. These have been developed since the 1960s and could be improved still further. Lofting them into orbit "cold"—that is, before turning on the nuclear portion—may well erase the environmental issues. Fuel fluids can be flown up separately, for attachment to the actual rocket drive. Then the nuclear segment can heat the fuel to very high temperatures. Economically this seems the most promising way to develop interplanetary economics for the benefit of humanity.

The Nuclear Rocket: Workhorse of the Solar System
Geoffrey A. Landis

The Problem of Scale

Possibly the greatest achievement of our era, the Apollo mission to send men to the moon and return them safely to Earth was achieved using the power of chemical rockets.

And chemical rockets are impressive! Launching with an Earthshaking roar, with a trail of brilliant orange flame pushing a thousand-ton rocket away from Earth at twice the acceleration of gravity: no one who has ever seen a Saturn-V launch could fail to be awestruck by the power and the barely-controlled violence of a chemical rocket launching into the sky.

Yes, rockets are impressive. They took us to the moon; a feat unrivaled in human history. But chemical rockets aren't good enough.

The moon is 400,000 kilometers away—call it an even quarter million miles. Mars, at its closest approach, is about fifty million miles (depending slightly on where it is in its elliptical orbit)—but a trajectory to Mars would follow an orbit, not a straight line, and so a journey to Mars would be more like a quarter of a billion miles. Call it a thousand times longer journey than a trip to the moon.

But Mars is still, in terms of the solar system, Earth's neighbor. If we go all the way out to Pluto, it's a distance more like four billion miles, and its myriad cousins in the Kuiper belt perhaps as far as ten billion miles from Earth. So we can say that a trip to the edge of the solar system is perhaps 200 times farther than the distance to Mars at closest approach.

(The solar system doesn't really have an "edge," of course. The actual edge of the solar system is much farther away, if we call the Oort cloud part of the solar system—now we will be talking in trillion of miles, not mere billions).

In science fiction, where the heirs of Flash Gordon zip around from the asteroid belt to the moons of Jupiter in the time it takes to turn to a new chapter, this is a fact that can sometimes be overlooked: The solar system is big.

The stars are farther away yet.

To reach the stars, we will need more than merely a planetary-scale effort. I don't yet know which of many technologies we will use, when finally we choose to assail that endless ocean of night between the stars. Some of the smaller plans call for fusion-powered starships built in space, of a size that would dwarf the Empire State building; some require making lenses larger than the size of Earth to focus laser beams that use more power than has been generated by all of human civilization; some suggest mining fusion fuel from stations floating in the atmosphere of Jupiter or Saturn. Even the smallest of generation starships would have to be a self-sufficient space colony of the kind envisioned by Gerard O'Neill, miles in diameter and with thousands or tens of thousands of inhabitants, built in space from the resources harvested from the moon or the asteroid belt

By almost every plausible path, an interstellar mission will require us to employ not merely the resources of Earth, but to utilize the resources of the solar system.

To go to the stars, first we will industrialize the solar system.

The Fires of Apollo

Chemical rockets are not good enough.

Chemical rockets took us to the moon, but to do that, most of what was launched from Earth was fuel and the tanks to hold the fuel. Of the 3,000 tons launched on the trip to the moon, after all the fuel was burned and the unneeded systems jettisoned, the mass of the spacecraft returned to Earth was barely six tons.

Chemical fuels run up against a limitation known as the *rocket equation*. The mass of fuel must be carried by the rocket, which takes more fuel to carry, which in turn adds more mass. The result is an exponential: as you go faster, it takes exponentially more fuel. Specifically, if mi is the initial mass of the fully fueled rocket, and mf the final mass of the empty rocket with all the fuel expended, to achieve a mission velocity of ΔV, the rocket equation is:

$$m_f/m_i = \exp(\Delta V/V_e)$$ where Ve is the rocket's exhaust velocity.

(The efficiency of a rocket is often specified as a specific impulse, with the symbol I_{sp} instead of exhaust velocity. Specific impulse is measured in units of seconds. This corresponds to how long a rocket producing one pound of thrust will take to burn through one pound of fuel. By definition, specific impulse equals exhaust velocity divided by Earth's gravity, g, so both units measure the same thing, and rocket scientists easily flip back and forth from one to the other.)

The higher the exhaust velocity—the higher the specific impulse—the longer the rocket can produce a given amount of thrust on the same amount of fuel, and so the more fuel-efficient it is. The exhaust velocity of a rocket is thus the parameter that characterizes how fast you can go.

The exhaust velocity of the best chemical propellants, liquid hydrogen and oxygen, is about five kilometers per second. The ultimate limit is set by the amount of energy in chemical bonds. You really can't do much better than this—if you use chemical propellants.

But if you use something more energetic than chemical fuels?

One way to do this would be to use solar energy: instead of storing energy in the fuel, use the energy of the sun (or, better yet, beam power to the rocket!). I've spent much of my career looking at solar-electric propulsion, and at even more efficient propulsion using light-pushed sails, which are indeed one way of improving efficiency using the rocket equation. These can be quite superb technology for the missions for which they are best suited—but I'm not going to talk about these here. Solar propulsion systems, in general, are methods that use very little fuel, but have extremely low thrust, and in consequence tend to be rather slow, possibly more suitable for cargo missions than as a propulsion system for humans to explore, settle, and industrialize the solar system.

An alternate method of getting past the limitations of chemical rockets is to use a power source that has vastly more energy than the energy stored in chemical bonds: nuclear power.

If we want to go to the stars—if we want to settle and industrialize the solar system—we need nuclear rockets.

A Pickup Truck for Space

What we need is a spaceship that is not built on the principles of a hotrod, designed for maximum efficiency and minimum weight, overpowered but capable of carrying little more than the driver. What we need is the spaceship equivalent of a pickup truck: a simple, rugged, and basic transportation system capable of hauling a load of cargo for a wide variety of jobs.

A nuclear rocket is fundamentally extremely simple. A nuclear reactor produces heat. If you pass a gas over that heated reactor core, the gas heats up—and you can expand heated gas out a nozzle to produce thrust. (Very often we call this a *nuclear thermal rocket,* to distinguish it from other ways to use nuclear power in a rocket.) A nuclear thermal rocket is so simple a design that you could build one with the technology of ancient Rome, had a nuclear reactor happened to be available to the Romans.

Actually, a nuclear thermal rocket is far simpler than the nuclear generators used to produce electrical power on Earth; most of the parts needed to generate electricity aren't even needed. A nuclear thermal rocket consists of a tank, a pump, a nuclear reactor, and a nozzle—that's it.

In fact, space is the ideal place for nuclear power: between the trapped high-energy particles of the Van Allen belts, cosmic rays, and solar flares; space already is radioactive. You wouldn't want to crash one down on Earth (and most proposals for nuclear rockets propose operating regulations that don't allow nuclear rockets in orbits that would decay in any time shorter than 100,000 years)—but assuming you don't allow them on Earth, their natural home is in space.

Nuclear rockets are so effective, of course, because nuclear bonds are so much more energetic than chemical bonds. In a nuclear rocket, the energy source and the reaction mass are separate. In principle, then, you could use almost anything for reaction mass: heat it to a plasma, and let it expand out the bell in back. In practice, though, there's another law of physics that you have

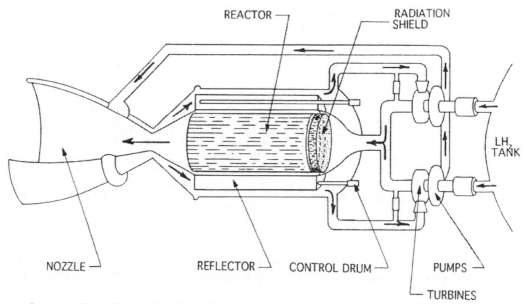

REACTOR

RADIATION
SHIELD

LH₂
TANK

NOZZLE

REFLECTOR

CONTROL DRUM

PUMPS

TURBINES

***Schematic of a nuclear rocket: Liquid hydrogen stored in the LH2 tank is used to cool the nozzle
of the engine, and then is pumped into the core of the nuclear reactor, where it is heated to several
thousand degrees, and expanded out through the nozzle, where it produces thrust.***

to pay attention to. For any thermal rocket (nuclear heated, or otherwise), the exhaust velocity is proportional to the square root of the temperature.

So, the higher the exhaust velocity you need—which is to say, the more oomph you need from a given mass of propellant—the higher temperature you have to run the core. How hot can you go? Ultimately, that's a materials science question. The early test rockets of the 1960s and early 1970s heated the hydrogen to about 2,200° C (about 4,100° F), achieving an exhaust velocity for the hydrogen propellant of about 8.5 kilometers per second, almost twice that which is achieved by the best chemical rockets.

The exhaust velocity decreases as the molecular mass of exhaust increases, though. Hydrogen was used because hydrogen atoms are lightest, and thus move fastest when they're heated.

In fact, exhaust velocity decreases as the square root of the atomic mass. So, if you used water for the reaction mass, with a molecular mass of eighteen, rather than hydrogen, with a molecular mass of two, the exhaust velocity is only one third as high.

(In the interest of simplicity here, I'm ignoring a lot of complicated physics regarding whether the exhaust is ionized, dissociated, or molecular, and whether frozen flow conditions apply. The factor of three mentioned is a quick approximation.)

Unfortunately, if you divide that exhaust velocity down by a factor of three, your nuclear rocket doesn't even do quite as good as the 4.5 kilometers per second exhaust velocity you'd get if you electrolyzed the hydrogen and oxygen of the water and used them in a conventional rocket.

So there's a good reason to use hydrogen as the fuel—but complications as well. Hydrogen has extremely low density—liquid hydrogen is fifteen times lighter than water; less dense then

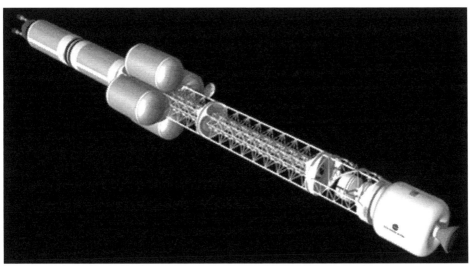

Artist's visualization of a spaceship propelled by a nuclear rocket engine. The nuclear rocket engine is at the far left. The crew compartment, on the right, is separated from the engine by the fuel tanks and by a long boom, providing the crew protection from stray radiation (primarily reactor-generated neutrons) from the engine.

Styrofoam. You need big tanks to hold hydrogen fuel, and you have to keep it cold, about twenty degrees above absolute zero.

Water is abundant in the outer solar system, however. After all, once you get far from the sun—roughly past the middle of the asteroid belt—water ice is just another kind of rock. Closer in to the sun, many asteroids have water as part of their composition in the form of water of hydration. So you can refuel your nuclear rocket using hydrogen generated from harvesting rocks in the Trojan asteroids, or short period comets—and, since you have a transportation system, you can haul fuel wherever you need it in the inner solar system as well. You can either

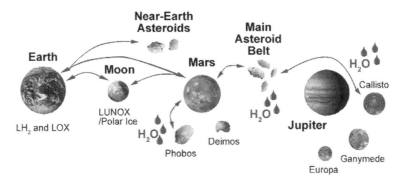

The water road: water, the most useful resource in the outer solar system, could be the trade-goods for a future "silk road" of space, as nuclear freighters fetch water from the icy moons of the outer solar system and main asteroid belt and bring it in to the inner solar system.

use electrolysis to split up out the hydrogen for reaction mass—or, if you want higher thrust but can afford lower performance, you can use water (or ammonia—another ice common in the outer reaches of the solar system) as the reaction mass.

So water could serve as the core for a future space-based economy—the "silk road" trading routes of space may very well go to the outer solar system, gathering water to split into hydrogen and oxygen, usable as fuel, for life support, and for industrial processing; and bringing it in to the inner solar system, in exchange for goods taken outbound.

Back to the Future?

Nuclear rockets are not just theory. Back in the 1960s, NASA had a program called NERVA ("Nuclear Energy for Rocket Vehicle Applications") to develop nuclear rockets. The Russians also had their own development project. In 1971, when Nixon decided not to follow the Apollo missions to the moon with a program to send humans to Mars, the NERVA program was canceled for lack of an application—but not before more than twenty nuclear rockets were designed, developed, and even tested. Along with the related "Rover" program, NASA did twenty-eight full-power test firings of nuclear rocket engines at the Jackass Flats test site in Nevada, with thrust levels ranging from 5,000 pounds (Kiwi-A) to a quarter of a million pounds (Phoebus 2) including running one engine for as long as 109 minutes, and restarting one engine twenty-eight times. It's been forty years, and they never were used in an actual flight, but nuclear rockets definitely are a technology that's been developed and tested.

From 1987 through 1991 the US developed nuclear rocketry further in Project Timberwind. It made advances in high-temperature metals, high-strength fibers and nuclear engineering. This gave dramatically improved performance, achieving *specific impulses* up to 1,000 seconds.

Beyond NERVA

But can't we do better?

As I stated, the exhaust velocity of a nuclear rocket goes up as the temperature to which you heat the exhaust gas goes up. The record temperature used by the NERVA and Rover rockets tested in the 1960s was at a core temperature of 2,500° C, heating the gas up to about 2,300° C. So the secret to higher performance is simple: how hot can you get?

Well, it's all a question of materials technology. You don't want to melt your reactor, and you have to bolt it onto the vehicle somehow. The NERVA rockets had a core of uranium carbide and zirconium carbide, in a graphite matrix. This was pretty high tech for 1969. Another serious limitation is the thermal gradients, since the core, at the thousands of degrees, has to be somehow attached to the vehicle, at somewhere near room temperature, and is cooled by the flow of liquid hydrogen, at 250 below zero. Still, better materials are under development. The "Model T" nuclear rocket may operate at 2,400° C, but advanced tricarbide materials might go up as high as 3,000. Beyond that, it becomes extremely difficult to find materials that remain strong.

But, still, that might not be the end of the road. One approach to higher temperatures is the "particle bed" nuclear reactor, where the uranium-oxide fuel is mixed with the high-temperature carbide and formed into pellets, which are put into a rotating reactor drum. This maximizes

Nuclear rocket developed for Project NERVA

the area exposed to flowing the hydrogen, and the individual pellets need not be structurally strong, but can be taken nearly to the melting point. Beyond that is the possibility of a liquid-core reactor, and one step beyond that is the gas- or even plasma-core reactor, in which the uranium fuel is held in a magnetic fields. These may be beyond the range of current nuclear design, but the reaction physics is still the same—so, in a fundamental sense, "it's an engineering problem."

But there's another approach: for higher exhaust velocity, we can get beyond the thermal limit by avoiding using thermal rockets.

Instead of using the nuclear reaction to heat hydrogen, we can generate electricity. This can power an ion engine, and/or one of the ion engines' many cousins and descendants, such as stationary-plasma thrusters, magnetoplasmadynamic thrusters, microwave thrusters, pulsed-inductive thrusters and variants and competitors to these. The fundamental feature of these is to use electric and magnetic fields to manipulate and accelerate the exhaust, and thus get beyond the square-root-of-temperature limit of thermal thrusters. Since they're not limited by temperature, these rockets now have a much wider range of fuels that they can use as reaction mass. Many of the ion engine types use Xenon, a gas that is easy to liquefy, easy to ionize, and dense; others use easily liquefied metals such as lithium or cesium.

But there's a downside to the high impulse of ion engines: the higher the exhaust velocity, the more power it needs, and the less thrust it produces per unit power. Electrically-powered rocket engines are power hogs, and rather than the quarter-million-pound thrust of a NERVA engine, produce thrust of a few pounds, or even less. Over a long time, this can build up—typical trajectories involve thrusting for months, or even years—but these high-impulse rockets are the very opposite of hotrods. This is a technology for long-haul cargo transport, not for jetting around the solar system.

Possibly the best solution, then, would be to do both: a "bimodal" nuclear rocket, able to directly use the heat of the nuclear core when high thrust is needed, and then able to generate electrical power as well to run a highly-efficient, low-thrust ion engine for months at a time. This would be the best of both worlds, an engine that can accelerate when needed, and be fuel-efficient the rest of the time. Here's a rocket we can use to colonize the solar system, to send humans to the moons of Jupiter and to mine the atmosphere of Saturn; to harvest asteroids to build colonies in space, and to bring us to the edge of the solar system, where we can look out at the stars.

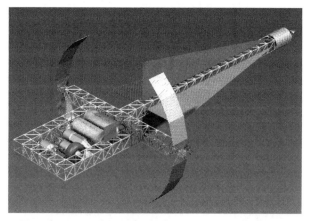

Visualization of a nuclear-electric cargo vehicle for a mission to Callisto. The nuclear engine is on the far right side. The fin-shaped thermal radiators, used for the thermal to electric conversion system, are in the shadow of a shield protecting them from reactor-generated neutrons.

Beyond Fusion: the Path to the Stars

Nuclear fission rockets may open up the solar system, and even take us into the Kuiper belt and the inner edge of the Oort cloud, but uranium fission is still not energetic enough to take us on the ultimate journey. For that, we need more efficient rockets yet. Fusion rockets offer the possibility of even higher performance. These require a level of technology a little higher yet, with plasma that is either compressed to fire in bursts, or a continuous reaction confined by magnetic fields.

But that's a topic for another day.

Acknowledgements:

All images courtesy of NASA. I'd like to thank Dr. Stanley Borowski of the NASA Glenn Research Center for helpful discussions, and for his many design studies and tireless advocacy of nuclear rocketry in space.

This story follows on the logic of Neal Stephenson's story, but over a longer timeline, as the interplanetary economy develops in Wild West fashion. Modeled somewhat upon the famous Robert A. Heinlein story, it treats a focused man who has one dream and devotes his life to it.

The Man Who Sold The Stars
Gregory Benford

Vain is the word of that philosopher which does not heal any suffering of man.
~Epicurus

It will take a thousand years for the frontier to reach the Pacific.
~Thomas Jefferson

2016

Harold Mann idled at a corner and watched an enormous guy come out of one of the adult movie houses and stride over to his Harley.

Harold was on his second job—at fifteen, using fake ID—driving a cab on South Jefferson Street of St. Louis. Business late on a sweltering night was slow. The big bearded guy's bike was under a lightpost next to a Honda Hawk. The man in black leather pants and a black T-shirt shouted at the whole street, "Who parked this turd next to my bike?"

He then grabbed the Honda Hawk, grunted as he lifted it, and threw it all the way across the street. It hit another Japanese bike, a yellow Kawasaki. The clanging smashup echoed in the moist night.

Some gasoline dripped from the Kawasaki and the man walked over, puffing on a brown Sherman's cigarette, and—dropped it. The gasoline *whomped* sending flames licking across the sidewalk. The biker glared at Harold and walked up to the cab window. He pulled out a big Bowie knife, grinning. Harold looked straight ahead and heard the tapping on the window.

"What ya think a dat?" He slurred the words and spat on the blacktop.

Harold rolled down the window and looked into the scowling sweaty face. "I don't think you threw that rice burner hard enough."

Glowering: "Yeah?"

"Man's got to throw long in this life."

The biker walked away laughing. The bikes burned. Harold finished his duty time, drove to the cab station, and quit. *Maybe not my best line of work,* he thought.

Five months later he had turned sixteen and had another fake ID saying he was twenty-one. He pitched a smartware app to a startup company in St. Louis, by walking in cold and asking to see the vice president. The app assisted robots with finding their footing and orientation while working in low Earth orbit. They could then assemble parts for the first orbital hotel.

The key to the app was using the new composite carbon girders with holes punched every half meter. Robots could count on having a dual-pivot purchase no more than fifty centimeters away, to torque or support a mechanical advantage. This increased their mobility and mass carrying capacity.

The vice president was intrigued. While his engineers looked over the app he asked Harold for his credentials. He gave them a certificate saying he had graduated from MIT with a degree in astronautics, remarking that it was the same program in which Buzz Aldrin had gotten his PhD. The e-certificate was authentic, though he had artfully hacked it to omit the detail that he had done the classes entirely online in three years without ever being in Boston.

The startup bought his app and he got a job. Within two years he was their CEO, and they issued an Initial Public Offering. His share was nearly a million, since he had worked mostly for stock options. Driving home that night, he saw the same biker guy coming out of a bar. Harold pulled over and bought the guy a drink, never saying why.

He recalled his first job as he watched the vids from the latest big satellite telescope. The deep resolution views were striking, and they brought back a moment when he was ten years old.

He had rented beach chairs to tourists down at Orange Beach, Alabama. All day long he let nobody get past him without a friendly, insistent, "A chair to make you more comfortable? The sand's hot. Just five bucks for the day."

The usual brush-off he eased by with, "Keeps you away from the sand flies, sir."—and that usually did the trick, especially if he had a woman with him. She would usually wrinkle her nose and badger the man into it.

Decent money, and he was only ten. His father thought it was good training and Harold did, too—the Great Recession was not yet in the rear view mirror. The tourists officially had the chairs till sundown but many stayed with their beer and got fried oysters from the stand down the way. He stayed late, reading used paperback science fiction novels under the fluorescents of the greasy burger stand. He was an addict; science fiction sold the sizzle of the science steak. Even when he got tired he remembered to be polite, smiling and using the *yes sir* a lot—and so he discovered tipping.

Some just left the chairs strewn around, so he had to drag them back, two in each hand, to the shed. He had just finished stacking chairs and was turning to plod down to the bus stop to ride home when he turned toward the surf and saw them.

Saw them truly, for the first time. The whole grand sprawl of jewels across the blue-black carpet, hovering above the salty tang of gulf waters like a commandment. The Milky Way spanned the sky, vanishing into the horizon, glows shimmering of emerald, ruby and hard diamond whites.

That's what we're part of, he thought. *The real, ultimate way the universe is, not just this moist curtain above a sandy stretch. Reality, big and strange and wonderful.*

2023

At twenty-two he decided to tint his black hair gray to appear on his first business panel, about resource extraction from asteroids. With dark glasses and long sideburns the tinting made him look older. The moderator was the famous Interplanetary Resources exec, Peter Diamandis, who deftly kept the talk flowing without digressing.

A steely looking woman in a stylish blue suit got into an argument with a panelist guy from NASA, saying, "Your main goal appears to be not to fail. In the bigger space companies and academia, the mission has to work. So you gold plate everything and your price soars."

"And you're about profit, period," the guy shot back.

"People give us money, their choice, we pay them with dividends. You take taxes with laws."

He said slowly, "Lockheed, Northrop Grumman, the rest—they have great track records."

"They're monuments, this is a movement."

The audience murmured and people started arguing with each other. Harold surveyed the discontent with a bemused smile.

He'd seen their likes before. Merit-driven products of the test-prep industry, capable cogs. And yet they did their jobs while thinking they were countercultural rebels. Their generation loved the Standard Storyline: insurgents fighting the true establishment, that distant dull group which was always somebody else. They were sharp and from Ivy Leagues, Stanford, Caltech. That unconscious attitude prevailed in corporate boardrooms, so they could rail against the establishment over cabernet in the evenings.

As he watched this woman he reflected that in a way he had accomplished the life goals his parents had taught him, mostly by example. He had found a good way to make a living, had started a business, and enjoyed it all. He got up each morning eager to get to the office. But this woman made him realize he had other goals left to achieve.

The focused woman said, "If you're young and lean, things can *fail* and you still keep going. For the big space companies the whole competition is just getting the government contract, then it's all risk aversion. It's not at all about doing something cool, first to market, then making money so you can do more. That's what I like: not playing it safe. To shift gears, to follow your nose."

She seemed startled when she got applause. Harold nodded and smiled at her. His talk was next. Fairly technical, about universal joints, AI linkages, and space applications—but she listened intently as he outlined a rock-prospector team of robots he had worked out and tried in the Arizona desert.

Moving on, Diamandis commented that for robots, deserts might be easier than space. Then he asked on a concluding note, where all this work would lead. "Prosperity!" the woman said.

Someone in the large audience called, "To the stars!" and another voice shouted "The stars? Impossible! Why do it anyway?"

She glanced at the moderator and said, "Why go to the stars? Because we are the descendents of those primates who chose to look over the next hill. Because we won't survive on this rock indefinitely. Because they're *there.*"

The panel met in the bar with Diamandis for drinks after. He could not take his eyes off her, even when she was talking to Diamandis. He learned her name was Sara-no-h—Sara Ernsberg. As the group broke up he said impulsively, "Do you dance?"

They got back to the hotel at three o'clock a.m. He slept in her arms till noon, and they missed the entire morning session, including his own talk.

2029

"You're going off the deep end," Sara said. They were in his office with a big view of Pike's Peak. The slender mountain had snagged a looming purple thundercloud on its slopes. Lightning flashed in its belly. *Mine too,* Harold thought.

"I can make a billion in a year if we can repeat the old Air Force test trials, make 'em work," Harold said. "It's a calculated risk."

"Look, the public's against nuclear rocketry."

"Has anybody really asked? The nuke flies up cold as a salmon. SpaceX can deliver it. We turn it on after we've flown a tank of hydrogen up and mated it to the nuclear thermal unit. That assembly flies my robot team to the candidate asteroid and runs a nuke power source for their exploration."

She twisted her mouth in a skeptical red-lipsticked torque that he loved. "It never comes back into low Earth orbit?"

"Never. We use it for smelting in orbit beyond the moon."

Sara said, "I prefer more conservative invest—"

"In five years this *will be* conservative. It'll be raining soup and we'll have a bucket."

"So this is gambling on a certainty."

"Launching a nuke rocket core, piggybacking on a two-stage to orbit, it makes economic sense."

"Nukes. The UN can block you."

"Elon says he can launch us from mid-Pacific. His platform's not a UN member—or subject to nation-state controls." Elon Musk, Jeff Bezos, Richard Branson—they had been like the Carnegie and Rockefellers of high vacuum, just a generation before, and were still big players.

"Any skeptical legal advisor will say"—she frowned, did a bass growl—"You could be sued for every double-yoked egg a hen lays after launch."

Harold flicked on a complex legal flow chart. "Ah, that old mink case as 'proximate cause'? Nope, we'd legally have to be sued on the moon. The parent company is Blue Sky Nuclear, incorporated as of today on Luna."

"Luna? There's no—"

"I and some investor friends sent a robot office clerk there. They'll have to argue for years over whether that's a legal registry. I'm happy to let them go trotting down that alley."

"Sorry. You've still got enough minimum contacts with the planet for them to get personal jurisdiction. It just might take a while—or you might end up losing that Laguna mansion."

Harold shrugged. "Keegan over at Consolidated will be after us for sure."

"Then there's the Outer Space Treaty—the US is a signatory."

"Yeah, Consolidated will work that. That's how the Microsoft competitors shamelessly worked with the Clinton administration to take down Microsoft."

"You've got it all figured out."

"Nope, just the fun parts. The rest I leave to you."

"Me!?"

"I need a pretty face with a razor mind to present our target asteroids to the excited investing public."

"I haven't said I'll put anything into this swan dive off a skyscraper—"

"Take your pick." He flashed a list onto the wall using his thump command:

Most Cost Effective: 2000 BM19, a very small O-type asteroid (diameter less than 1 km). Makes several close approaches to Earth. Estimated value is $18.50 trillion and an estimated profit of $3.55 trillion (USD).

Most Accessible: 2009 WY7, small asteroid with regular close approaches of less than 1 AU. A silicaceous or "stony" object that has a high accessibility score on Asterank of 7.6577. Some hints of water content.

Most Valuable + Most Profitable: 253 Mathilde, a 52.8 km-diameter carbonaceous asteroid. Estimated mineral/metal value of over $100 trillion. After robot mining and transport, estimated profit $9.53 trillion

She shook her head. "Your audience is venture capital guys, not astronomers."

"I want to snag their economic ambition, not their intellect. You can charmingly smooth over the distinction. Now, come with me into the Board meeting." He spread his arms wide, gave her a sunny smile.

"What?!" She looked terrified, which was actually rather attractive.

Five minutes later he was gesturing at a 3D PowerPoint slide without looking at it. *Always face the audience.*

The nuke fission rocket idea he had prepared the Board for, so it took only some cheerleading from Sara and a fast finish after the tech details: "Look, remember the 1990s and the 2000s? Computers toppled and then rebuilt industries. Retail sales—Wal-Mart and Amazon. Banking–ATMs, online services. Finance—high-speed online investing down to milliseconds, global markets you needed a stop watch to work in. Entertainment—Web streaming, downloads,

YouTube, streaming. Publishing—e-books took down old-style print media, aggregators did in paper news, so nobody goes to 'press' anymore. The march of time."

A well-known savvy techno-skeptic sniffed and gave them all a bemused scowl. Harold had invited him on the Board, knowing the name alone would raise stock value. "Thanks for the history lesson. How's that tell us what comes next?

"The big demand now is for raw material." Harold stood up and spread his hands expansively. "So we butt into the old mining company territories by bringing in new sources—the asteroids. Next we do communications—we provide cheap, easy repair to geosynchronous comsats that have gone dead. Energy—we go get the Helium-3 that's just sitting in the first few meters of the moon's surface, along with plenty of rare earths as a bonus."

This provoked a full hour of discussion, some heated. The skeptic stormed out, angry—only to return fifteen minutes later, red-faced, claiming he'd just gone to the bathroom. Sara got in some telling jabs, even some laughs. Finally the Board turned to a plausible scenario for the mining operation. As usual, many tradeoffs and unknowns, as always in blue sky investing—"or black sky," one Board member joked. Harold grinned and decided to go radical on them.

"Friends, a chunk of high-quality metal twenty meters across generates an impact explosion on the scale of megatons, granted. But!—without radioactive fallout, and it doesn't spray much after impact, especially if it hits sand." He flicked a wrist and a satellite view of Earth rotated on the wall screen. The Board leaned forward, studying sites picked out in livid color. A graphic showed a fiery dot skating down through the atmosphere, diving in suddenly, and spiking into a bleak tan Mexican desert.

"There are plenty of remote deserts we can hit to make an impact safe. All the miners need do is to find a mostly metal asteroid on the appropriate orbit, give it a calculated nudge. Let our robots work metal and rare earths out of its crust. Save that, bundle the sludge with some orbital debris, and fuse it as a shell over the refined metal. Do your smelting where the robots can blow it off. The solar wind picks up the waste, takes it to interstellar space. Do this while the rock's in transit. Get the robots off with an unmanned shuttle. Pilot the metal into orbit so it can skim along the upper atmosphere, maybe skip it like a rock on a lake. Then bring it down, slam it in, mine it."

He finished with a 3D budget, detailing axes: profit and loss and cost. "I'll take you the Board to, let's say, Libya to watch the fireworks."

Mermin raised his big hand. "Reality check. You'll never get clearance."

"Your reality check will bounce. The U.N. Security Council is a debating society. Current ideas evaporate in societies struggling to feed their people. If a desert country can gain a new industry by letting rocks drop into their land, they will. Libya already says it will."

An investor said mildly, "Genuine realpolitik trumps wishful-thinking-realpolitik these days."

Bo Duc Anton, CEO of *Astroprospects,* smoothed her classic Chinese cheongsam and said, "Start small."

"How?" Harold asked.

"My company has a Spacefarer contract to find and de-orbit debris. I can compact some of the space junk we collect with robots: booster rockets, lost gear, urine-icicles, dead satellites.

We just grabbed Explorer I! Not typical—we'll bring it down to sell to collectors. Real junk we can bundle and drop back in as a trial, into a desert."

"Isn't all that supposed to go into the ocean?" someone shot back.

Lo Duc shrugged. "Sometimes we miss." Laughter. She blinked; it had just popped out, before she had a chance to really think it through. Harold knew how to get momentum going. Another objection about bad publicity made him shrug, saying, "Some days you're the pigeon, and some days you're the statue."

"Make it a small mass," Sara said. "You need to test targeting anyway."

Another woman said, "Better to start out clean, have a contract with Libya. They need the cash and the mining business."

The corporate lawyer began, "Risk is too high to allow—"

"Committees don't open frontiers—people do," Harold said. "We've got to look beyond the Official Future people learn in business school. It's risky to *not* do this."

2031

Their trial package came screaming down like an orange arrow. It hit within 200 meters of their bulls-eye and Sara shouted as the shock wave hit them in a rolling roar, from two kilometers away. Sand rose in a tan wave above the Libyan desert and crashed in front of them like dirty foam.

She and Harold were the only ones who dared to stand this close. His simulation group were confident that with the cladding Sara's company provided, made from orbital junk, it would shed red-hot debris and follow a predictable path. Unlike the earlier satellites that had come flaming down, this was dense and hard and so kept to its programmed path.

The shock wave blew her hair straight back in the thunderclap roar and she laughed as he kissed her.

He realized then that while others had sports like golf, or rich-guy risky hobbies like flying experimental aircraft, or just ran through a series of ever more expensive women…and he didn't. He liked to work and Sara was just the right balanced love he needed, no more.

Whenever he needed to feel his inspiration, he could turn from his meetings and engineering details and deal-making…and simply look up into the night sky. The frosty splendor of stars glimmered there, eternally beckoning.

2032

His company had an executive retreat of a rather different sort in the high Sierras. They all went backpacking out from Mammoth Lakes and after some grunting and swearing got camped around Deer Lake at 10,000 feet. Harold liked to see how people did when not behind a desk. Some seemed to see the natural world as an exotic, hostile place. One groused about no cell phone service.

Lying outside and staring lazily into the crisp clear night sky, he recalled the same sensation of awe at the grand silvery sweep of the Milky Way that he had felt as a boy. He couldn't see Alpha Centauri from here. Could a star be closer, yet discovered? Could he get there? The planets he would reach, sure, but *stars….*

The Voyagers were out there too, swimming slowly across interstellar space. Later that night he dreamed about them in some far future, visited by astronauts in a silvery ship. They recharged its nuclear generators and left a beacon for later travelers. Later, Kuiper-Belt settlers considered it a must-visit for space-suited family poses, until preservationists started rationing tourist visits. When he awoke to a ruddy dawn he realized that by then it would be ancient, like the Hagia Sophia in Constantinople and the Pantheon in Rome—deep time monuments. The dream had been wonderful, as crystal clear as the high Sierra air.

As a boy in 2013 he had read a book about starships and how they might be built, following a century of burgeoning interplanetary commerce. After all, Columbus had found the New World in caravels designed for warm, tranquil Mediterranean waters, not the Atlantic. But within a generation the promise of gold and land spurred rugged craft that grew steadily larger.

That distant dreaming boy had gone on to read novels about starships reaching strange worlds, of generation ships whose crews forgot in the laboring bulk of the ship what the purpose of it all was, or even that they were in a ship. He found that unlikely, since people were naturally curious and poked at their boundaries, always alight to the notion of new territory. No, they would *know*.

Going to the stars would be harder than anything ever done, maybe more of a struggle than laboring up out of an ocean and onto land. The true vast extent of the problem was the charm of it, too.

The next night around their snapping campfire along the John Muir trail he talked about building capability to reach the stars. It would be a step-by-step process, a side effect of developing the solar system economy.

His execs rolled their eyes and looked at their CEO as if he was going loopy on them. *Well, maybe I am,* he thought. *For now.*

2033

Two years later the Chinese tested a nuke rocket engine in the open air, just as the USA had in the 1960s and 1970s. There was little escaped radiation. They even blew up a nuke configuration with implanted explosives, and found very low residual radiation—also as the Americans had. Intelligence found the Chinese had simply bought the entire program developed by the Soviets at Semipalatinsk; the Russian nation was having a fire sale. Now the Chinese were rapidly improving on the old designs.

This alerted the world and became the next "Sputnik moment." It was almost as if the new Chinese semi-democracy, struggling to pacify its nationalist faction, was trying to deflect internal stresses into an external competition. Soon enough, it quite obviously was.

Suddenly capital rained from the skies. Harold was ready with his bucket. He had made his deals to get a program under way for his company, FarVoyager, and their test-bed results were good. They orbited the first test nuke half a year later.

The Chinese were still reorganizing after the overthrow of the old Red regime, their politics a factional scramble. They decided to not launch a nuke rocket from the ground, even though there was little escaped residual radiation. So they converged on the American design of assembly of the cold nuke rocket and command module with regularly supplied fuel tubes that inserted into the big liquid pods. This became the standard method for fission transport.

Harold was a private sort of man who made a point of letting no one at all, even Sara, call him Harry. Still, at the celebration of the first nuke burn from low Earth orbit, he let himself go. It had gone perfectly, the hydrogen jetting out in luminous fury to boost a major tonnage of commsats and a few working zero-grav laboratories to high orbits. Proof of principle. Now the whole solar system seemed to yawn open at last.

Committees don't open frontiers—people do.

Somebody on his staff had put it up without asking. *Good,* he thought. Then he burst into a song he had written:

"A fact without a theory is like a ship without a sail.
Is like a boat without a rudder.
Is like a kite without a tail.
A fact without a theory is as sad as sad can be.
But if there's one thing worse
in this universe,
A fact you just can't hack—
it's a theory
without a *fact*."

Though he hit every note perfectly, he never sang in public again.

Soon SpaceX lofted another long gray tube of liquid hydrogen to fly alongside the FarVoyager engine. Specially designed Astrominer bots mated fuel to engines. The craft ignited and left orbit, bound for a near-Earth asteroid to carry out a prospector mission. It visited five water-rich asteroids that were energetically easier to reach than the surface of the moon. Orbital Sciences and Virgin Galactic, deeply into their orbital hotel businesses, offered their services for a second flight.

The Chinese had built an industrial empire with autocratic methods, but they had never opened a frontier before. They found it hard to let enterprises compete. The instincts of Late Marxism were hard to overcome. On the other hand they were an eager market for products from space. Toward the end of the decade, Chinese billionaires competed to top the latest lavish champagne party in the Spherical Hilton.

2037

Harold and Sara took their first flight to the big new Spherical Hilton together. The two-stage-to-orbit shuttle was smooth, a big rugged composite airframe. Out the porthole they could see Phoenix, now more like Jericho or Ur, with its shriveled relics of golf courses and the dusty hulls of swimming pools beside abandoned homes.

The Spherical was ornate in its lightweight way. They had a suite, which meant a big bubble window they could float within and see the universe passing in review. Hanging there in an ocean of night, they made love. Each time with her he felt a new depth, an unexpected flavoring. This one came where all parts of her converged, his head between her thighs, the zero-G making every angle easily realized amid the salt musk. She was moist and cool and diamond-sharp stars drifted behind her. New heat rose between them as she fluttered her tongue. Their bodies said what their words could not. Energy rippled along their skins, somehow liberated by the weightlessness. He felt his own knotted confusions somehow focus in a convulsed thrust into her slick mouth. Yes, here was their center.

They had preference for the spherical swimming drop. He plunged into the ten-meter diameter shimmering cool ball and felt it coil around him in a way water on Earth did not. He hung suspended and kissed her foot, grinning madly. Kick, stroke, and he was back in air, barely in time, gasping.

The hotel furniture was as light as air. Carbon fibers with diamond struts, all puffed into a foamy shape. Their stick-to chairs faced the huge window where Earth slid by. The sights somehow loosened their tongues as they sipped from wine spheres. Lightly, Sara revealed that before him she had been the happy hypotenuse of a triangular domestic ménage.

"You gave that up for me?"

"No, for me. That way *I* got *you.*"

Over dinner she finally got him to talk about his odd childhood. Maybe the altitude gave perspective? He grimaced. "Look, the information age rewards people who mature fast, are verbally and socially sophisticated, can control their impulses. Girls were way better at that. Schools praised diversity but were culturally the same. Different skin color, same opinions. The girls ran my senior class and I slept through classes."

"So you got out." A gyrating hurricane, the fifteenth already this summer, churned across the Atlantic. He knew what she was thinking. It was easier to get him to talk if she didn't look at him, and this sight gave her a perfect reason.

"They wanted to teach me how to share. I wanted a curriculum that taught how to win." He grinned. "And how to lose. I've done my share of that, too."

"You got that in startups?"

"And Internet training competitions. My Mom had her way of teaching, too."

"Like?" The African drought was a stark gray expanse, dotted with brown oases.

"To be grateful, f'instance." His voice became a thin soprano. "*There are millions of unfortunate children in this world who don't have wonderful parents like me.*"

When she laughed he went on, "Religion, too." The high voice: "*You better pray that comes out of the carpet. And when I talked about my plans of going to the planets, then the stars, I told you once, I've told you a million times. Don't exaggerate!*"

She floated over the table and kissed him. He nuzzled her throat and murmured, "Okay, now your turn."

She had to look at him for this, and luckily they were now chasing a ruby sunset over India. "Teachers wrote my parent that I was 'a fiercely rambunctious girl.' After a while they found out they couldn't tame me by assigning some of those exquisitely sensitive Newbery-award-winning novellas."

"Yeah. Social engineering just ain't that easy. You're always going to get malcontents who like money and movement more than contentment."

"You're after the money?"

He noticed that she had deftly turned the talk back to him, but let it go. "Oh, you mean those investments in nanotech and 3D printing? Sure, to get cash for the Botworks orbiter up here."

"What about all that opposition to nanotech? They teamed up with that failed campaign to stop your launches of cold nuke rockets."

"Marches don't stop markets. So maybe nanotech pushes noses out of joint. Big social impacts, sure. Life's like that. The world's getting stranger and I'm going to get stranger right along with it."

When they took the trans-orbital shuttle to the Botworks in higher orbit, Australia was burning again. An angry black shroud north of Melbourne cloaked the already parched lands. Geoengineering with aerosols in the stratosphere was redistributing rainfall to counter this, but imperfectly. At least it was cooling the world and avoiding the terrible droughts of the early 2030s. Meanwhile chemical plants steadily worked to offset the rising alkalinity of the oceans.

Harold immediately used his phone to call through to his company office, Astromines AU, in Melbourne. "Give extra compensations to staff in the burn area," he said to his operations manager. "And make a corporate donation, say a hundred million, to the disaster zone."

He could tell from her expression that she was surprised. "Y'know what I miss most in this warming-up world—the north pole." He pointed to the south, where Antarctica was smaller but still brilliant white. To the north no sea ice remained.

She nodded. "Maybe we should spread more aerosols, bring the summer sea ice back?"

"The Russians want the sea lanes open wide. It's a tradeoff." He sighed.

The Botworks were served by the big nuke complex and the attendant rocket, floating like a shark beside a clumpy whale. Astromines had snagged into high orbit a nominal working asteroid fifty meters across. Bots swarmed over it, directed by an octagonal manager craft secured to the dark rock by a carbon fiber tower. Bots walked on this, their thick legs swinging from one toe-hole to the next with startling speed. They then spread over a mesh that grasped the asteroid, helping the smelting units grind and heat. The nuke power plant gave electricity and hot smelting fluids through a flexible, stark white piping array.

"Looks like a spider web around its prey," Sara said.

"Not far wrong," Harold said. "The Chinese have one like this, but many more people. See that hab on top of the command ship?" he pointed at the busy octagonal assembly. "Has three crew, all women. They work together better out here, seems like."

"The *Journal* says they're undercutting your price."

Harold shrugged. "They want to win not a space race, but a space marathon. They learned the big lesson from the 2020s. Manufacturing is all that matters in the long run. Service economies only shuttle money from person to person. Manufacturing creates wealth, services distribute it."

"The Euros want to keep space public—something that every citizen feels like they're involved in, a public utility rather than a private playground."

"Yeah," Harold said, "I saw that speech. 'Not just billionaires doing their bungee jumping in reverse.' Good line, bad logic."

The three women aboard were all agog that the CEO had come out to see them. Sara and Harold got into skinsuits and 'the gals' tugged them around the asteroid, pointing out bots as they performed complex tasks. Bots all had AIs that knew kinetic tasks down to the millimeter, and bots never tired. "They don't get retirement plans," Harold remarked.

Their departing gift was a chunk of the ore. Sara sniffed it. "Smells like gunpowder."

"Smells like the future," Harold said.

After they left, Sara said, "My, they prettied themselves up for you. Skirts in space?"

He shrugged. "A nice gesture."

"They didn't know I was coming, did they?"

"I may have neglected to mention it."

"I almost laughed at how their faces fell. Didn't your staff send a manifest?"

"I don't use staff much. Like to work hands-on."

"Obsessive, the *Journal* said."

"People exaggerate."

That evening, their last in orbit, she said, "I'm not going to marry you."

He blinked just once. "How did you know?"

"Private pod dinner, stunning view, waiter who says nothing, gleam in eye—little clues."

"And—no?"

"I'm CEO of three companies, you of two. We have cross-lateral contracts with constraints under that damned ExoLegal legislation. Even a contract marriage—"

"I was going for the whole thing."

"Trad marriage? I'm not an antique."

He returned the already-palmed ring back to his jacket pocket while she sipped from her wine bulb. "Okay, but a contract marriage of five years—"

She leaned over and kissed him slowly. "It's a wonderful thought, and we love each other, but—look at our lives. Always on the move. Eighteen-hour days. Legal thickets."

"At least we couldn't testify against each other."

"Now there's an incentive! Just what a girl wants to hear." She smiled into some distant future.

"So long as we don't commit any crimes, it'll keep the lawyers off us."

"No, I like it better this way. I'm here because I want to be. And you realize that Marquand fellow I also see isn't going to hedge you out, yes?"

He raised eyebrows, pursed his lips. "I was kinda hoping."

"I don't need a ring to stay with you. Or you me. We see others, we live intersecting lives. Great lives."

He nodded. Most of the way down from orbit they were relaxed, discussing how Harold could arrange to designate countries as "commissionaires" rather than wholesalers of raw metal in cylindrical ingots. Because commissionaires never took possession of ingots, which would mean pay taxes on them, an agent in France could sell them on behalf of a subsidiary in low-tax Singapore, people he had never even seen. The ingots themselves never saw Singapore, either. They came down in capsules that parachuted directly onto the grounds of assembly factories. Once in the landing yard, robots took ingots to processors.

Years later, he remembered three things from that first orbital trip. The crew in skirts, yes. Sara's rejection, certainly. But what really paid off were the profits that stacked up in Singapore. Harold joined others in lobbying for a "repatriation holiday" that let them bring profits into the US or Europe at reduced tax rates.

"Reminds me of some lyrics from that old pop song, 'Humanity's Way,' remember?" he asked Sara one evening, after a long bout in bed.

Where there is need
Some will bleed
Call it greed
But all will feed.

From that commissionaires maneuver he got money to pay for more astronomy studies, for finding new asteroids to visit, prospect, mine, for new engineering designs. Plus he could intervene where governments were slow to act, as with the Australian fires. Capital could move faster.

A few months later, exhausted with work, he needed some down time. He spent a week on Maui having his longevity looked after, using a service LifeCode had just started.

Most of aging was failure to do maintenance. The latest innovation used genetic information to spot-repair genes, then stimulate them into doing better. While a team poked and prodded him, fed him odd, thick, slurpy drinks and got him into lab-rat style exercises, he let his mind roam, thinking about his boyhood dreams.

It was time to fund them.

His excuse for spending the money was a celebration. Sara had finally agreed to a five-year contract marriage. The stock in his companies leaped over 10 percent.

2041

Dr. Katherine Amani looked up from her work. Harold said, "I made an appoint—"

"You actually *are* Harold Mann. I thought it might be a joke."

"Uh, that's what I said, Ma'am—"

"I thought it was a…boast."

"More like a confession."

He knew her wry look. Didn't this man know that in the '30s no sophisticated person divulged wealth or talent or ability? It was unseemly, unsettling, and rude.

"Or a prank—that was my second guess."

"Your friends masquerading?"

"Last week my boyfriend…never mind. What do you want?"

"I want to know about the follow-on to the Wide-Field Infrared Survey Explorer satellite."

"What? Oh…WISE."

"Oh, should've known. Nobody uses full names; I really am out of it."

"WISE, that's ancient…and there was no follow-on."

"My people tell me one was built, but not launched."

"Ah, that. Too expensive back in the Drawback era. Still is. It's probably sitting on a shelf in Maryland."

"You agreed with the conclusion—no brown dwarfs closer than five light-years?"

She got up and made a graphic display open on the well. "Here's the final 3D plot. We surveyed the sky—well, nearly all—"

"How much did you miss?"

"About 2 percent, as I recall. Some cycling glitch with the 'scope."

"How sure are you of the five light-year number?"

She grinned. "You mean, the contradiction with the average density of brown dwarfs? It was indeed a puzzle. There should've been several closer, but we didn't see them?"

He decided to use a method that worked in business planning. "Suppose I told you there was one? What would explain why you missed it?"

"Um…I always wondered if our method was right. See, we surveyed in three spectral bands, and used the ratio of luminosity between those bands to see if it was a dwarf."

"How did you know the ratio test?"

"Modeling of dwarf atmospheres. See, dark stars have methane clouds at the top of their cold atmospheres. We used metalicity plus methane and water absorption bands to select out the millions of faint dots we had."

"How about a pure hydrogen dwarf?"

She gave him the wry look again. "You've read up on this. A pure hydrogen star would be undetected in these kind of model-driven observations."

"Good answer. How much would it cost me to have all the data examined again, using a less narrow band model?"

She blinked. "Funding like that—maybe a million, even."

"You and a team could do it? I'll make it a gift to the university."

"Uh, yes, we could. And…wow. We never see such grants anymore."

"Hard times. Sad. And stupid." He leaned against a filing cabinet. "How does an infrared astronomer like you get new data?"

"Mostly we don't."

Harold allowed himself a smile. She was still trying to take in his offer. He liked keeping people off balance; they told you more that way. "How much would it cost to pull that old WISE satellite off that shelf, update all the electronics, launch?"

Another blink, eyes to the ceiling, lips moving. "Um…we haven't gone much farther in e-development, and launch costs are down thanks to—" she stopped. "To you guys, I guess."

"And a hundred million others."

"With people living fulltime offworld, too." He could see the question in her eyes. *A tycoon comes visiting an astronomer? What's up?*

"Estimate the cost of a new WISE with broader modeling in the analysis?"

"Maybe 200 million."

"Okay, let's redo the old data first. And work up a sharper cost estimate for a re-launch."

She was still trying to take it all in when he left for a business appointment, feeling the old dazzle of opportunity lifting his eyebrows. The university got out the paperwork in three days; they were feeling the long-term research pinch. He delivered the check himself, and then took her out to dinner with her post-docs. She got a bit tipsy.

2040

The several-year recession—even worse than the Great Recession—took its toll on his enterprises. Trimming programs and people were part of business, he knew, just like gardening. Indeed, he took up gardening to get his mind off the failures.

Chief of these was the inevitable collapse of the Muslim countries that had run for a century on oil revenues. Their reserves were depleting, green tech was lowering consumption, and their vast overpopulation was getting ugly in the streets. The Arab states began their collapse, plus Indonesia. Of course that didn't affect Pakistan and the other Stans, or the coming Muslim majority in Sweden. Older countries stiffened their immigration, letting few in but the very skilled. Financial markets turned turtle, heads in, going nowhere.

Plus he put some of his own money into a startup study of building a space elevator. It seemed a great idea, a crucial component to get into space cheaply and effectively. There were super-strong composites, part diamond, and his self-managing AI bots made it seem plausible.

His engineers used a junk debris mass left over from mining to build the first lengths, with multiple redundant cables with cross-strapping to distribute loads around a break. When a meteorite or dead satellite inevitably hits, you would only need to replace a small part.

But the first high launch tower went into rapid launch mode then, a big one that had been building for years. At twenty kilometers height it provided myriad services and could throw capsules up its central stalk. They came out in thin air at about a kilometer per second and then their rockets took over. They chopped small package lift costs considerably below even the traditional two-stage to orbit.

The space elevator looked less plausible all the while. Plus there were tech failures and as well the corporate treasurer turned out to have his hand in the till. That was the big, final hit. *Engineering was hard, but people were harder.* It became his mantra. Harold could not for the life of him hold together the investment coalition to build farther.

He had to fold and take a considerable haircut. The unfinished cable became an embarrassment. He finally sold the useful parts for orbital scrap use, just to stop people taking pictures of it and laughing.

For three years he took a dollar a year salary and sent his stock option profits to the astronomers for their infrared work. Sara offered to help but he thought it safer to keep their assets separate. She got miffed at that and they didn't see each other for two months.

Then she appeared at his skyscraper office at quitting time, in a fine new red dress. Red was always a signal between them, better than roses any day.

"That damned treasurer," he said to Sara that evening. "Takes all kinds but—"

"It must." She smiled as she shed the red. "They're all here, like 'em or not."

At the burial of his mother he felt anew the deep sadness that had settled over him at his father's death two years before. Now he was on his own in a way that did not bring joy but rather a steady commitment. They had taught him how to live and now he had to carry forward what they had also taught without words but with deeds. Somehow.

In this and many other ways he learned to put defeats behind him. And to press on, as well.

2043

His first target back in the 2020s had been an asteroid rich in rare earths. That paid back all startup costs in a single shipment.

Spin could throw the robots off, so he paid rockhounds, with their satellite telescopes and fast-flight sail craft, to find one with a tiny spin and good elements. The key innovation was a rock filter that worked best in microgravity, and not at all on Earth. It separated molecules by charge and mass, using tribocharging induced by simple friction, the same process that yielded sparks on doorknobs after walking across a rug. His first half dozen asteroids made him rich, and then he got started on truly adventurous ideas.

As soon as he had robots that could manage well even on the surface of rocks that spun around several axes, he made spin an asset. There was Board opposition, but he called in favors and won a straight up/down vote with the Board.

A new class of robots spun out long, rigid armatures of carbon fiber, creating a throwing arm. Cargo got hoisted out on the arm, and then released. Careful calculation of spin phase and orientation made it possible to send packages on precise trajectories to high Earth orbit. With a midcourse correction nudge, they nosed into nets waiting to receive them—free transport, paid for by a slight decrease in the asteroid spin. Fuel depots, orbital factories and hab colonies got their goods delivered, for the minor trouble of snagging them as they passed by.

The "right to mine" industry took off then. Brokers traded those rights and gathered capital. Only when shipments started did he fully own the rock he had paid options on. There were plenty of tricky accounting riffs to play, especially since the rules kept changing. The U.S. Geological Survey, which originally had been formed in 1879 to discover ore and stimulate the mining industry, was still in the Department of the Interior. It soon became an interplanetary agency.

Then under the North American Community rules, taxes and deductibles became even more fraught with peril

Plus a new one: hijackers could grab the cargos in flight. It did not take long for other companies operating in near-Earth space to figure this out. Such pirates then inadvertently supplied a defense against the main danger—big masses missing their targets and slamming down through Earth's atmosphere. But with better beacons and tracking, piracy dropped off.

Three multiseason series, about the exotic gangs of 'roid pirates, had run on worldwide 3D—all before Harold ordered arming of his robot 'roid escorts.

2045

He got to Katherine Amani's office within an hour. "Where is it?"

She grinned. "Less than a light-year away. Small, cold, but there."

"How?"

She blinked, used to his abrupt questioning style by now. "We had those suspicions, recall? Early estimates predicted as many brown dwarfs as typical stars, but the WISE survey showed just one brown dwarf for every six stars. So how could we have missed some? If they were close to us, they could move enough between our two surveys. Not showing up again eliminated them."

He nodded eagerly. "So you looked at them again."

She nodded and showed him a star map of many small blotches. One she circled. "It took a while. Atmosphere temperature is a tad above this room's."

He pursed his lips and leaned forward. "Wow, lower end of the Y dwarf range. And close!"

"As you wished." She smiled.

Harold got up, started pacing, then looked at her intently. "You saw last month's discovery—an Earthlike with an ozone line?"

"Yes, great, clearly a biosphere. Nearly a hundred light-years away. All attention's focused on it. Both the Chinese and USA/Euro want to put up new satellites to pick up as many pixels as they can, analyze the atmosphere, maybe get a picture."

He stopped pacing, sat down. "It'll get all the attention. Let's keep this quiet for now. How about putting WISE 2 up?"

Her eyes widened. "Withhold—? Ah, I see. Let them ignore infrared studies while we get more data."

"If you don't mind." *What's my action item here?* he always asked himself. "I like springing surprises, but they have to be substantiated."

"There's not much chance anyone's going to revisit this old data soon. If you have the money to put up another, better WISE…I suppose so…."

There. "Done, then." She blinked. Her mouth opened and nothing came out. *She's being handed a hundred million bucks, after all. I know CEOs who can keep their cool at moments like this, but they are few. So…keep it mellow….* "Let's keep all this quiet for now. Meanwhile, I want to put some money behind a way to get there."

"But our rockets would take—"

"Better than rockets. We can leave the engine on the ground."

2052

He discovered a useful rule: If you want to know what's going on, don't ask the person in charge. But to get the truth, especially from the edgy government bodies that regulated space industries, you had to come in from the top management, and then drill down. So Harold learned that to score a big-name dinner guest or a favor from a V.I.P. in Washington, there was no point messing around with official channels or wasting time with midlevel functionaries. Invite the Big Exec up to one of the orbiting hotels, then take him to the Botworks.

This meant he and Sara often returned to what they referred to as their honeymoon suite. Any deductible adventure that got him into space—*closer to the stars* he often called it—was golden.

It worked. Underlings fear for their careers and are more likely to examine new acquaintances for potential peril. But there's an unexpected naïveté among the truly powerful; they assume that anyone who has arrived at their desk has survived the scrutiny of handlers. They liked anything that helped get ahead of the Chinese. Especially if they got a junket—"learning tour"—out of it.

That was how he learned of a new profit angle. The craters near the moon's north and south poles were like the dusty attic of the solar system—an attic in a deep freeze. In a hundred square kilometer area there were a billion gallons of water in the top meter of dirt—and even better, the same load of mercury. Water made the place livable for the few shivering humans who had to run the robot teams mining the metals. Pure hydrogen poured out when a reactor's waste heat warmed the soil, capturing the harvest in big balloons inflated by the gases. This rocket fuel spread throughout the swelling fleets of mining craft.

All this wealth squatted in the dark craters. After exhausting the several hundred asteroids that were energetically easier to reach than Luna, the frozen poles were the latest economic hotspots.

To his surprise he won an Environmentalist Of The Year award. The citation read, "For shifting mining and refining of metals and rare earths to space, helping in the ongoing clean-up of Earth's surface." He attended the ceremony and found the crowd to be delighted that he had "come so far." They apparently thought he lived permanently in orbit. Indeed, terrestrial mining for metals was a very dirty business, and metal ore mines tended to leave a lot of poisonous waste behind. So the costs of terrestrial metals rose as people grew more environmentally demanding, making asteroid mining and smelting even more rewarding. He pointed out to the crowd that they had increased his profitability and got applause. On the spot he gave the Environmentalist Agenda cause a billion-dollar gift.

2055

By now Harold Mann was one of the *ultras,* the chummy though distant club of the trillionaires. Some said there were mysterious others, the transcendental rich, or "transrich" but Harold didn't think they existed. If they did, they left no signs among the vast and fast trading markets. The constrained AIs who governed those provinces would not say if any transrich existed, but then, they were coy.

Definitions didn't interest him. He was of the *determined elderly* now, rich and harboring the ambition of those who knew they had little time to accomplish more…much more.

Sara said, "You're exercising those stock options close to the line, given all this new legislation." They were swimming off Maui, so the subject seemed odd. "I need cash for R&D."

"I've picked up some legal sniffers around my operations," she said. "The North American Community needs cash so—"

"They always do. Hey, see if we can body surf this wave!"

His advisors told him not to discuss his ambitions so much, and certainly not his intricate finances, what with all the suffering in the world. The vast differences between economic levels had led to the fashionable view humans had messed up enough worlds already. So if life were detected on a distant Earthlike world, humanity had best leave it alone.

The same argument had arisen over the subsurface life discovered on Mars decades before. That life, organized through microbial plants, was remarkably strange and showed clear signs of consciousness. It was entirely anaerobic, oxygen-free, and of separate and earlier origin than Earth's. Many scientists thought that the undeniable connections between Martian and Earthly DNA, going back to the Archea ages, proved that we were Martians. Thus we were not damaging an entirely separate phylum of life. There was no damage anyway, since human Martian colonies had no biological transactions with the true Martians at all, which were far below ground.

Still, Harold made no secret that he had plans for future exploration. He wouldn't say what they were, ever. He funded propulsion studies by exercising stock options in several minor companies, taking his profits, then plowing it into secretive companies pursuing low probability/high payoff technologies.

Secrets created their own fandoms, in the sprawling, intensively interacting solar economy. He was quite surprised when the mysterious aura around his name made the public like him more; people wanted intriguing puzzles now, a sense of things coming.

2059

Some asteroids were icy, with up to 20 percent water and frozen carbon dioxide; miners called them iceteroids. Melt the 'roid rock with circulating nuke- heated fluids and the water comes off first. Condense and separate it out, squirt it into expando spheres for packaging and let it freeze in free space. Hang the spheres on frames holding a bare nuke engine, with no shielding needed. Then robo-ship the whole unlovely contraption to near-Earth habitats for life support or, with the CO_2, for propellants.

The rocky, metal-rich asteroids got teams of mining craft that deployed smart minebots, which could siphon off metals by weight and fluidity. Platinum was the biggest prize, so prospector bots sought it first when they touched down on a new rock. "Fat plat" was pure strain metal that could go straight into Earthside catalytic converters. Auto-facs and 3D printers made electronics or even jewelry. High-value ore shipped in low-energy orbits arrived at Working Earth Orbit space with its market value already set—never less than $50,000 New Bucks a kilogram—because it had been mass spectrum-sorted by bots along the way. Those rugged devices could take all the time they needed to get the measures right. They were slaved to the MarketWatch integrators, beyond question more honest than a human could even pretend to be. The lesser stuff—iron, copper, aluminum—got fed into orbital factories to make spacecraft fittings and hulls in vacuum-dry foundries. Behind all this was the laser comm Net that kept bots coordinated and standards aligned.

Yields accelerated in what became known as the Astro Moore's Law, though in fact the similarity was superficial. The true driver was the plentitude of free fresh mass, coasting out there among the planets.

His was not the first mining company to go out into the main asteroid belt. It wasn't even the tenth. But it lasted.

So did Harold's one, personal R&D budget. He was surprised to find one morning at a news story calling him the biggest research funder except for China, the USA and Europe, in that order.

2060

Dr. Katherine Amani handled the press well. Harold sat in the back and watched her proclaim discovery of not one but two dwarf stars nearer than Alpha Centauri. Yes, she said, she had taken the years of study essential to be quite sure these stars were truly there. One of the virtues of not reporting to government panels, he thought, but said nothing.

Press attention was still focused on the distant, Earthlike world called Glory by the public. Of course no expedition was feasible, but reporters immediately asked about this new star. Did Redstar, the nearest, have planets?

Dr. Amani demurred. It was too early to tell, but "anonymous donors" were readying a far more sensitive infrared study of the region close around Redstar.

The possibility of going there got little attention in the press. The current worldwide depression had bled most of the sense of opening possibilities from the general class who paid attention to more than just getting through their difficult days.

And who named this dull dwarf Redstar, anyway? Surely the International Astronomical Union had naming rights?

Dr. Amani opened her mouth and looked at the back of the room, but Harold was already gone. You get better coverage if the media uncover the story themselves, he had learned—then feed their eager faces.

2063

Harold sighed. "Did you ever think that we're just stuff, the odd sort of stuff that comes into consciousness, reproduces, swims through this universe, and dies, that's it?"

Sara frowned. "You don't believe that."

"No, I don't. But I could."

They were inside a tight capsule of Mooncrete, heavy shielding against a solar storm of great lancing ferocity. This first trip to the L1 resort was not turning out very well and Harold felt claustrophobic—the most common affliction among deep-space travelers.

She kissed him. "Have another glass of wine."

2069

Often when he was in an immersion tank having his body scanned, inspected and improved, he would reflect that the meeting about WISE 2 was where his life began to accelerate. The sensation of time collapsing along its own axis was common with aging, of course. It arose from the lack of novelty in later life. Travel, new friends, fresh hobbies—these helped. But he had been to every country he wondered about, 87 of them when he stopped. Friends were fine, too, though he never had hobbies. Intense interests outside of work, yes—but they were always pointed at the sky, the solar system, the stars.

First came the sails. Entrepreneurs had already developed the fundamentals of solar sailing, a thrifty way to survey and prospect myriad asteroids. The sun's photons were free, but skimpy. Better to focus a microwave or laser beam on a sail and shoot it out of low Earth orbit, saving it years of climbing up Earth's gravity well. Better still, coat its inner face with a designer paint that, heated by the beam, would blow off—an induced rocket effect.

But the real kick came from diving deep. *Sundiver I* had already plunged to within a few radii of the sun, shed the asteroid that shielded it, and unfurled before the furnace star. Its gossamer disk carried an intricately designed burden that now warmed furiously. Painted on, this layer blew off under high temperatures, a momentary rocket exploiting a bit of Newtonian physics: change velocities when the craft has high speed, and the boost gets amplified. A blue-white jet arced for tens of minutes. Once gone, the painted fuel revealed a blazing white sail. That intense hour sent it shooting outward at speeds rocketeers only dreamed about.

Sundiver I had entered the far reaches of the Kuiper Belt by the time Harold assembled the Inner Network System of microwave and laser beams. This INS made considerable profit by lowering commcosts among mining communities, asteroid habitats, bases on Mars and even the new exploration teams around Jupiter's moon, Ganymede.

Sail development accelerated, making quick exploration of the outer solar system efficient. Harold bought into several small sail startups. By the time they had found some ripe iceteroids ready for steering into the inner worlds, and be harvested, he exercised his options and extracted profits on the expectations-bounce tech stocks often get—for a while.

By the time *Sundiver XI* came speeding out from its searing solar encounter, the INS system was ready to pour radiance on its kilometer-wide sail whenever it passed nearby. Timing was exquisite, intricate, a marvel equated in the media to ballet. *The INS goes OUT* one headline proclaimed. Headed for Redstar.

2073

Fusion nuke rockets had become the Conestoga wagons of the solar system, and the lunar poles were their frigid watering station. The reversed-magnetic-field configuration had finally made big-bore fusion rocket chambers practical, and they had higher thrust per kilogram than fission. Shipping got cheaper.

Harold had a piece of polar development, mostly because he wanted to drive toward ever-larger nukes. The Chinese and Arabs who threw in with them tried the same approach, but proved to be slower to respond to the myriad problems that arose. The poles were *cold* and gear refused to work unless warmed by the waste of big operating and distilling nuke plants. Harold had investments in breeder reactors and recycling reactor wastes, so he benefitted from both ends of the enterprise.

Soon AI-piloted nukes ran whole mining parties to asteroids. Complexes grew, which demanded humans to supervise the smart but limited AIs who had narrow intuition and common sense.

On carbon-restricted Earth, such work cut back on wasteful, inefficient, and polluting processes to mine and smelt. There was much less digging, grinding, and greenhouse gas emission. Social benefits rebounded, wealth spread through largess, and the work week fell. Workalcoholics emigrated into space, where opportunity and 100-hour weeks abounded. All that fervor and wealth came from spinning habitats and solar mirrors melting rocks way out in space.

2074

Harold imposed strict rules on how supplies and parts for his asteroid habitats got delivered. The specs laid out exact sizes of the storage canisters, plus where the securing bolts went, how latches fitted, corner configurations and thickness down to the millimeter. His suppliers grimaced but complied, wondering why he was so exact.

"Your competitors aren't so damn picky," one said.

"But I pay more for it," Harold said. "You want the contract?" He smiled at the grudging nod.

He told no one the reason. His work crews dutifully unloaded the supplies and parts in zero-grav, handing them to the bots who did the assembly of borers and smelters. Only then did the next bot crew appear, taking the shipping canisters apart, clicking together the light carbon-fiber walls—and producing the actual outer walls of the habitats. Spun up, the tight joins held full atmospheric pressures. He had gotten them delivered at the suppliers' cost, not his.

2081

"We can hold them off for a while but not forever, Lin said.

"You've done that for decades now," Harold said, hands clasped in front of him at the conference table. Sara sat beside him, shaking her head.

"Decades?" Sara said. "I forget details… Writing all the development costs off—"

"As business investment, yes," Lin finished for her. A 3D lattice condensed in the air over their table. It mapped in axes of time, dollars and legal avenues. A spaghetti of multicolored strands connected big orange dots. *Like a nightmare medusa,* Harold thought.

"All this was legal, deductible back when—"

"You started, yes." Lin gave him a wan smile. "Not any more."

Sara looked startled as she followed the info-dense tangles. "These topo maps in cash and progress indices—there's a whole development line just in ion engines!"

Lin grimaced. "That doesn't get us off the hook any more."

Harold nodded. "Our poor old planet has seen a lot. Environ damage, the greenhouse not going away as fast as we thought, resource scarcity, big collective regimes. So they go after offworld cash."

"And change the rules in the middle of the game," Lin said. "You've been in the game a long time, so you have to change the most."

"So…" Sara was still entranced by the luminous spaghetti and its projected territories. "…you can pay some back taxes?"

"Sure, if I strip myself of most that I own," Harold said. "Or—"

"You can fight," Lin finished for him. She obviously knew her boss well, Harold saw, and was one jump ahead of him. "Legal dodges, evidence of sequestering funds offworld, not available for testimony—"

"The whole suite of dances every business learns the hard way," Sara finished. "Let me guess—back taxes doubling every three years, retroactive penalties, the usual post facto nightmare."

"Can I go back to Earth?" Harold asked, eyes veiled.

"I don't advise it."

Sara jerked as though she had just come awake. "What?!"

"They can arrest you." Lin gave them a *that's how it is shrug*. "Not likely maybe, but it depends on who's got the helm of the North American Community when you land. You have enemies in the Mexican faction."

Harold remembered the friction surrounding the adoption of the the Community Constitution. Yes, the Mexicans wanted a Right of Confiscation and he spent a lot to block them. Mexican Constitution from over a century ago stopped foreigners from owning land in Mexico. Now there was a French-style tax on net worth, too. He sighed. "It doesn't take a very big person to carry a grudge."

Lin said softly, "The sitting Chairman called you out publically for using corporate funds to further your 'hobbies' yesterday."

"I thought hobbies were supposed to broaden you," Harold said. "Mine is R&D."

Sara's mouth twisted into a cautious tilt. "Remember that silence is sometimes the best answer."

Lin said, "I started the legal defenses running."

"How about my reserves—and Sara's—in Earthbound accounts?"

"I got them posted free to Lunar Security last night. It caused a drop in several markets this morning."

Sara said, "No hiding when you're this big, I guess."

Lin nodded. "I had to sacrifice some transactions in progress. Resource plays, a convertible debenture or two."

Harold got to his feet in the 0.4 Grav. "I smell Keegan over at Consolidated."

Lin said, "I do too. Rumor says he hates you. He wants to weaken you, maybe create opposition on our Board."

"I'll work the Board some." Harold shrugged. "Sometimes you get, and sometimes you get got. Let's go for a swim."

Their new swim sphere was forty meters in diameter. For safety a black cable lanced through the middle, so swimmers too far from air could haul themselves out quickly. He made a point of never using it and lapped furiously around the perimeter, letting drops scatter everywhere. The blue-green water cohered, enhanced surface tensions gathering up the drops amid the air currents. Sara was not so proud; she swam subsurface most of the time, using an oxy enhancer.

When she surfaced he was there with a frown and after her first gasp for air she said, "Don't worry, you've faced troublemakers like this before."

One of his signs of anxiety was a slight lapsing back into his southern accent. "The biggest troublemaker you'll probably ever have to deal with watches you from the mirror every mornin."

2085

In principle it was simple: stay in orbit, use the centrifugal 0.4 grav advantage that the Mars Effect people had shown did indeed lessen damage to the general neuro and cardio systems. But some were shy of orbiting at all, so they lost old friends. Harold and Sara too tired of orbital life, of running their far-flung businesses electronically—so they had battalions of lawyers fight the Community edicts. This allowed them some immunity to prosecution and seizure while on the ground. Still, they made visits short, mostly to see the latest work on robotics and catch up with the red dwarf star scientists. A hobby of sorts, he still maintained for the lawyers.

2093

A reporter accosted him and Sara in their home overlooking Kings Canyon. It was an "arranged surprise opportunity" as his media advisors called it, so he feigned being startled.

"You're certainly the man behind Dr. Amani's announcement of Redstar's discovery, Mr. Mann. Now you're launching a probe to look at it." The reporter feigned astonishment, as required. "So you knew about it for years!"

"Decades, actually." Deadpan. *Hold Sara close*, he thought. *And smile, dammit*. After all, it was their umpteenth five-year contract anniversary. She smiled, also obligatory. He couldn't hold his smile for long. He focused on the view. The distant pines and elegant granite peaks weren't clear and sharp like the old days, since the warming gave a heat ripple everywhere.

"Why did you and your scientists not announce—"

"I wanted to get my ducks in a row. Now the whole world is training their 'scopes on Redstar, to tell us what they can. I'll give us all a close-up."

"That's arrogant!"

"I suppose so. This isn't a hobby, it's a lifelong obsession. I wanted to do it my way."

"The entire world scientific establishment—"

"Is just that, an establishment. Funded by governments—that is, by you citizens of so many nations. I wanted to move faster than that. And make a profit while I built the INS."

"What will *Sundiver XI* find?" The reporter seemed to be getting desperate.

"I don't know. That's the point, yes?"

As they walked on and Sara said, "Y'know, people are saying that, after all this, we'd decide to retire. We've finally accomplished everything we set out to do, so…." She gave his a raised-eyebrow glance.

"I don't think we're safe, just retired."

"Oh—the Universal Rights rules?"

The inevitable collision between a stressed, overrun planet and lots of retired but vigorous well-to-dos was looking like a train wreck. The central political message seemed to be, *You have that cash, see, and I have this gun…and these lawyers.*

"So they'll come after us for more taxes," Sara said. "We can live with that."

"It's getting harder to stay Earthbound. There's no government space program anymore; can't afford it or get out from under their bureaucracies. That Exceptional Needs Tax alone—"

She jostled him, kissed him. "Enough worries! Don't let tomorrow use up too much of today. Let's go for a hike on the John Muir trail."

She was right. He went. They had friends to meet later, a good cabernet tasting—life was good. He should lap it up. Even so, in the back of his mind on the trail that afternoon he thought of *Sundiver XI* and mulled, *Even better to see it in person. No, that's silly. Not that I can. Too damn old.*

2100

He and Sara arrived at the robo-control station in a high arc ship. The zero gravs were a blessing to some of his joints. He and Sara had spent as much time as they could in the 0.4 grav floor of the Cylindrical, but there were limits. Whether 0.4 was optimal was still a raging debate, but optimizing life always had been—just usually without statistics and multivariate analysis.

As he swam through the transfer bubble, looking out through UV filter walls, he saw a waiting array of ceramic and carbon fiber bots. The gliding serenity of deep space made a slow, artful ballet of even routine industrial processes.

He knew their designs well, had even worked on some of the coupling joints. Stubby, mostly arms and counter-levered hold joints highlighted in yellow, they sat dutifully in a fiber rack array, drawing power from the idling nuke a few meters away at their backs.

Zero-grav bots had no front/back bias at all. They could spin their heads to bring digital eyes or metal sensors to bear, all housed in a rotating platform that sported microwave antennas and a laser feed, too. Omni-able, the industry pundits called it. He was proud to be one of the 59 inventors listed on the patent—whose fraction he had given to Sara, when they'd celebrated one of the five-year contract marriages.

The rest of the bodies were chunky mech pistons and grapplers, driven by fluid hydraulics. No beauty, but they were yellow and red and blue as suited their jobs, so the control bot could tell them apart by mere visuals. People in space made their habitats and work areas gaudy, a reflex against the surrounding black. The color splashes would have driven an Earthside decorator to violence.

And here came the orange transfer pressure gate at the end of the tube. The door dilated.

"Officer aboard!" a young woman declared firmly as he popped through the dilating lock. Her name badge woven into her flexsuit said *Nguyen* and she gave him a grin. "It's protocol, sir—I know you're not an officer."

"Daddy Spacebucks aboard would be better," he said.

The man next to her laughed but it seemed unlikely anyone here knew the reference. "Yessir." He wore a T-shirt that proclaimed SAME SHIRT DIFFERENT DAY. Harold sniffed and was thankful habitat air cleanup was well developed. "Sorry you're not an actual daddy," Sara said beside him.

"Um, what? Oh…I never wanted to be. I'd have been terrible at it."

"I felt the same, for me. We're better at making things work."

"Too bad about the genes not getting passed on."

She chuckled. "What makes you think that?"

He couldn't suppress his grin. "OK, we've all got stuff stored. We live quantified lives with backed-up selves. Your eggs…?"

"I might want to use them someday. Or another lady."

"Way things are going, could be another, well, guy."

"So be it." She shrugged.

"We're here as a luxury to the species. We expand horizons, most people live inside them. Hey!" Harold looked around, uncomfortable with abstractions, as usual. "I'm here to catch that new rock work."

"I got that from the squirt," Nguyen said. "That's why we vectored you here. The bots are about to eat it."

"What's the problem?"

"Mech coherence," the man said. His name was Frakto, he had purple skin, and he waved Harold through a side bend into a narrow dilation lock. "They're not catching the flyout right."

As Harold settled into a tight shuttle bore he saw the bot boat flare with a long ivory jet driving it. Against the stars the steam wake billowed like a wavy porcelain flag, then vanished as it expanded and cooled.

Nguyen piloted their shuttle so it stayed away from the jet, Harold and Sara hanging onto a grip in the short accelerations. They coasted alongside the bot boat. The sun swam across the blackness like a glaring eye, though blotted by the smart walls.

For a moment the sun passed behind the bot's jet and a shimmering rainbow appeared, framing the long snarl of pearly steam in ovals of blue and soft red. *A sundog,* Harold thought. The phenomena on Earth framed circles around the sun, tracking it. Here the circles faded on both sides of the steam plume, as the jet density fell off.

This surprise made him smile. He had never thought of this effect acting in high vacuum. Once more he had the sensation of discovering something in the obvious-once-you-see-it category that never ceased to delight him.

Later, when they cast an entire iceball cylinder into high velocity, he felt the slow pleasure of delight again. It picked up considerable speed in the magcat, a big lumbering beer can with a snow white cargo. Then the catapult sleeve fell away as the cylinder's ion jets cut in. It shrank with bewildering speed, faster than any takeoff he had ever seen.

"It can rendezvous up to what speed?" Sara asked Nguyen.

"A hundred klicks a sec," Nguyen said. "Maybe better."

"Um, what's the commercial app?"

"Emergency supplies needed in a habitat," Harold said crisply. "Funded it myself."

She smiled. "You belong out here, you know."

"I've been here, one way or the other, since I was driving a cab."

Man's got to throw long in this life, he thought.

2102

At first he had noticed the years accelerating. Now it was the decades.

He and Sara did all the new techy health things plus taking a special Lifecode series of molecules targeted on their own genes. They up-regulated their repairs and kept their bodies fixing the innumerable insults of advancing age. Tedious, sometimes, but it worked. They wanted to be around when *Sundiver XI* arrived. Harold didn't say so, not even to Sara, but he wanted even more.

He had to tell his advisors repeatedly: *Become a mere steward of your own assets? Boring!*

A rich bank account did not mean rich ideas; in fact, often the reverse. The bigger your ass, the more you want to cover it.

So the second step was the comet-grabbers.

Investors care a lot more about return on invested capital than about optimized hardware or new technology. But give them dividends and they will give you your technologies. Start with an easy consumer come-on first.

Luna was in many ways a pleasant place—right distance from the sun, quite nearby, light gravity. But it was bone dry; if ordinary Earthside sidewalks had been there, miners would have leaped at them to suck out moisture.

With air and water, people would visit the moon's striking plains and mountains. For now, bubbles blown in excavated cavities would suffice for flying, the best sport of all. The retaining bubble was actually the new flux-diamond, a carbon liquid that condensed into a rigorous firm seal. So he invested in the new Lunatic Hotels and luck intervened. Their first cavity struck a totally unexpected lode of ice, so they had no water problems—the enterprise went profitable. Contrary to the conventional wisdom, Luna was a soft, stony sponge.

He paid off the support loans at the earliest possible no-penalty date. When the comet nuclei he had ordered sent in from beyond Uranus by robot crews finally arrived, they weren't essential, thanks to the lake discovery—and he ordered a swimming sea made of them. They crashed onto the surface but succulent machines captured a lot of their moisture.

Nobody knew it, but this was the beginning of the terraforming process that would take many decades before a filmy skin of moist air clung to the Lunar craters. That shallow gravitational well could still hold for 10,000 years the atmosphere slammed into it by a cascade of comets. But a full atmosphere could wait. Soft winds would blow across those ancient lava fields, eventually, but the Loonie Hotel business made profits *now*.

So did throwing ice and hydrocarbons sunward, the real point. The robot teams who snagged floating mountains of iceteroids and steered them inward had their troubles, but bots worked 24/7 for little maintenance cost, no vacation time or health care, and with no retirement plan. They benefitted humans who loved the interplanetary splendor they saw on their screens, then went to the sunny sandy beach after their twenty-hour work weeks.

Profit made his own research teams possible. They developed a way of shaping iceteroids and torpedoing them at very high speeds into the hoppers of passing nuke rockets. Vapor flared in the nuclear chambers and forked out at furious speeds. The magnetic catapult system led to efficiencies in nudging comet-candidates onward. In turn this increased the profitability of supplying light molecules in shimmering spheroids to the inner solar system colonies.

But such distant wonders were scarcely the whole point. The shareholders thought so, and Harold obligingly said so. He owned the largest share fraction but ruled through a coalition on the Board. Still, he ran his latest corporation, Farscape, with bigger goals in mind.

"I sold my debris companies," Sara said with a sigh. "We've done most of the job. Mop-up is boring, casting big sheets to grab small stuff. Time to get out."

"Good. I have a trip planned, we'll be free to go."

He lounged back while a nurse did the daily IV. He did not allow her to joggle his hand holding a vintage red wine; it was part of the medication, too. Deductible, even, though they had stopped thinking about such accounting matters decades ago. Then a good swim in the spherical pool, yes. They were on the 0.4 g level of the Great Cylinder Hotel, which seemed to be the right gravitation for longevity. Turned out the studies using lab mice and pigs had been right. Sara especially was aging very slowly; he had to hustle to keep up with her.

She smiled. Time had been kind to her wrinkles; indeed, she had few. They had just finished celebrating their 100th birthdays, three months apart, and even a thousand kilometers from Earth the media hammered on the walls, metaphorically. It was hard to escape the grasping Earth.

"Where?" she asked.

"To the whatever end there is."

2103

Sundiver XI plunged by Redstar in less than a day. After getting its first delta-V near the sun, then long-term boosts from microwave beamers and lasers—the Inner Network System—it had turned on a high efficiency, nuke-driven ion drive that gathered molecules from in front of the sail. The combination shaved many centuries from the voyage. It also meant the big sail was a relativistic missile streaking through, swiveling its cameras and sensors to grab torrents of data.

Harold watched intently with Sara at his side and Katherine Amani on the other. The smart sail was tacking as well as it could, zooming by the star. All this had happened nearly a year ago, of course, delayed by the light travel time.

The team below them studied their screens and abruptly a startled shout came up from them. On the big screen appeared a reddish world with a shimmering ivory haze of atmosphere surrounding it. Harold saw an unmistakable glint of sunlight from the left edge. Polar caps were a dirty gray.

"An ocean," he said. "On a tide-locked planet."

Katherine Amani said, "We picked it up earlier but this is the freshest image. There are plenty of spectral signatures. There's an odd cutoff in the far infrared, something like the edge we see in light reflecting from Earth plants. This spectrum is really dark red, down to reddish-brown to black."

The next image showed clear continents and somber seas, even big lakes and some white-capped mountains. Reams of data slid down screens around the room as the sail did its job.

Katherine said, "There's going to be so much to study—"

"I'm going," Harold said. "Now."

"No," Sara said. "*We* are."

2104

Lin was in a hurry. "You're not safe here anymore. The legal walls around you have come down."

Harold said, "It's that bad?"

"The North American Community reached a midnight agreement with the Euros. They got new laws passed, to let them seize assets and argue their case later. It applies to anyone offworld, too. No matter how big."

Lin had worked for Harold for decades, so he knew she was holding the worst back for last. "And…personally?"

"There are criminal charges available." She said it calmly but he could tell she wasn't.

"Could they come up here with warrants?"

"I'm afraid it's explicitly allowed for you two, plus about fifty others. The ones on Luna will get more warning—"

"But we're just an hour away," Sara said.

"Afraid so."

"Could this be just intimidation?" Sara asked.

"It doesn't smell that way," Lin said, presenting a projected summary of the legislation. Harold had read enough bad prose to know the bomb was buried in the footnotes, and with a tap Lin brought up their names.

Harold Mann, indicted.

Sara Ernsberg, indicted.

Subject to precautionary arrest.

Sara seemed undisturbed. "We're packed for a long boost. I've got our vital memorabilia in a carry case. Clothes, meds, the rest."

Harold watched the two women get the staff moving—hustle, bustle, rustle. He didn't regret growing older, it was a privilege denied to many. But he had trouble grasping the unspoken assumptions behind these new societies. Boundaries got redrawn at the point of a sword, and the legal frame followed. When he was growing up the paradigm had been *with liberty and justice for all*—but now on a world stage jammed with swarming masses in desperate need, it seemed to be *three hots and a cot and whatever you got.*

"You were right, back decades ago," Lin said in passing. "Going for high-efficiency boost. And building that development complex out in the asteroids, where bot teams could do the assembly."

"Bots don't blab," Sara said.

Harold smiled and nodded. "This is going to be more fun than retirement to a prison."

2105

Harold had started letting people call him Harry, now that he was over 100.

They pushed him into the ship on a zero-grav gurney. It was massive with med devices and monitors, all wrapped around a lean though not frail body.

Sara smiled beside him "Are we going first class?"

"The only class. It's just us."

"Crew?"

"They bail out at Neptune. After that, bots."

"You planned it that way? From the first?"

"This was a worst-case option. The bots can do human coldsleep tending in flight. Our genomes and specs are already run in a lot of simulations and some lab trials." He shrugged. "Best I could do."

In the run-up preps he had rejected a mix of genes derived from naked moles and eagles to improve his hearing and vision. He also shook off, with an irritated snort, suggestions that for a trifling sum he could have his multiracial brown skin suffused with the fashionable golden scarlet. He had scrawled across the memo, *As is!*

"My, these look more like coffins than I'd like," was Sara's only remark when the immersion team helped them try on the big blue sleeves for the Sleep Crucibles. Harold's LongSleep company had tested for decades the induced-sleep pods he and Sara would use. With hydrogen sulfide bleed-ins and low temperatures, long cruise ships had been carrying slumbering passengers to the outer solar system for over a decade. This would be a logical but untried extension of those, by an order of magnitude.

The entire solar system media maze was now intent on Harold's "quixotic indulgence" but he had left all that behind. Some of it was amusing, though. Already crackpots were baying that Harold's fast probe would announce our presence to unknown alien bullies, who would come steaming in to trounce us. Their detailed explanations of why were useful diagnostics of the crackpot's problems, often amusing. The most common was that aliens with the right ontological bias would read out the state of our technology, our vulnerabilities, and deepest secrets. Then they would come for our riches.

"Maybe the Redstar carnivores will eat us," Harold mused.

"They'll have to catch us first," Sara said lightly. "If they exist."

They were lying in the crucibles while the team worked on their hookups.

It all seemed dreamy to Harold, yet he did not want to miss any of it. After all, he might not wake up. Still, when the team left and Katherine Amani leaned over them, he felt a longing for the world—no, worlds—they were leaving behind. "I will never forget you," Katherine said, and kissed each of them. Now she would go back Earthside to help his staff and companies face the consequences. For once he did not know what to say.

He fell asleep as the long burn began.

The automatic systems did revive them for system checks as they cleared Jupiter's orbit. When they passed Saturn he insisted on getting up and looking at the big screen nearby. Saturn was just a dim glimmer, with the sun a glaring white coin. He held up his hand at arm's length and realized he was covering the entire orbit that Earth swung through. All the great acts of human history had played out on that scale.

Then their long sleep began, with the sour, stinky tinge of hydrogen sulfide cocktail swarming into their nostrils as the cold seeped in.

2192

Brown dwarf and planet (Jon Lomberg)

He heard, "—hydrogen sulfide bind to cytochrome oxidase reversing complete. All blood oxygen returned to normal. Beginning neuro—alert!—is awake."

"You bet…I am," he croaked. "Sara…?"

"Reviving." The voice was precise and melodious and of course a mech—an animated AI that seemed to swim in his blurred vision. It had eyes and a grill speaker mouth but that was the only concession to human-like appearance.

"We made it." He had harbored doubts, of course. Now a great feeling of triumph swelled in him.

"With conditioning, we will escort you to the world below," a mech said. "But you must undergo the restoration treatments."

"Anything the autodoc wants," Harold said. "Anything. Try to make it fun."

What had he used to say, back as a kid? *Committees don't open frontiers—people do. With smart machines.*

The extravagant ruddy sun painted them pomegranate. Redstar had banded clouds of methane that echoed Jupiter, in a constant slow swirl.

"Let us see it better," he said to the mechs. "Tune helmet filters to our eyes." *A strange new world,* he thought.

"Full sensorium, please," added Sara.

Even though their helmets amped the visible spectrum, the effect was eerie. Stars shone in pale gray here against the inky black. The huge hull of Redstar hung as a burgundy disk cut off by the sea. Here and there across the long panorama of perpetual twilight, slanting rays of a deep Indian red showed floating plants, lapping on the waves in a somber sprawl. Everything glowed with infernal incandescence.

"Good to sit," Sara said. The mechs had brought them low sloping chairs from the descent vehicle. The 1.34 gravs here made walking odd and harder, although they had been reviving their muscles for five days aboard the main ship. He had learned to edit out the joint pain, too.

Down from the desolate slope to his left came an echoing cry, long and slow. In the thick air a thing like a huge orange gossamer butterfly fluttered on a thin wind. It swooped across a sky peppered with amber clouds and vanished with deliberate, long flaps of its enormous wings, vanishing behind a low eroded hill.

"Thick atmosphere, something that big can fly," Harold said, still trying to take it all in. This was far stranger than Mars, with an entire ecology on ready view. Most of the vegetation was low slimy growth, hugging the land.

"I wonder…," Sara said and stopped. Something moved on the beach.

It looked like a reddish rock at first until he could see the legs articulating with a sluggish grace. A huge crab-like creature with long antennae waving. Now that his eyes adjusted to the odd light he could see other small forms. A big thing broke surface out on the oily swell, then slid away.

Slow-stirring life abounded on this stony beach. "Well chosen landing site," Sara said to the mech crawler nearby. The machines kept their silence, as if understanding the importance of this moment.

He recalled that in the media storm before they departed, a faction had argued that some ethical imperative made ever visiting other living worlds morally reprehensible. Such people thought alien life should go undisturbed, never realizing that their notions would kill the very impulse behind astronomical curiosity. Why find it if you can't study it further?

They would never see such wonders as this. Still…something tickled at the back of his mind.

Then he felt it. They were sitting in seats that sloped back like loungers, and the old memory came fresh again. This subdued crimson landscape recalled Orange Beach, Alabama, where he had rented chairs to tourists. That night his boyish eyes had truly seen the stars for the first time and yes, the ideas had mingled. Learning to do a job, making money…and at work's end the sudden huge perspective of the galaxy itself, a sprawl of stars like jewels on velvet above the salty waters. He could see the design of his life in a single scene, leading to this alien beach.

"My life has come full circle," he said. Sara just grasped his hand through their suit gloves.

Her analytical gaze swept across the view. The crablike thing was still lumbering slowly down the shoreline, apparently looking for food in the slight surf. "I wonder how long we can live out here."

The nearby mech said crisply, "We will have to harvest supplies from this world, begin expansion of living quarters. I assume you wish to live on the ship, of course."

"For now," Harold said. "But to live down here, yes, that's my goal."

Sara studied him for a long time. "You know, we're nearing 200 years old."

"Not done yet," he said.

He and Sara had decided to rest, just a quiet doze in their suits while sitting on the beach. Somehow the scene was restful.

So Harold was surprised when the mech called him up from a muzzy sleep.

"There is news," it said. "A signal."

"What? Um. Pop it through."

In oddly accented English a relayed voice said "—hailing the Mann vehicle, we are approaching orbit. Harold Mann expedition there? In middle of doing a delta-V around the planet to lose last of our velocity."

"What?" Harold said, suddenly alert. "Who are you?"

"Translight expedition eleven, sir," the soft high woman's voice said. "Are you human or mech?"

"Human. Harold Mann."

"What? Then you survived! We did not expect—never mind, this is great news." The voice rose, elated.

Sara said, "How—?"

"We hoped to get here when your expedition was to arrive, but your transmissions were sometimes confusing. So you're there! Looks like—on the surface?"

Harold realized there must be video on this link too. "Yes, a relaxing day at the beach. Recliner chairs. Cocktail hour coming up."

"We're not equipped for landfall. This is just a compact carrier, experimental. We'll have to meet you—"

"Wait, how the hell did you get here so fast?"

"Translight, sir. It's a relativistic warp effect, been working on it for decades. Our teams have made some jaunts into the Oort, now this. I admit it was a long shot, trying to catch up to you. We're funded by one of your own companies, Galaxy Nautics."

"So…." Harold was having trouble following this woman's fast, odd accent. "Looks like I left the right people in charge."

"Some of them still are—we live much longer now. Mr. Mann, you're the first. You beat us here by maybe a few weeks. We're still getting the translight calibrated."

Sara broke in, her voice slow from just waking up. "So there's a… new method?"

"New physics, straight out of the new Insight. Is that—Mrs. Mann?"

Sara managed a dry laugh. "Ms. Ernsberg. Our five-year marriage contract ran out about a century ago."

"You *both* survived! Wonderful news. Estimated probabilities were less than 10 percent, so— sorry, we came expecting to be talking to just the mechs. We'll have to rethink our procedures—"

Harold said firmly, "I believe you work for me, Ma'am."

"I…suppose I do." The woman's voice hesitated and came back a bit subdued. "I can understand your Anglish! You're…the oldest people, ever. Not even our Optimals have gotten this many years logged. We didn't really think—well, anyway, we brought modern gear for your body upgrades. Our CEO insisted."

"Is that Mark Martin?"

"Well, no, he's Chairman of the Board now. But it may have been his idea, yes. We can do micro-repairs, sort out your accumulated epigenetic effects from the long-sleep—"

"Good, start getting your gear up to speed for us," Harold said. "My joints are aching and I need a high-mileage checkup."

The woman laughed. It felt good to be back in business.

Harold stretched against this planet's strong gravs. "Y'know, the formal name of this was the Forward Expedition. Let's call it that, OK?"

Sara glanced over at him and said to the general comm, "Pretty soon now, we'll have our mechs lift off this beach and get into orbit. Say, you don't have a warrant for us, eh?"

"What? I don't—oh, our AI shows the doc history…." They could hear the woman muttering to someone. "No, that regime killed a lot of people. No real records. No wonder you left in a hurry! They didn't last long."

Sara laughed and said, "When we left, Earthside had its same problem—too many humans on the planet using a destructive technology to live by. Has that changed?'

"Oh, plenty. Fewer of us now, getting the climate punched up, importing plenty from offworld."

Harold wanted to know everything that a century had unfolded, but restrained himself to, "Keegan over at Consolidated—what happened?"

A long pause. "Checked my infold mind. Both of them went under in some scandal."

"Ah." Perhaps it was small of him, but Harold grinned. "Good. Aim to rendezvous within two orbits."

"Uh, yes sir. We're modifying our delta-V now. As soon—"

"Just get it done. I want to see your system specs as soon as we're aboard."

"Oh, yes, I think—My Lord, there are statues to you back home! I don't know how to—"

"Yeah, well, this statue talks. No pigeons here, either. Make it snappy. We've got work to do here."

***Starship orbits a planet of Barnard's Star. Nemesis* (Don Dixon)**

BUILDING STARSHIPS

Pioneer 10 heads into the galaxy **(Courtesy of NASA, nasaimages.org)**

Across the Dark, the Pioneers
Geoffrey A. Landis

The ships first sent across the dark ocean,
pebbles flung into the universe vast,
rocket-propelled, a flash of motion
past Jupiter, Saturn, the Kuiper cloud:
    they glide outward to the stars

    now silent, dead, pitted by dust

    a voyage of a hundred thousand years:

    the Voyagers and Pioneers.

The next probes sent out across the dark
the swiftest ships yet made by man
ion-engined craft, faster by far
with nuclear reactors making power
speed past the planets, and brave the dark
    and distant silence between the stars;

    and dwindling in their rear-view mirrors:

    the Earth, the sun, and Pioneers.

The light-sail probes soon follow on
huge sails that dive down toward the sun
and outward thrust by just the force of light.
They need no fuel to challenge the sea of the night.
    The mirrors reflect the dwindling sun

    pass past all planets, one by one

    they see reflected in their vast mirrors

    the silent coasting Pioneers.

James Benford & Gregory Benford

And faster sails, faster far,
pushed not by light from our feeble star
but focused beams of laser light;
or pushed by microwaves in flight
    pass the ion-engine ships,

    prior sails reflecting now but dark.

    They'll leave behind in their rear-view mirrors

    Earth, the sun, and Pioneers.

Then fusion probes, massive and fast
with exhaust bright as a thousand suns
flickering diamonds in the sky
dwindle in the darkness as they fly
    past sail ships already on the way

    past the laser craft launched after

    and far away, left in the rear

    the Earth, the sun, and Pioneers.

And we wait at home, listening intent
for messages from the probes we've sent
signals nearly too faint for us to hear
attenuated by transit across light years
    the first to reach a distant sun

    that tells of wondrous worlds unknown,

    the glory reflected in distant mirrors

    the voyage begun with Pioneers.

And so we fly, through centuries
faster and farther across the emptiness;
we send out probes, our robot selves
On voyages of decades across the darkness
    and dream one day humans too will go

    the ultimate voyage, which has no end.

    Behind us, in our rear-view mirrors

    we'll see the sun, and Pioneers.

Dreams of reaching for the stars date back to antiquity. Plans for how to do it have benefitted from even the simple plans of the early twentieth century, and promise to become more focused as we realize the enormity of the task.

Starship Pioneers
Adam Crowl

Early starship designers were hopeful that the vast distances between stars could be crossed at great speeds. Their designs pushed at the limits of physics and engineering understanding, only to have serious flaws emerge upon further development of their basic concepts. This history of Starship Pioneers teaches us that ideas tried and rejected never really remain forgotten but frequently lead to better ideas. Present-day starship concepts are challenging, due to the extreme scale of the problem. They represent the latest evolution of early concepts that were explored, found wanting, then evolved, and improved. On this historical foundation twenty-first-century starship designers stand.

In ancient times, the distance to the stars was a hazy "much farther away than the planets," and had been since the earliest discussion of their distance in Archimedes's *"The Sand Reckoner,"* in the third-century BC. Archimedes interpreted heliocentric astronomer Aristarchus to imply that the distance to the stars, compared to the sun, was proportionally as far away as the ratio of the radius of Earth to the distance to the sun—about 10,000 times. Thus the sphere of the stars was an unthinkable 100 million Earth radii away, leading many to settle for the closer 20,000 Earth radii of second-century AD geocentric astronomer Ptolemy's estimate. Satirists such as Lucian of Samosata wrote of trips to the stars, but it wasn't until after the sixteenth-century's Copernican Revolution that the stars become possible destinations. Johannes Kepler, the German astronomer who first determined the laws of motion of the planets, wrote to Galileo Galilei, the Italian scientist who had just discovered the four largest moons of Jupiter, the following:

But as soon as somebody demonstrates the art of flying, settlers from our species of man will not be lacking. Who would once have thought that the crossing of the wide ocean was calmer and safer than of the narrow Adriatic Sea, Baltic Sea, or English Channel? Given ships or sails adapted to the breezes of heaven, there will be those who will not shrink from even that vast expanse. Therefore, for the sake of those who, as it were, will presently be on hand to attempt this voyage, let us establish the astronomy, Galileo, you of Jupiter, and me of the moon....

By the nineteenth century, telescopic techniques finally become precise enough to observe the annual parallax motion of the nearer stars. Using this, in 1838, Friedrich Wilhelm Bessel measured the distance to the nearby star, 61 Cygni. At the time, the largest astronomical distance measure was the distance between Earth and the sun—the aptly named Astronomical Unit or AU. Bessel measured 61 Cygni to be 660,000 AU distant, a figure so large that he invented a new unit, which he dubbed *the light-year*. He computed 61 Cygni to be 10.3 light-years distant—a value not far removed from its current measured distance of 11.4 light-years.

What nineteenth-century scientists finally measured, nineteenth-century science fiction soon imagined crossing. Jules Verne's adventurous Frenchman, Michel Ardan, in *"From Earth to the Moon"* (1863) prophesized the means for flying beyond the moon to the distant stars in his first speech to the American public:

Well, the projectile is the vehicle of the future, and the planets themselves are nothing else! Now some of you, gentlemen, may imagine that the velocity we propose to impart to it is extravagant. It is nothing of the kind. All the stars exceed it in rapidity, and the earth herself is at this moment carrying us round the sun at three times as rapid a rate, and yet she is a mere lounger on the way compared with many others of the planets! And her velocity is constantly decreasing. Is it not evident, then, I ask you, that there will someday appear velocities far greater than these, of which light or electricity will probably be the mechanical agent?

As fleet as the planets were, as dramatized by Verne, flying to the stars would require millennia at merely interplanetary speeds. The experience of nineteenth-century telegraphy showed perceptive writers such as Verne that "electricity" was perhaps a thousand times faster, and light even more so. Yet Verne's insight as to the means for flying to the stars remained fiction for the rest of the nineteenth century. Anti-gravity forces and gravity-proof materials vied with gigantic cannons in the pages of late nineteenth-century science fiction, epitomized by the interplanetary hydrogen gas guns in H.G. Wells's *"War of the Worlds"* (1898). Around the turn of that century, physics transmuted from classical mechanics based on fields in absolute space and time, into twentieth-century physics based on atoms and quanta subject to relativity and quantum mechanics. The discovery of radioactivity hinted of energies within the atom sufficient to reach the stars.

The Astronauticists and Their Starships

The first realistic means of achieving interplanetary, and interstellar, travel was developed in parallel by Konstantin Tsiolkovsky, Hermann Oberth, Robert Esnault-Pelterie, and Robert Goddard in the early twentieth century. Their theoretical and experimental work in the first quarter of the century was continued and expanded enthusiastically in Germany and Soviet Russia in the 1920s, but officially ignored in French- and English-speaking countries despite the establishment of Interplanetary Societies in the UK and USA in the late 1920s. Groups of enthusiasts and science-fiction writers, in particular, kept apace of the progress quietly being made in laboratories around the world. After the V-2 bombing of England in 1942 and the atomic bombing of Hiroshima and Nagasaki in 1945, neither government nor the man in the street could ignore astronautics and nuclear physics.

Konstantin Tsiolkovsky was first to publish, in 1903, the *Rocket Equation*, from which all modern day astronautics derives. A rocket is a reaction vehicle—propelled by Newton's "for every action there is an equal and opposite reaction" as propellant mass is expelled at high speed, and the rocket vehicle is thrust away by reaction in the opposite direction. Unlike propeller or jet aircraft, rockets must contain all of the mass they expel as propellant. Tsiolkovsky's *Rocket Equation* tells us that any reaction vehicle containing all its propellant on board gains speed linearly as its starting mass of propellant rises exponentially. To double a rocket's final speed, the ratio of the starting mass to the vehicle's final mass (its so-called *mass-ratio*) must be squared. To triple the final velocity, the mass-ratio must be cubed. This is a steep curve to climb.

By the 1920s Tsiolkovsky had designed a series of ever more capable rocket-propelled vehicles as part of a master plan that sketched out a logical progression for humanity's inexorable expansion into the galaxy. His reason for leaving the solar system became a common rationale for interstellar travel—escape of the sun's inevitable decline and death. When Tsiolkovsky first wrote, early twentieth-century astrophysical discussions predicted a lifetime of the sun of about twenty million years, as dramatized in H.G. Wells's finale of *The Time Machine* on a near-frozen Earth under a fading sun thirty million years in the future. This embarrassingly short solar life span was replaced in the 1920s by estimates of many trillions of years of life based on the relativistic mass-energy relationship. This was in turn replaced by the discovery of nuclear fusion as the chief stellar power-source in the 1930s, by Hans Bethe, which gave the sun a multibillion-year life span.

Tsiolkovsky knew that known energy sources, chemical fuels, were insufficient for the task of interstellar flight, so he imagined storing up solar energy by some as-yet unknown means. He wrote of multimillennial interstellar voyages in what would later be known as *generation ships*, gigantic vessels—artificial planets really—filled with the means of sustaining a sizable population for the duration of the journey.

Tsiolkovsky was the first true astronauticist, but outside of Tsarist, then Soviet, Russia's scientific establishment, he was at first unknown. By the 1920s, as Tsiolkovsky's work became better known, Hermann Oberth, a German researcher, had already captured the imagination of the broader European and English-speaking public with his books on interplanetary travel. Oberth eventually moved to America and lived to see the first moon landings. In France, another pioneer astronauticist's career was not so fortunate. Robert Esnault-Pelterie, the inventor and aeronautical pioneer who had built and flown the first monoplane, independently derived and published the rocket equation in 1913. He extensively developed the theory of space flight and coined the term "astronautics." By 1927 he was lecturing on the equations of relativistic rocket motion, publishing the first paper on relativistic interstellar flight in 1928. However, the multiyear flight-times and vast energies to reach even the nearest stars caused him to declare interstellar travel improbable, though he did predict mass-energy conversion propulsion would make the outermost planets readily accessible. Esnault-Pelterie tried to stir up interest in his native country of France, and though his lectures were instrumental in creating interest across the Atlantic in the US, he eventually gave up astronautics research in despair and moved to Switzerland, dying there in 1957.

In the United States, Robert Goddard, independent of Tsiolkovsky and the Europeans, developed the equations of rocket motion in the early 1900s. His 1919 Smithsonian monograph, *A Method of Reaching Extreme Altitudes*, was his first paper to be noticed by a wider audience. Unfortunately, the attention came from an infamous *New York Times* editorial in 1920. Contrary to myth, Goddard's critic didn't see anything wrong with his suggestion of a rocket reaching the moon, but did (erroneously) think Newton's Laws meant the rocket needed an atmosphere to react against in order to accelerate, and criticized Goddard's work on that basis. Though Goddard's more knowledgeable supporters pointed out the error, the resulting notoriety caused Goddard to shun publicity. His space travel concepts, as outlined in letters to the Smithsonian, went much farther than the moon. He discussed the feasibility of automatic photographic fly-by probes to the planets, the use of liquid hydrogen/oxygen propellant, and use of a solar-concentrator to power an early design for an ion-drive he had developed experimentally.

Goddard kept his starship discussion private, putting them in a 1918 sealed letter titled *The Ultimate Migration*, recently republished by the British Interplanetary Society. Goddard saw two options for reaching the stars, based on the best physics of the day. Option one, if "intra-atomic" energy could be released, would be to make starships from asteroids with their interiors hollowed out for living room, illuminated by an artificial sun, their outer layers used for insulation and protection against meteoroids. Option two, if intra-atomic energy proved impossible to release, Goddard imagined multimillennial journeys carried out by crews bred to be able to undergo reversible mummification, revived for course-corrections. Eventually all the ship's passengers would be revived after perhaps a million years of slowly crossing the galaxy. If either option proved impossible, then humanity might send forth cells able to re-evolve humanity after flying through the void for billennia to suitable planets.

Philosopher Olaf Stapledon repeated these concepts in fiction in his twin multibillion-year histories, *Last and First Men* (1930) and *Starmaker* (1937). In the former, he imagines the Fifth Human Species developing "sub-atomic" propulsion 250 million years in the future. Humanity uses this to escape the moon falling to Earth, by migrating to a terraformed Venus. A later human species then avoids the sun's sudden brightening from a cosmic collision by migrating to the now torrid planet Neptune 500 million years later. Ultimately humanity discovers how to change planetary orbits via its sub-atomic engines, but failed interstellar expeditions mean that when the last Human Species, the Eighteenth, is confronted by the rapid rebrightening of the sun, they are trapped. As a desperate last measure, they launch forth designed cells to seed other star systems just as Goddard had imagined a decade before. In *Starmaker* Stapledon goes further, imagining interstellar travel by other species throughout the galaxy, employing sub-atomic engines to propel artificial planets at up to half light-speed.

Asteroid Starships went on to become a standard SF trope, achieving an apotheosis in George Zebrowski's *Macrolife* (1979). Inspired by Dandridge Cole's early 1960s discussion of asteroid habitats as super-organism-like stages in human evolution, Zebrowski, in the spirit of Tsiolkovsky, went farther than either Goddard or Stapledon, imagining that asteroid starships would eventually escape the demise of the entire cosmos.

Herman Potočnik (aka Hermann Noordung), in his influential work *The Problem of Space Travel—The Rocket Motor,* discussed interstellar travel and examined continuous acceleration trajectories to the nearer stars, computing the timescale for such flights at tolerable levels of acceleration (1.5 g) He concluded that, even without relativity complicating matters, up to thirty years would be required for round-trip missions. While he mentioned relativity's restrictions, he left exploration of the relativistic rocket for a future discussion that he would never have as he died of a pulmonary disorder the year his only book was published.

The early 1930s saw voyages to the planets designed in considerable detail. The Second World War brought about what many had not believed possible the decade before—rockets capable of leaving the atmosphere and atomic power.

Light as a Mechanical Agent: The Photon Rocket

Photon rockets had been discussed in fiction ever since people realized that radiation had "push." Beams of radiation produced by "sub-atomic energy" propelled Olaf Stapledon's "ether-vessels" of the *Fifth Men*, but the physical process for converting atomic mass into pure energy was only revealed by the discovery of antimatter.

In 1946 Swiss aeronautical engineer Jakob Akeret, unaware of Esnault-Pelterie's work, redeveloped the relativistic rocket equations. Twenty years of progress in nuclear physics since the early work of Esnault-Pelterie and Potočnik made the conversion of mass into energy no longer seem farfetched. Science-fiction writers, as they did with spaceflight and the atomic bomb, were the first to apply the new understanding. Robert Heinlein's fictional invention of the mass-conversion powered "Torch-ship" is an example, able to undertake high sub-light speeds to the nearer stars in his novel *Time for the Stars* (1951). Heinlein's torch-ships appear in several other novels, but the hoped-for mass-annihilation reaction has proven elusive.

Les Shepherd's 1952 JBIS paper, *Interstellar Flight,* was the first serious engineering study of interstellar travel. He pointed to the most severe design challenge for interstellar rockets—handling the immense power loads required. A photon rocket must radiate 300 megawatts of light power for every newton of thrust, meaning a "small" 1,000-ton vehicle accelerating at one-G would need to generate and direct 3,000 terawatts, the equivalent of a 0.75 megaton nuclear explosion, every second. A tiny fraction absorbed as waste heat would vaporize the vehicle. Shepherd reluctantly concluded that slower interstellar arks, propelled by ions and taking centuries to travel between the stars, would be necessary.

The difficulties didn't stop another engineer from promoting the incredible performance of matter-annihilation photon rockets. Eugen Saenger, the wartime designer of the *Silverbird* sub-orbital rocket-bomber, published several papers on the mechanics of photon-rockets in the 1950s, though he had been working on the idea since the 1930s. He proposed using matter-antimatter annihilation reactions to make gamma-ray photons, speculating that a dense electron gas could be the perfect mirror for reflecting them in a useful beam. He computed the effect of time dilation while undergoing continuous acceleration, leading him to conclude that flights to the Andromeda galaxy (then thought to be just 750,000 light-years away) could be achieved in twenty-five years of on-board ship-time, while circumnavigating the three billion light-years of the then known universe would take merely forty-two years.

Every light-year travelled while accelerating at one-G increases the time-dilation factor by an extra 1.0323—halfway to Saenger's Andromeda meant a time-dilation factor of almost 400,000. However, the difficulty for any rocket is the need to carry *all* the propellant—to accelerate and slow down. Deceleration requires a total velocity change that's double the required top speed, which squares the mass-ratio. Pure photon rockets approaching the speed of light have a mass-ratio that is the square of twice the time-dilation factor at the halfway point. Thus to travel the 0.75 million light-years to Andromeda's M31 and slow to a halt again requires a mass-ratio of 600 *billion*. Modern distance measurements to M31 of 2.5 million light-years only inflate the problem. Saenger's photon rocket would inspire future interstellar designs, but several miracles of applied physics and engineering are needed for it to achieve its promise.

Project Orion: Atomic Propulsion for Real

The 1950s saw early breakthroughs in nuclear fusion and the start of both internal and external nuclear rocket programs in the USA and the USSR. *Internal nuclear rockets* use a fission reactor as a heat source to heat to high speed any suitable volatile; hydrogen gives the best performance, due to its low atomic mass and its ability to soak up heat at high temperatures. Nuclear fission reactors require maintaining and containing a sufficient density of fissioning material, while exposing the engine to a high heat load, which limits the internal nuclear rocket from attaining high exhaust velocities. A variant, known as the *fission-fragment rocket*, can theoretically propel interstellar missions and has been revived in recent years. In essence, the fission-fragment rocket uses as its propellant the fragments produced by nuclear fission of uranium or some other fissile element. Such fission-fragments can achieve speeds of over 10,000 km/s, but a dense gas of fragments can lose energy via x-rays, causing heating problems for the engine. To reduce the heat loads, efficient reflection and direction of x-rays need to be developed, something that might not be unreasonable with the recent discovery of x-ray reflection by diamond crystals.

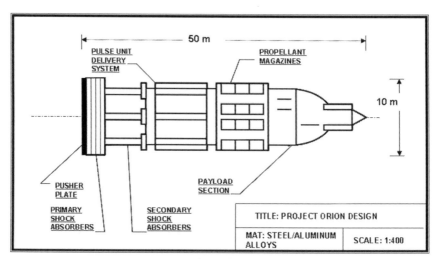

Orion External Nuclear Pulse Rocket

The *external nuclear rocket* can achieve interstellar exhaust velocities, in theory, because it doesn't try to contain the heat. "Project Orion," a semi-military design study, first explored the concept in 1957-1965. Famously the idea for Orion came from the observation that large graphite-coated metal spheres placed near ground zero of a nuclear test were propelled by the expanding fireball and left largely unharmed.

The idea of external nuclear propulsion is to explode a sequence of small fission devices behind the ship. The shockfront pushes on a "pusher plate" and propels the ship forward. Several experimental rigs, dubbed "Putt-Putt," were propelled by chemical explosives and demonstrated that explosions acting on a shock absorber equipped pusher plate could propel a vehicle at high accelerations. These experiments led to intensive research and an audacious design for a multithousand-ton interplanetary vehicle. As originally envisaged, Orion could have launched itself from Earth's surface, but calculations of the level of fallout created cooled several Project Orion physicists on the idea. When NASA took Orion over as a civilian spaceflight program it was then imagined to be lofted into orbit via a Saturn V. Such a "Baby Orion" was still capable of launching large payloads to Mars, or even Jupiter and Saturn.

However the original Orion was never meant to be an interstellar-capable design, as it used fission devices. Freeman Dyson, one of Project Orion's physicists, adapted Orion to the task of interstellar travel in his seminal 1968 *Physics Today* popularisation, "Interstellar Transport." Dyson couldn't reveal classified aspects of his work on thermonuclear weapons, and had to estimate the performance of nuclear fusion devices from publically available information. Dyson sketched two designs, one that absorbed and dissipated the heat of the blast passively, and the other that operated at much higher power levels thanks to low observed heat transference when riding an explosion's plasma shockwave. The theoretical top performance, using pure deuterium fusion devices, with 25 percent efficiency of energy transfer between explosion and spacecraft, was a mission speed of 10,000 km/s (3.3 percent of light-speed) able to reach Alpha Centauri in 130 years, including braking to a halt there. Dyson's computed exhaust velocity was 15,000 km/s and he argued for a mass-ratio of four, based on efficiency. In principle, there was no reason why mission velocities couldn't be made higher by higher mass-ratios and staging.

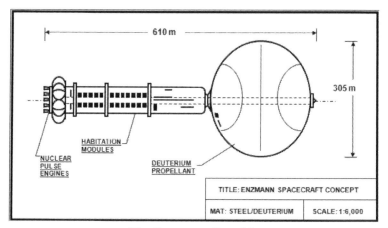

The Enzmann Starship

Orion as an interstellar vehicle was given its ultimate expression by Robert Duncan-Enzmann in the form of the distinctive *Enzmann Starship*. This design, essentially a *world-ship*, was popularized by G. Harry Stine in a 1973 *Analog* article. It was a thousand-foot-long cylindrical habitat mounted on twelve to twenty-four Orion pulse engines, with a thousand-foot mirror-finished spherical fuel tank out front, relative to the direction of thrust. Stine described a fleet of Enzmann starships travelling to the nearer stars in a century or so, carrying up to 2,000 colonists each. Such century long journeys made it a milder generation ship concept. As a single stage rocket vehicle the Enzmann starship could only achieve lower fractions of the speed of light, thanks to the limitations of the rocket equation. It's a milder generation ship, because journeys are decades, not centuries.

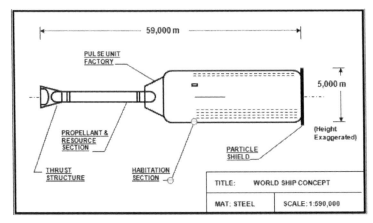

Bond-Martin Gigantic World-Ship

Coincidentally the 1970s saw the birth of the *space colony*, a potential justification of a massive space-based economy. Gerald K. O'Neill, a physicist, was the first to argue for immense orbital cities to be built to house a workforce to be employed in building thousands of solar-power satellites. His grand plan promised to wean Earth civilization off fossil fuels, solve the energy crisis, and defuse the population bomb via a single concept. The juxtaposition of fusion engines and space colonies quickly led Alan Bond & Anthony Martin, of "Project Daedalus," to the idea of the *interstellar ark* or *world-ship*, essentially a revival of Tsiolkovsky's idea for crossing the interstellar void. Bond and Martin's concept called for gigantic vehicles, large enough to transport a sample of terrestrial society and ecology in a vehicle many kilometres in size. The cruise speed was envisaged to be 0.5 percent of the speed of light, meaning multimillennium voyages. Ignition of the propulsion units was by the kinetic energy of very high-speed projectiles, able to ignite 7,000-ton fusion detonations, a system that would be prohibitively large on smaller vehicles. Acceleration time to 1,500 km/s was estimated to be 100 years. Such a grandiose scale would require a mature solar-systemwide economy, achieved after some 700 years of economic development in the estimates of the authors.

Fusion rockets are capable of higher speeds, in principle. Duane Spencer and Leonard Jaffe, in a 1963 JPL study, showed that staged fusion rockets could achieve speeds arbitrarily close to the speed of light given a sufficient numbers of stages. The challenge would be engineering suitable fusion engines. Using reasonable parameters for known fusion fuels and basic features of fusion reactor design led Spencer and Jaffe to conclude that fifty-year missions to five light-years would require handling challenging waste-heat loads. Throughout the 1960s large mass-ratio fusion rockets were studied. One study concluded that a mission to sixteen light-years in forty years of ship-time was possible with a mass-ratio of 10,000—presumably via a multistage vehicle. Higher speeds, in the relativistic speed range to take advantage of time dilation would require even higher mass-ratios, achievable thanks to the mass-ratio multiplying effect of staging. However, such multistaged vehicles would mass many millions of tons, making the prospect remote. Fusion rockets would be very large, very slow, or both.

Fusion Powered Ramjets: Realising Saenger's Dream

As fusion rockets seemed limited to short-range or centuries-long missions, another fusion concept received far more attention. Robert Bussard was a nuclear rocket designer before he wrote his classic "Galactic Matter and Interstellar Flight," first published in 1960, in which he sketched the concept of the interstellar ramjet. A ramjet avoids the fundamental limitation of rockets—the exponential growth of their mass-ratio for a linear increase in velocity—by scooping up the interstellar medium and using that for fuelling a fusion propulsion reactor. Not carrying fuel meant the required propellant mass only increased linearly with speed. But only if sufficient mass could be scooped up. That turned out to be the problem.

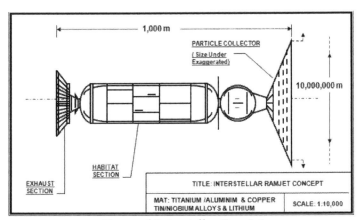

Bussard Interstellar Ramjet

Carl Sagan eagerly promoted Bussard's ramjet as the means for reaching the distant stars in his 1963 paper "Direct Contact Among Galactic Civilizations by Relativistic Interstellar Spaceflight." Sagan concluded that ramjets would allow Saenger's extreme relativistic time dilation, which would enable galactic-scale voyages by starship crews. He computed that if galactic civilizations launched a starship every year, then an average life span of ten million years for a civilization would allow every star in the galaxy to be visited every few thousand years. In science fiction, the Bussard ramjet became a standard part of the technological furniture, fictionalized in Larry Niven's "*Known Space*" short-stories and novels, and appearing on TV as the swirling lights on the engine nacelles of the most famous starship of all, the *USS Enterprise*. The ramjet's fictional apotheosis was Poul Anderson's *Tau Zero* (1970), which depicted a ramjet crew outrunning the universe's gravitational implosion in a Big Crunch, still one of the most audacious hard-SF scenarios yet imagined.

Bussard identified a number of issues before ramjets could realize Saenger's dream, the chief problem being the choice of fusion fuel. The interstellar medium is mostly the lightest isotope of hydrogen—a single proton with an electron attached—and this is the most difficult hydrogen to fuse. Protium, aka regular hydrogen, doesn't really undergo nuclear fusion proper with itself, as most fusion reactions involve the strong nuclear force. Instead the proton-proton fusion reaction involves the transformation of a proton into a neutron, a weak force reaction, which results in the formation of deuterium. Even at its peak reactivity, at about 10 billion degrees, the marriage of protons happens very slowly and a sustained reaction is only possible inside the sun and the stars by quantum tunnelling. Their union's product, deuterium, and its further reactions with itself and more protons, provides most of the energy inside the sun. But because of its relative willingness to fuse, it's cosmically rare. This poses a fuel-choice dilemma for a would-be ramjet builder—the most abundant hydrogen is the least fusion-friendly. The easiest to fuse, deuterium, numbers about one for every ten thousand protium atoms.

To collect sufficient hydrogen to sustain one-G acceleration, a ramjet needed to have a large collection area but very low mass. Bussard suggested a magnetic field could form a vast immaterial scoop over a very wide area. Larry Niven's *Known Space* stories fictionalised one possible difficulty of magnetic scoops—lethal field strengths—though this is dubious. Another objection came from a paper in the late 1960s that determined that after travelling for several million light-years, the required magnetic pressure in the field coils for generating the scoop-field would be impossible to contain using even the strongest known materials, such as diamond. A later analysis by Anthony Martin, co-designer of World-ships with Alan Bond, suggested the materials issue wasn't as severe as imagined. A more serious problem, discussed first by Bussard, was the apparent region of under-density in the interstellar medium within which the sun and the nearby stars exist. More useful densities of fuel are much farther away, found near star clusters and star-forming regions.

An apparently critical issue for ramjet operation was raised in the early 70s. Pure proton fusion reactions were far too slow to produce reasonable thrust and deuterium too rare to fuel the ramjet. Daniel Whitmire proposed a possible solution in the form of catalysed nuclear fusion. In the hotter stars, protons fuse with trace elements of carbon, nitrogen, and oxygen, undergoing what is known as the CNO Catalytic cycle—four protons are added serially to carbon-12, which is fused into nitrogen, then fused into oxygen, before finally a helium ion is emitted, leaving the original carbon-12. Of course, in real stars the densities and pressures are much higher than what can be achieved in a light-weight starship reactor, posing serious challenges for future engineers, but in principle the CNO cycle can burn protons sufficiently quickly for useful thrust. Robert Bussard, mindful of the many critiques of his concept, believed that the CNO cycle would allow his ramjet to work, once the many engineering challenges of fusion power were overcome.

Alternatives to a pure ramjet soon emerged. UK propulsion engineer Alan Bond proposed the *Ram-Augmented Interstellar Rocket*, which avoided the fuel issue by carrying the fuel needed to energise the much greater mass-stream being scooped from the interstellar void. In theory the *RAIR* could multiply the final velocity by three- or fourfold over a pure rocket alone, though the challenge of controlled fusion remained. Daniel Whitmire and NASA scientist Al Jackson collaborated on an alternative, the *Laser-Powered Interstellar Ramjet*, which avoided the fusion complication by using lasers to energise an ion-engine using scooped-up interstellar gas for reaction mass. Alternatively, lasers might be used to power a pure rocket, leading the authors to design and describe the *Laser-Powered Interstellar Rocket* as well, and comparing its performance against the laser-powered ramjet.

Given that fusion could be controlled, then the rocket equation might be avoided by another ramjet concept—the *Ramjet Runway*. Imagine a very long track-way enriched with fusion fuel pellets. A ramjet then first ionizes the pellets via laser, and then scoops up the ions up via a magnetic field. The track-way is long enough for the acceleration phase of flight, but much shorter than interstellar distances. A much slower vehicle could be used to lay out the fuel. Conceivably some other motive force could be used, for example solar-sails. To decelerate when approaching the destination, the ram-scoop would again be activated, to brake without expelling rocket propellant. It can be used to gather interstellar gas to blast against a *braking sail*. Gregory Matloff, who used drag on a high-temperature boron-sail to dissipate the ship's kinetic energy, developed this concept. Within a stellar system, a magnetic-scoop might also be used to gather the ionized material of the stellar wind and charged dust particles of the interplanetary medium, conceivably acting as a refueling system for a return journey.

Light as a Mechanical Agent: Pushing with Lasers

The discovery of sunlight's momentum in the nineteenth century led naturally to the idea of the solar-sail for interplanetary travel. Light, because it is reflected from an object and has momentum, exchanges some of that momentum to the object impacted, thanks to Newton's Third Law. As a result sunlight exerts an appreciable pressure on all exposed surfaces in free space, and a sufficiently large, lightweight reflective surface can use that pressure for propulsion. Solar-sail concepts had been discussed by the earliest astronautics pioneers, and fictionalized as early as the 1920s, by for example J.B.S. Haldane's *The Last Judgement* (1927) as the means of interplanetary travel by Venusian colonists, and the means of interstellar travel for Pierre Boule's space travelers in the original *Planet of the Apes* (1953) and the earliest starfarers of Cordwainer Smith's fictional cosmos. However, the faintness of sunlight makes this a slow-paced prospect at best.

A long-time science-fiction trope, seen in E.E. Smith's *Spacehounds of IPC* (1931), was propelling spacecraft via beams of energy mounted on a planet. Though attractive because it put the heavy power plant off the space vehicle it wasn't theoretically possible until the invention of a means of directing light as intense coherent electromagnetic waves—first with microwaves ("maser") in the 1950s, and the laser, first demonstrated in 1960. In an early interstellar application, Eugen Saenger proposed (1963) using a ship-board lasing system to convert matter-antimatter annihilation gamma-ray energy into lower frequencies that could be reflected. Robert Forward, as a young physicist in the early 1960s, was first to propose pushing vehicles via laser, but soon he wasn't alone. George Marx published an influential paper in the journal *Nature* in 1966, which made the topic scientifically respectable. Marx analysed the energy efficiency of laser propulsion, coming to the startling conclusion that efficiency approached 100 percent as the speed of light was approached. This claim was heavily debated, and eventually shown to be not quite correct. But an important fact remains that laser efficiency is relatively low at sub-relativistic speeds below about 0.2c. John Bloomer proposed, in 1966, a laser-powered star probe able to reach 0.25c and fly-by Alpha Centauri in just seventeen years, using the energy collected by a sixty-mile-wide liquid metal mirror focused onto a solar-powered laser one mile across. In a 1993 revision of his ideas, Bloomer has extended the concept, proposing a way to decelerate larger crewed vehicles at Alpha Centauri via much larger collector mirrors.

After Bloomer's proposal, several NASA papers from the 1970s also pointed out the potential for fast fly-by probes propelled to relativistic speeds by laser, but the concept didn't seem applicable to rendezvous missions because the push of the laser was one-way, implying no way of stopping. However P.C. Norem and Robert Forward had developed a way of turning a laser starship, using large charged wires to interact with the galactic magnetic field and undergo deflection by the Lorentz force. Conceivably this would allow a sail to turn around to allow the propulsion laser to slow it down, though with the penalty of tripling the journey time. More recent analyses have shown that such a concept requires very large turning circles, taking centuries and light-years to produce a significant deflection. Yet the idea of interacting with the interstellar medium to brake is physically sound and has been revived by several later concepts.

In a 1979 discussion on SETI, Freeman Dyson proposed a propellantless means of braking from interstellar speeds—using what he called *Alfven Braking*, based on the observation that very large inflatable satellites of the early 1960s, the Echo series, underwent a magnetic interaction with the plasma environment around them, falling from orbit faster than expected. Starships using such Alfven Braking should be detectable from the electromagnetic energy released, thus Dyson proposed searching for advanced extraterrestrial space vehicles by looking for their skid-marks. An Alfven Brake would need a very large area, but a very low mass to be practical. A potentially better magnetic-braking option did appear in the late 1980s, the mag-sail.

External Nuclear Revival

A 1971 paper by nuclear physicist Friedwardt Winterberg, based on declassified work into fusion ignition, suggested fusion could be triggered in small fuel pellets via electron beams, instead of the large and inefficient fission-bomb triggers of 1960s hydrogen bombs. This led to a flurry of designs for "pure fusion" external nuclear rockets, using both electron beam and laser triggers, some of which were computed to produce interstellar exhaust velocities of 10,000 km/s or more. In 1973 a British Interplanetary Society team, led by Alan Bond and Anthony Martin, began Project Daedalus—a design study of an interstellar probe deliberately using realistic short-term technological advances. Concluding in 1978 after more than 10,000 hours of effort, Project Daedalus led many to conclude that interstellar travel was possible, perhaps within few decades.

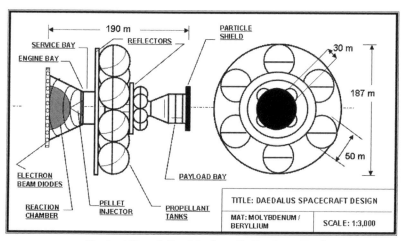

Project Daedalus Nuclear Pulse Star-Probe

However a few features of the probe design would require a substantial space-based industrial capacity and decades to implement, chief of which was the 30,000 tons of the exotic fusion fuel He-3. This required mass was more than the known quantity in Earth's atmosphere. Getting it would mean that a substantial extraction and refining industry would need to be developed in the helium-rich atmosphere of the outer gas giants. Project Daedalus proposed mining Jupiter, but reaching orbit while struggling against more than twice Earth's gravity demands very high performance nuclear rockets. Significantly easier access to orbit can be achieved from any of the other gas giants, where the gravity is significantly lower and their cooler atmospheres make chilling hydrogen/helium mixtures significantly easier. Robert Zubrin has proposed Saturn as the ideal location for He-3 mines, while other researchers suggest Uranus.

A more modest nuclear pulse starship was proposed by the NASA sponsored student project, Project Longshot, which examined a starship capable of a 100-year data-return time from Alpha Centauri B. The payload was a nuclear-powered probe for orbital and exploring the Alpha Centauri system, massing thirty tons—far less than the hundreds of tons of Project Daedalus.

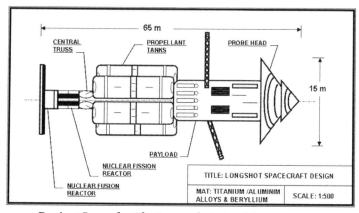

Project Longshot fusion rocket starship concept

Several propulsion options were studied, but the pulsed fusion pulse propulsion of Daedalus was chosen due to its perceived near-term development. The students proposed fuelling up the probe via a stop-over in orbit around Uranus. Although it has since been widely referenced, Project Longshot was a didactic exercise, not a serious NASA study, unlike another fusion space-vehicle concept from the 1980s, VISTA.

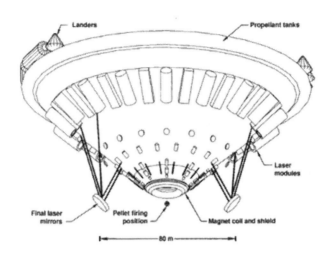

VISTA—A Vehicle for Interplanetary Space Transport Application

While not strictly designed as an interstellar vehicle, Charles Orth's detailed work on VISTA since the 1980s provides a baseline for fusion-propelled vehicles, and was adapted for interstellar missions by Raymond Halyard in the 1990s. Using multiple high-powered lasers, some reflected back to the detonation point via mirrors, the VISTA ignition system is akin to Daedalus, but with a higher compression symmetry. The vehicle's unusual shape is due to the necessity of minimising the area directly exposed to the neutron radiation produced by the fusion reactions. Orth has argued, based on more advanced fusion physics modeling than the original Daedalus study, that a deuterium-helium-3 fuel mix would be less effective than pure deuterium, as both would produce similar amounts of neutrons, but He-3 is much harder to acquire. Like Daedalus, VISTA uses a deuterium-tritium fusion trigger in the core of the fuel pellets, so to be used for an interstellar orbital mission, where the vehicle cruises for decades or centuries between the stars, a system of liquid lithium neutron absorbers would be required to manufacture tritium due to its relatively short half-life. Halyard's interstellar VISTA would take several centuries to reach the nearest stars.

Light as a Mechanical Agent: The Laser-Sail

Robert Forward, as inventive in the 1980s as he was in the 1960s, produced a series of new laser and maser sail concepts. His first breakthrough was the use of an immense Fresnel zone lens to increase the useful range of laser light. Due to diffraction, which causes the laser wave-front to spread out, the range of optical lasers would be limited to just astronomical units of distance without a lens to improve the focus. Useful range was inversely proportional to the wavelength of the laser-light, so x-ray wavelength lasers would be needed to reach across light-years without corrective optics. Thus Forward's lens allowed the more tractable optical wavelengths to be used.

Perhaps more significantly, Forward suggested a means of solving the braking problem of laser-sails with his invention of the stage-sail—a laser-sail composed of successively smaller sails to be pushed via light reflected from the larger sail surrounding it. In Forward's analysis, this allowed for the exciting prospect of reaching nearby stars, like Epsilon Eridani, at 0.5c—and returning the vehicle to the solar system. Laser-sails could thus be used for fast roundtrip missions, something that couldn't be said of their more conventional fusion rocket competitors.

To send laser-light usefully concentrated across ten light-years or more, the beam would need to be focused by an immensely large Fresnel zone lens—about 500,000 kilometres in diameter. The lens itself, which shaped the wave-front of the beam via alternating rings of light and dark, need only be made of thin space-capable material, but such an immensity would still mass a half million tons. The laser stage-sail itself would mass 75,000 tons and be 1,000 kilometres in diameter, but its gossamer areal density is only 0.075 grams per square meter. Accelerating it requires a pushing power of many thousands of terawatts. Clearly not a near-term prospect, but a design requiring a massive solar-system-based infrastructure capable of producing lenses massing hundreds of thousands of tons and millions of tons of energy collectors and laser generators in orbits close to the sun.

Like all good ideas, Forward's laser-sail systems had immediate critics. Many of the criticisms hit the mark for good, yet subtle, physics reasons. For example, NASA scientist Geoffrey Landis pointed out that Forward's Starwisp microwave propelled sail (massing mere grams, but accelerated at 115-Gs to 0.2c for a fly-by mission of Alpha Centauri in twenty-two years) would be vaporised by the proposed power-levels, due to a miscalculation in the absorption of microwave energy by the proposed aluminium sail.

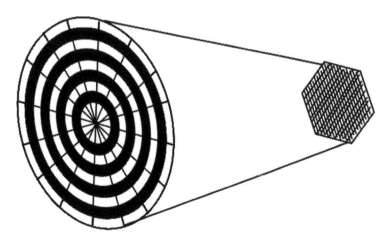

Bob Forward's Fresnel Lens focusing microwaves on a Starwisp Microwave Sail

All was not lost for the laser-sail, however. By the late 1980s Forward's work had inspired application of newer materials and newer concepts to the task of interstellar travel. In a series of studies improving on Forward's laser-sails, Geoffrey Landis used high-temperature materials such as boron and carbon to allow much higher accelerations. He also proposed a means of markedly reducing the mass of the Fresnel lens. Instead of a single, very large lens, a string of lenses could be deployed, much smaller in size, and much lighter in mass. The immense power requirements of laser-sails would still require a large space-based infrastructure, but more modest probes might be feasible within a few decades of development of in-space solar-power satellites.

Other materials advances inspired a different kind of sail concept. The discovery in 1986 of high temperature superconductors allowed designers to contemplate use of superconductivity for creating low-mass magnets on a grand scale. One such application was the magnetic-sail, which had begun life as a failed design for an interstellar ramjet. The inventors, Jonathan Vos Post, Dana Andrews and Robert Zubrin, had discovered to their dismay that their ram-scoop design created more drag than the proposed ion-rocket attached to it could produce in thrust. They quickly realised that it could be used as a braking system from interstellar speeds. The "mag-sail" would save large amounts of fuel. Coupled to a laser-sail, a mag-sail would allow deceleration from relativistic speed in just a couple of decades, without the problems of focusing laser-light across light-years in order to brake. High temperature laser-sail materials would allow much shorter acceleration distance, reducing the scale of the focusing optics to more manageable sizes. In one analysis a focusing mirror and mag-sail-equipped laser-sail as "small" as fifty kilometres across would achieve the same performance as Forward's 1,000 kilometre Fresnel lens/light-sail combination for a mission to Epsilon Eridani.

Electricity as a Mechanical Agent

Alternatively the lasers could be entirely done away with and the magnetic-sail could be pushed via a particle beam—a near-literal realisation of Verne's dream of using electricity to propel projectiles to the stars. However, charged particle beams expand due to their charge, as well as being deflected by the ambient magnetic fields around them. A beam composed of heavier neutral particles would deflect less, which leads to a rather different concept entirely, Clifford Singer's interstellar pellet-launcher.

At roughly the same time as Forward was proposing a new way of launching vehicles to the stars via lasers, Singer, in the pages of the *Journal of the British Interplanetary Society*, was proposing the use of small pellets to project momentum to a starship. His concept used a very large magnetic accelerator, spread out over a very large track-way, to propel small particles to relativistic speeds and then push against the magnetic-field of an interstellar vehicle. The advantage of the pellets was that they were too big to deflect via the feeble magnetic fields of interstellar space or the random impacts of interstellar gas and cosmic rays. Being so numerous, the loss of a few pellets to random dust collisions wouldn't impact performance.

Singer refined his concept over a series of papers, but the idea has languished, outshone by the enthusiasm of researchers for laser-sails and antimatter propulsion. But in the 1990s, as nano-technology was becoming a part of the conceptual landscape, a new pellet-propulsion

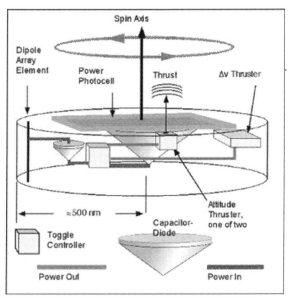

Nordley Smart-Pellet

concept emerged, championed by USAF researcher Gerald D. Nordley. Instead of "dumb pellets," externally steered and shot toward a starship, Nordley proposed nano-technological "Smart-Pellets," which could maintain themselves in the beam path to a starship via a simple artificial intelligence and tiny single-atom rockets for course correction.

To power either system would require immense solar power-collection systems, which Nordley proposed to be built via self-replicating machines. Optimistically assuming a single self-replicating power-satellite that supplies one gigawatt of power that copies itself in a year, then within mere decades sufficient power would be available to propel a 1,000-ton starship to 0.86c at five-Gs, and a decade later a thousand such starships could be propelled per year.

Light and Electricity Combined: Antimatter Pion Rockets

By the 1980s, antimatter creation and handling had advanced technologically to the point where serious engineering designs could be developed and explored. Robert Forward was a vocal advocate of antimatter propulsion, especially after his realisation that the mutual annihilation of protons and anti-protons didn't result in the immediate conversion of mass into a random spread of gamma rays. Protons are composite objects, composed of a variety of quarks, joined together via the so-called Colour Force. When protons and their anti-proton nemeses met, they reduce each other into a series of short-lived charged quark-pairs known as pions. Being charged, a series of magnetic-field coils can direct pions into a rocket exhaust and the back-reaction of the pions on the magnetic-field coils would provide a means of mechanically coupling their departure into a thrust for the rocket. Thus the Forward antimatter pion rocket was born.

However modeling of the interaction of pions with magnetic-field coils using the computers of the late 1980s gave rather unpromising rocket efficiency. The departing pions would transfer to the rocket only a fraction of the energy released by the matter-antimatter reaction, the rest

Antimatter Pion Rocket (after Frisbee)

being lost as hazardous gamma rays. Using standard rocket-design techniques, NASA's Robert Frisbee computed that four-stage matter-antimatter rockets able to reach a forty light-year target in 100 years would mass many millions of tons, and be immensely long. This was partly due to the hazardous gamma rays emitted, but also due to the difficulty of storing antimatter. To avoid annihilating with the normal matter tanks holding it, antimatter needs to be stored frozen within a fraction of a degree of absolute zero (minus 273 C), then levitated via a magnetic field, possible only because of the paramagnetism of frozen hydrogen. Such a low temperature can only be sustained by heroic refrigeration and immense heat radiators, resulting in a rocket design metres wide, yet thousands of kilometres in length.

An alternative antimatter starship design is seen in fictional form, spectacularly visualised as part of the drive system of the ISV "Venture Star" in the 2009 global blockbuster *Avatar*. The Venture Star is arguably the first realistic starship design produced by Hollywood and is a spectacular example of *poly-propulsion,* use of two or more drive systems with different performance characteristics for different stages of a mission—the other half of the Venture Star's propulsion system being a Forward laser-sail.

Light and Electricity Today

In Verne's *From Earth to the Moon*, Michel Ardan was a French adventurer who proposed to ride to the moon on a cannon shell. Until he so dared, the Baltimore Gun Club designers had been content to launch the projectile as a mere message to the possible inhabitants of the moon.

Starship design today still needs adventurers. As technology has developed what was once theoretical and broadly defined has become ever more detailed engineering designs. There is a technical base from which serious work is advancing against the starship problem.

There is now a resurgence of interest in starship designs, applying twenty-first-century knowledge to ascertain which concepts appear more practical.

Icarus Interstellar is an umbrella organization developing conceptual designs for starships using several technologies mentioned here, including nuclear rockets, world ships, the Bussard ramjets and beam-driven sails. The most advanced part of Icarus Interstellar is Project Icarus, which has a goal of designing a starship to launch within the century. The oldest organization is the Tau Zero Foundation, named after the Poul Anderson novel. The TZF maintains the premier interstellar news website, *Centauri Dreams*. The newest organization is the Institute for Interstellar Studies in the UK. *"Exploring Further"* contains online contact information for present organizations interested in research toward development of starships, as well as links to online versions of many of the sources mentioned herein.

The current generation of starship designers are refining and building on the concepts of Tsiolkovsky, Goddard, Saenger, Forward, and Bussard. Many starship designers mentioned here are still active—Freeman Dyson, Friedwardt Winterberg, Daniel Whitmire, Al Jackson, Gregory Matloff, Geoffrey Landis, and Gerald Nordley—but the task is big enough for anyone and everyone to be involved. One day, the designs will be built and the adventure will begin a new phase, needing *real* Michel Ardans to ride them to the stars.

Exploring Further

Many of the works cited in the footnotes of the chapter are available online. Some are even free. Older works may exist in multiple editions, and translations, of varying readability and quality.

Archimedes, *The Sand Reckoner*, http://4dlab.info/archimedessandreckoner.pdf

Also for a more mathematical discussion: http://www.calstatela.edu/faculty/hmendel/Ancient%20Mathematics/Archimedes/SandReckoner/SandReckoner.html

Johannes Kepler, *Conversation with the Sidereal Messenger*, [online: https://eee.uci.edu/clients/bjbecker/ExploringtheCosmos/week4e.html]

Jules Verne, De la Terre à la Lune, Available online courtesy of Project Gutenberg: http://www.gutenberg.org/ebooks/8986

Nikolai Rynin, "Interplanetary flight and communication. Volume 3, no. 7: K. E. Tsiolkovskii, life, writings, and rockets," Available: http://ntrs.nasa.gov/search.jsp?R=19720013307

Mike Gruntman, "From Astronautics to Cosmonautics." Chapter 1 discusses the life of Robert Esnault-Pelterie. Available: http://astronauticsnow.com/astrocosmo/index.html

Robert Esnault-Pelterie, "Astronautics and the Theory of Relativity," in Nikolai Rynin, "Interplanetary flight and communication." Volume 3, no. 8: *Theory of Space Flight*, 1932. This volume contains classic papers by several early astronauticists mentioned. Available: http://ntrs.nasa.gov/search.jsp?R=19720015159

Robert H. Goddard, "A Method of Reaching Extreme Altitudes," The Smithsonian Miscellaneous Collections, Vol. 71, No. 2. (1919) Available: http://www.clarku.edu/research/archives/pdf/ext_altitudes.pdf

The New York Times, January 13, 1920; page 12, column 5. Available: http://en.wikisource.org/wiki/The_New_York_Times/Robert_Goddard

Robert H. Goddard, "Report Concerning Further Developments," The Smithsonian Institution, (1920). Available: http://siarchives.si.edu/history/exhibits/stories/march-1920-report-concerning-further-developments-space-travel

Goddard's "The Ultimate Migration" online courtesy of the British Interplanetary Society - http://www.bis-space.com/2012/03/23/4110/the-ultimate-migration

http://www.bis-space.com/2012/03/23/4110/the-ultimate-migration

Olaf Stapledon, "Last and First Men: A Story of the Near and Far Future," Methuen (1930). Available: http://gutenberg.net.au/ebooks06/0601101h.html

Olaf Stapledon, *Star Maker*, Methuen (1937). Available: http://gutenberg.net.au/ebooks06/0601841h.html

George Zebrowski, *Macrolife*, Harper & Row (1979). Any good online bookstore.

Herman Potočnik, *The Problem of Space Travel—The Rocket Motor*, Richard Carl Schmidt & Co., Berlin W62, (1929). The online Museum of Potočnik's life is well worth visiting. http://www.ksevt.eu/eng_index_herman.html

His book: http://www.noordung.info/NoordungEng.pdf

Jakob Ackeret, "Zur Theorie der Raketen," *Helvetica Physica Acta*, Vol. 19, 1947, pp. 103–112 (in German).

Les Shepherd, "Interstellar Flight," *Journal of the British Interplanetary Society*, Vol. 11, 149-167, July 1952. http://www.bis-space.com/products-page/magazines-and-journals/jbis-journal-of-the-british-interplanetary-society/

E. Saenger, "Flight mechanics of photon rockets," 1961. Available: http://ntrs.nasa.gov/search.jsp?R=19670081918

Pavel Tsvetkov, Ron Hart, Don King, Gary Rochau, "Planetary Surface Power and Interstellar Propulsion Using Fission Fragment Magnetic Collimator Reactor," AIP Conf. Proc. 813, pp. 803-810 (2006). Available: http://link.aip.org/link/doi/10.1063/1.2169262

Nuclear pulse space vehicle study. volume i- summary, NASA-CR-60655 (1964). Available: http://ntrs.nasa.gov/search.jsp?R=19650058729

Freeman J. Dyson, "Interstellar Transport," *Physics Today*, pp. 41-45 (1968). Available: http://galileo.phys.virginia.edu/classes/109.jvn.spring00/nuc_rocket/Dyson.pdf

G.Harry Stine, "A Program for Starflight," *Analog Science Fiction & Science Fact*, Oct 1973. Summarised: http://enzmannstarship.blogspot.com/

Gerald K. O'Neill, "The colonization of space," *Physics Today*, 27 (9), 32 (1974). Available: http://www.nss.org/settlement/physicstoday.htm

Anthony R. Martin, "Worldships—concept, cause, cost, construction and colonisation," *Journal of the British Interplanetary Society*, Vol.37, pp. 243-253 (1984). http://www.bis-space.com/products-page/magazines-and-journals/jbis-journal-of-the-british-interplanetary-society/

Alan Bond & Anthony R. Martin, "Worldships—an assessment of the engineering feasibility," *Journal of the British Interplanetary Society*, Vol.37, pp. 254-266 (1984). http://www.bis-space.com/products-page/magazines-and-journals/jbis-journal-of-the-british-interplanetary-society/

L.D.Jaffe & D.F.Spencer, "Feasibility of interstellar travel," NASA Report: JPL-TR-32-233 (1963). Available: http://ntrs.nasa.gov/search.jsp?R=19640006154

Eugene F. Mallove & Gregory L. Matloff, The Starflight Handbook: a Pioneers Guide to Interstellar Travel, Wiley Science Editions, New York (1989). Available: http://www.amazon.com/Starflight-Handbook-Pioneers-Interstellar-Editions/dp/0471619124

Norman R. Schulze, Fusion Energy for Space Missions in the 21st Century, NASA Technical Memorandum 4298, pp. 7-22 (August 1991). Available: http://ntrs.nasa.gov/search.jsp?R=19920002565

Robert W. Bussard, "Galactic Matter and Interstellar Flight," *Acta Astronautica*, 6, pp.179-194 (1960). Available: http://www.askmar.com/Robert%20Bussard/Galactic%20Matter%20and%20Interstellar%20Flight.pdf

Carl Sagan, "Direct contact among galactic civilizations by relativistic interstellar spaceflight," *Planetary and Space Science*, Vol. 11, p.485-498 (1963). Available: http://www.sciencedirect.com/science/article/pii/0032063363900722

Poul Anderson, *Tau Zero*, Lancer Books, New York (1970). Any good online book-store.

John Ford Fishback, "Relativistic Interstellar Spaceflight," *Acta Astronautica*, 15 pp.25-35 (1969). http://www.sciencedirect.com/science/journal/00945765

Anthony R. Martin, "Structural Limitations on Interstellar Space-flight," *Acta Astronautica*, 16, pp.353-357 (1971). http://www.sciencedirect.com/science/journal/00945765

Daniel P. Whitmire; "Relativistic Spaceflight and the Catalytic Nuclear Ramjet," *Acta Astronautica*, 2 pp.497-509 (1975). Available: http://www.askmar.com/Robert%20Bussard/Catalytic%20Nuclear%20Ramjet.pdf

Alan Bond, "An Analysis of the Potential Performance of the Ram Augmented Interstellar Rocket," *Journal of the British Interplanetary Society*, 27, 674-688, (1974). http://www.bis-space.com/products-page/magazines-and-journals/jbis-journal-of-the-british-interplanetary-society/

Daniel P. Whitmire & A. A. Jackson, "Laser Powered Interstellar Ramjet," *Journal of the British Interplanetary Society*, 30, pp.223-226 (1977). http://www.bis-space.com/products-page/magazines-and-journals/jbis-journal-of-the-british-interplanetary-society/

A.A. Jackson & Daniel P. Whitmire, "Laser Powered Interstellar Rocket," *Journal of the British Interplanetary Society*, 31, pp.335-337 (1978). http://www.bis-space.com/products-page/magazines-and-journals/jbis-journal-of-the-british-interplanetary-society/

Bond, *op. cit.* p.226.

Gregory L. Matloff, "The Interstellar Ramjet Acceleration Runway," *Journal of the British Interplanetary Society*, 32, pp.219-220 (1979). http://www.bis-space.com/products-page/magazines-and-journals/jbis-journal-of-the-british-interplanetary-society/

Gregory L. Matloff, "Utilization of O'Neill's Model I Lagrange Point Colony as an Interstellar Ark," *Journal of the British Interplanetary Society*, 29 p.775-785 (1976). http://www.bis-space.com/products-page/magazines-and-journals/jbis-journal-of-the-british-interplanetary-society/

Gregory L. Matloff & Alphonsus J. Fennelly, "A Superconducting Ion Scoop and its Application to Interstellar Flight," *Journal of the British Interplanetary Society*, 27, pp.663-673 (1974). http://www.bis-space.com/products-page/magazines-and-journals/jbis-journal-of-the-british-interplanetary-society/

Max M. Michaelis & Andrew Forbes, "Laser propulsion: a review," *South African Journal of Science*, 102, pp.289-295 (2006). Available: http://content.ebscohost.com/pdf19_22/pdf/2006/SJS/01Jul06/24261285.pdf

Robert L.Forward, "Pluto—The Gateway to the Stars," *Missiles and Rockets*, Vol. 10, pp. 26-28 (1962). Unfortunately a defunct Journal title.

George Marx, "Interstellar Vehicle Propelled by Terrestrial Laser Beam," *Nature*, Vol. 211, pp. 22-23, (1966). http://www.nature.com/nature/journal/v213/n5076/abs/213588a0.html

Michaelis & Forbes, *op. cit.* p.294.

John H.Bloomer, In Proceedings of 17th International Astronautical Federation Congress, 1966, IAF, 3-5 Rue Mario-Nikis, 75015 Paris, France.

John H. Bloomer, "Earthly Millennium Energy and Interstellar Shuttle Propulsion Potentials of Liquid Space Optics," 28th Intersociety Energy Conversion Engineering Conference, Atlanta, GA (1993). Available from: http://conradosalas.info/Earthly%20Millennium%20Energy%20 and%20Interstellar%20Shuttle%20Propulsion....doc

W.E. Moeckel, "Comparison of advanced propulsion concepts for deep space exploration," *J. Spacecraft and Rockets*, 9, pp.863–868 (1972). Available: http://ntrs.nasa.gov/search. jsp?R=19720024129

P.C.Norem,"Interstellar Travel, A Round Trip Propulsion System with Relativistic Velocity Capabilities," American Astronautical Society, Paper No. 69-388, (1969).

Gregory L. Matloff, Deep Space Probes, 2nd Edition, Springer-Verlag Berlin And Heidelberg Gmbh & Co. K, New York (2005).

Freeman J. Dyson, "Interstellar propulsion systems," in Extraterrestrials: where are they? eds. M.H. Hart & B. Zuckerman, pp.41-45, Pergamon Press, Oxford (1982).

Friedwardt Winterberg, "Rocket Propulsion by Thermonuclear Micro-Bombs Ignited with Intense Relativistic Electron Beams," *Raumfahrtforschung*, 15, pp.208-217 (1971).

A.R. Martin (ed.), "Project Daedalus," *Journal of the British Interplanetary Society Interstellar Studies*, Supplement, Final Report of the BIS Starship Study, (1978). http://www.bis-space. com/products-page/magazines-and-journals/jbis-journal-of-the-british-interplanetary-society/

R.C. Parkinson, "Project Daedalus: Propellant Acquisition Techniques," *Journal of the British Interplanetary Society Interstellar Studies*, Supplement, Final Report of the BIS Starship Study, (1978). http://www.bis-space.com/products-page/magazines-and-journals/jbis-journal-of-the-british-interplanetary-society/

Robert Zubrin, Entering Space: Creating a Spacefaring civilization, Jeremy P. Tarcher/Putnam, a member of Penguin Putnam Inc., New York (1999).

Curtis West, Sally Chamberlain, Neftali Pagan & Robert Steves, "Project Longshot: A mission to Alpha Centauri," NASA-CR-186052, NASA (1989). Available: http://ntrs.nasa.gov/search. jsp?R=19900014101

Charles D. Orth, VISTA—A Vehicle for Interplanetary. Space Transport Application Powered by. Inertial Confinement Fusion, UCRL-TR-110500, Lawrence Livermore National Laboratory, (2003). Available: https://e-reports-ext.llnl.gov/pdf/318478.pdf

Raymond J. Halyard, "Comparison of Proposed Inertial Confinement Fusion Propulsion Systems for Interstellar Travel," *Journal of the British Interplanetary Society*, 50, pp.129-136, (1997). http://www.bis-space.com/products-page/magazines-and-journals/jbis-journal-of-the-british-interplanetary-society/

Robert L. Forward, "Roundtrip Interstellar Travel Using Laser-Pushed Lightsails," *J. Spacecraft and Rockets*, Vol. 21, No.2, pp. 187-195 (1984). http://arc.aiaa.org/doi/abs/10.2514/3.8632

Robert L. Forward, "Starwisp: an Ultralight Interstellar Probe," *J. Spacecraft and Rockets*, Vol. 22, No.3, p. 345-350 (1985). http://arc.aiaa.org/doi/abs/10.2514/3.25754

Geoffrey A. Landis, "Microwave Pushed Interstellar Sail: Starwisp revisited," presented at 10th NASA/AIAA Advanced Space Propulsion Workshop, April 5-8, Huntsville AL (1999). http://arc.aiaa.org/doi/abs/10.2514/6.2000-3337

Geoffrey A. Landis, "Advanced Solar and Laser Pushed Lightsail Concepts," Final Report, NASA Institute for Advanced Concepts, Atlanta (1999). Available: http://www.niac.usra.edu/files/studies/final_report/4Landis.pdf

Dana G. Andrews, Robert M. Zubrin, "Magnetic Sails and Interstellar Travel," *Journal of the British Interplanetary Society*, Vol43, pp265-272, (1990). http://www.bis-space.com/products-page/magazines-and-journals/jbis-journal-of-the-british-interplanetary-society/

Dana G. Andrews, Robert M.Zubrin, "Use of magnetic sails for advanced exploration missions," in Vision-21: Space Travel for the Next Millennium; NASA Lewis Research Center, pp.202-210, (1990). Available: http://ntrs.nasa.gov/search.jsp?R=19910012840

Clifford E. Singer, "Interstellar Propulsion Using a Pellet Stream for Momentum Transfer," *Journal of the British Interplanetary Society*, Vol. 33, pp. 107-115, (1980). http://www.bis-space.com/products-page/magazines-and-journals/jbis-journal-of-the-british-interplanetary-society/

Gerald D. Nordley, "Interstellar Probes Propelled by Self-steering Momentum Transfer Particles," Paper IAA-01-IAA.4.1.05, 52nd International Astronautical Congress, (2001). More on Gerald's work can be found on his web-site: http://www.gdnordley.com/

Robert L. Forward, "Antimatter propulsion," *Journal of the British Interplanetary Society*, 35, pp.391–395 (1982). http://www.bis-space.com/products-page/magazines-and-journals/jbis-journal-of-the-british-interplanetary-society/

R.H. Frisbee, "How To Build An Antimatter Rocket For Interstellar Missions, Systems Level Considerations," in Designing Advanced Propulsion Technology Vehicles, 39th AIAA/ASME/SAE/ASEE Joint Propulsion Conference and Exhibit, Huntsville Alabama, 20-23 July 2003. AIAA-2003-4676. Available: http://hdl.handle.net/2014/38278

Icarus Interstellar: http://icarusinterstellar.org/

Tau Zero Foundation: http://www.tauzero.aero

Centauri Dreams: http://www.centauri-dreams.org/

Institute for Interstellar Studies: http://www.i4is.org/

Solar sailing is an idea nearly a century old. Imagined as vast films pushed by sunlight, they had an elegance that appeals, though not enough to lead to many missions. The Japanese sailed to Venus, but we have yet to use such technologies very much. With the advent of lasers and high power microwaves, driving sails with far higher powers than sunlight has gotten recent attention, and may be the conceptual breakthrough that makes sails take on new life.

Sailships
James Benford

I begin with a prediction: The first starship will be a sail—a *sailship*, driven by a beam of photons, attaining very high velocities and flying through the Kuiper belt, the Oort Cloud and into Deep Space. A sail driven by beamed energy is the twenty-first-century spacecraft that could genuinely achieve interstellar flight. The idea, which has been demonstrated experimentally over long range, is to exploit the ability of electromagnetic waves to transfer power through space, producing force at great distances.

I say this because sailships have a singular advantage: they leave the engine behind. So we can build a spacecraft that consists of only payload and structure—no fuel at all. The propellant is light itself, so sails reflect light waves, whether visible or microwave or laser produced, from a beam generated elsewhere. Sails can be made both light and smart, in the sense that control systems, sensors and computational ability can be embedded in the structure of the sail itself—a smart sail, with dispersed circuitry, and therefore far harder to damage by meteors or accident.

A century or two in the future, starships will be launched like this: A diaphanous sail, kilometers in diameter, deploys in space by slowly spinning up from an initial folded geometry. When all the embedded microcircuits are tested—those for command, control, sensor array, communications, and retro-reflective beam—the sailship is ready to fly.

The large aperture Beamer, which can be at laser, millimeter-wave or microwave frequencies, ramps up to high power, hits the sailship at its initially very close position. When it reflects from the sailship, it accelerates the sailship away. As the power comes up, the acceleration increases to a fraction of Earth gravity. In contrast to the fast burn of rockets, the beam stays on the sail for hours and days as acceleration continues, and velocity grows to interstellar speeds. Finally, as the sailship approaches the outer solar system, the beam spot size exceeds the area of the sail, the force on it drops and acceleration declines. At some point the beam shuts off and the sail is launched to the stars.

Why Sailships?

Starships need not be rockets. The problem with rockets is that for interstellar flight they will be very large, very inefficient and very costly. For example the kinetic energy of the Daedalus starship in its three-stage variety is enormous. With the final payload of 170 tons and a velocity of 13 percent of light, the kinetic energy of the starship is the equivalent of about ten times all the energy in the nuclear arsenals of the world, several hundred thousand nuclear weapons.

In contrast, a sailship can be extremely light, 100 kg, at a velocity of 13 percent of light it has a kinetic energy of 0.2 percent of the world nuclear arsenals. Now that's a lot of energy, about 50 average nuclear weapons. However, it's one-10,000th of the required energy for an interstellar fusion rocket.

Rockets are notoriously payload–inefficient. The Saturn 5 had a payload of only about 3 percent of its launch mass; the Daedalus starship concept had a payload less than 1 percent of its initial mass. In contrast, sailships have payload masses of at least 10 percent. On the other hand, sailships are energy-inefficient; only a few percent of the energy radiated ends up as kinetic energy of the spacecraft, due to the nature of momentum exchange, which drives them, instead of energy exchange, which drives rockets.

The Beamer—the source of the beam, including its powerful source and a large antenna or optic—projects a powerful laser, millimeter wave, or microwave beam onto a large sail. The sail reflects the beam, picking up momentum, accelerating away from the Beamer.

Beamers would look like the microwave dishes we currently see listening to satellites, only very much larger. The expensive part of this utility is the Beamer, which wouldn't be placed on Earth, but built in space from materials mined from the moon or asteroids. Beamers might be positioned close to the sun, where they could run on strong solar power.

The best feature of beaming energy is that the Beamer—with all its mass and complexity—is left behind, while the relatively simple ultra-light sail, carrying its payload, is driven far away. The Beamer can then be used for many sailship missions. Like the nineteenth-century railroads, once the track is laid, the train itself is a much smaller added expense. Another fundamental attraction of beamed power for space is simple: electromagnetic waves can carry energy and momentum (both linear and angular) over great distances with little loss.

I led the team that in 2000 demonstrated first flight of microwave-driven carbon sails using microwave beams to produce several-G accelerations as shown in the figure. Further experiments demonstrated *beam-riding*, stable flight of a sail propelled by a beam. Beam photon pressure will keep a concave shape sail in tension, and gives a sidewise restoring force. The beam can also carry angular momentum and communicate it to the sailship, spinning the craft. This effect can help control by stabilizing the sail against drift and yaw in flight.

For beam-driven sails, the physics is done. We have experimentally demonstrated flight, beam-riding, and beam-induced spin for stability. The engineering requirement is for large assemblies of both modular sources of the photons, and large antenna/optic arrays. We have vast experience in such sub-systems and progress is steady. There are clear paths forward to solving the engineering issues.

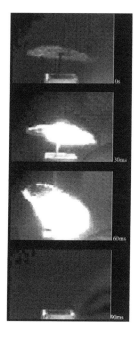

Carbon sail lifts off from end of waveguide at two gravities acceleration. Top frame is quiescent sail, in second frame, sail lights up from heat of microwave beam accelerating it, then tilts, moving upward, in fourth frame had flown out of view (30 msec frame interval).

We can think of the context of the interstellar propulsion challenge the same way we've always thought of fusion: first get the physics solved, then the engineering, and finally address the economic feasibility.

These issues are very different for the nuclear rocket. Nuclear fission physics and engineering are understood well enough, but I have seen no estimated cost of a nuclear thermal rocket. Fission-fusion hybrids are hypothetical until fusion gets out of the physics stage it's in now. Pure fusion, especially the aneutronic reactions that should be more desirable for rockets, is struggling with physics and engineering issues, both fundamentals and feasibility. And we know that nuclear is much more expensive to research and then develop, compared with beam-driven sails. And what would a fusion rocket cost? That's beyond this horizon at present.

Key Issues for Sailships

Deployment

First, an ultralightweight sailship must be deployed. The lightness and elasticity of new ultralight materials such as carbon fiber mats can tolerate virtually no external mechanical contact so that everything must be entirely "hands off." The practical requirements are to:

- Deploy and control the very light but very large structures with minimal mechanical contact.
- Deploy from a minimum stowed volume (maximize packing fraction).
- Provide for control after deployment.

One solution is to deploy by spinning up the sailship, which stabilizes the sail structure to both pitch and yaw. Spinning produces centrifugal force, which provides the required tension to hold the sail shape, preventing deformation and fluttering.

While spacecraft spinning up could be done with cold gas thrusters from unfurling to full deployment, it's attractive to remotely induce spin, entirely hands-off. The other virtues hands-off methods have are reversibility, should spin grow too fast, and real-time control.

We can do this with a purely electromagnetic torque exerted by a microwave beam. The directing microwave beam will carry angular momentum, if it is circularly polarized. A phenomenon that will do this is called geometric absorption and relies on a new understanding of angular momentum in classical electromagnetism. Spinning a conical carbon sail suspended above a microwave waveguide has demonstrated it. To optimize this efficient method of spinning the sail must be shaped so that currents induced on the conducting surface of the sail have a geometry that maximizes coupling of angular momentum. Making some cuts in the surface to change current flow paths does this.

Beam-Riding and Stability

Riding on the beam, i.e., stable flight of the sail propelled by beam momentum, places considerable demand upon the shape of the sail. Even if the beam is steady, a sail can wander off the beam if its shape should become deformed. Generally, sails without structural elements cannot be flown if they are convex toward the beam, as the beam pressure would cause them to collapse. On the other hand, beam pressure keeps concave shapes in tension, so conical concave shapes are the natural answer to sailship configuration questions, essentially a circular cone. These shapes resist sideways motion if the beam moves off-center, since a net sideways force restores the sail to its original position. Simulations show that this passive stability works quite well. And it is made more stable if the sail is spinning, as it would be if spin deployment is used. Therefore the natural configuration of a sailship is a conical spinning shape with the payload hanging from the apex along a flexible tether. It is shaped like a parachute. Suspending the payload below the sail is the more stable shape configuration. Stability also depends upon the ratio of the sail mass to payload mass.

Large-scale Space Construction

The sheer scale of the sailship, and the Beamer aperture which irradiates it, raises major issues. The Beamer aperture will be large numbers of independent radiators phase-locked to each other. Phase-locking is a well-understood phenomena widely used in radars for many decades. Generally, phase-locking uses amplifiers instead of oscillators, because amplifiers work better together in the large numbers of sources needed for power beaming. The Beamer is an assembly of replicated modules of power amplifiers and antennas, which can be something like the familiar microwave dish antennas or a laser output optics, is assembled, building the Beamer.

The sail will likewise be assembled from modular elements and connected together. Many types of materials have been suggested; many geometries have been tried for assembling sails. To date, the largest solar sail prototype is 100 feet on a side, deployed in a laboratory. The sixty-

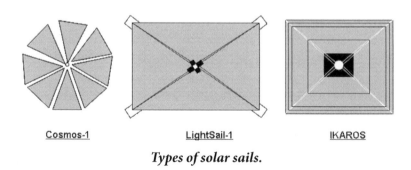

| Cosmos-1 | LightSail-1 | IKAROS |

Types of solar sails.

five-foot IKAROS sail has been deployed and flown to Venus. Nevertheless it's an engineering challenge to see how large diaphanous structures, perhaps miles in size, can be assembled, as interstellar sailcraft will require.

Pointing the Beam

A very precise control system keeps the laser beam on the sailship at large distances from the beam aperture. Tailored phase variations in the beam sources, which are amplifiers, will give control of the beam direction, aided by retro-reflective signals returned from the sail itself. But roundtrip transit delays will make such adjustments decline as the sailship flies away. Moreover, pointing and tracking systems have errors due to the system having imperfect optics, vibrations in the apparatus, and other factors. This is called jitter and varies the beam direction randomly. The beam center moves away from the sailship, and of course this increases with distance. The sailship acceleration should be done before this distance is reached. So there's a trade-off between the jitter and the distance over which the sail is accelerated: lower pointing accuracy, higher acceleration.

The present capability in angular precision pointing in space, for example the Hubble Space Telescope, is about ten nanoradians, which is about a millionth of a degree. The pointing and tracking accuracy needed for sailship missions will be of order microradians for interstellar precursors to picoradians for interstellar sailships. This is a difficult requirement, but in the past century, pointing accuracies increased steadily and rapidly. Pointing control and transmitter optics will likely improve substantially in future centuries, so future developments in astronomy may well make this requirement quite possible.

Slowing Down

Another fundamental issue of an interstellar sail ship will be deceleration as the sailship approaches its target star. Of course initial missions may be fly-by, although the rapid advance of space telescopes will make that short look unnecessary. This undercuts fly-bys, so a number of methods have been suggested, including magnetobreaking using an electrodynamic tether, as well as using drag in the interstellar medium and solar wind of the target star. Robert Forward suggested the trickiest method: subdivide the sail.

Forward proposed a way to both decelerate to the target star and even return back to the solar system: approach the target star, detach an inner part of the sail, the rendezvous sail, reflect a laser beam from Earth from the main sail, now a ring, to the rendezvous sail, decelerate it into the target system. The main sail flies on. For return to Earth, detach an inner part of the rendezvous sail. Send a laser beam from the solar system to the rendezvous sail, which bounces off it and accelerates the return stage toward home. Approaching the solar system, decelerate with the laser. The pointing and power requirements for this scenario are far beyond what we know how to do.

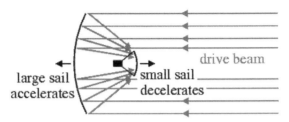

Forward's method of decelerating the sailship at the target star.

Developing Sailships

To eventually have a directed energy capability, space infrastructure must exist to build on. How do we get from where we are now to a future when directed energy can be used for fast missions, including interstellar? By developing in parallel both solar sails and other applications of directed energy.

Solar sail technology can lead to speeds on the scale of 100s of km/sec, perhaps even 0.1 percent of light-speed (300 km/sec), as shown by decades of work by pioneers such as Lou Friedman and Gregory Matloff. Such development enables directed energy sails:

- Sail engineering, especially materials: carbon nanotubes, carbon microtruss.
- Because sails are large-scale structures in space, they also influence the development of large transmitting antennas.
- This leads to: larger sails, lighter sails, faster sails.
- Fast solar sail missions will follow, for example to the Oort Cloud, Heliopause, and the interstellar medium.

This prepares the ground for beam propulsion.

Beaming power over great distances becomes economic only when it can move power from where it is cheap and accessible to places where it is hard to come by. It is often more economical to transmit power than to move the equipment to produce power locally. Modern power systems are complex, but if power for space can be located where it is easily accessible and adjacent to where the required skilled people are located, i.e., on Earth, then it becomes more practical.

Applications can be met by building up from existing technologies: microwave and millimeter wave antennas are already in use for astronomy, fusion has developed gyrotron sources at high frequencies (>100 GHz). The method is to build, stair-step-like, a sequence of applications of beaming power, as illustrated in the figure:

- Orbital debris mapping could be the first objective.
- Recharging of satellite batteries in Low Earth Orbit (LEO) could be economic, followed by recharging of satellites in GEO.
- Beam launch into orbit of 1,000 kg cargo-carrying supply modules can make industrial transport to LEO a reality at cost about an order of magnitude less than at present.

In today's frugal climate, it is important for technology development to be coupled to commercial applications. Several of the missions we've described are potentially commercial matters. Starting with orbital debris mapping, one can see an incremental commercial

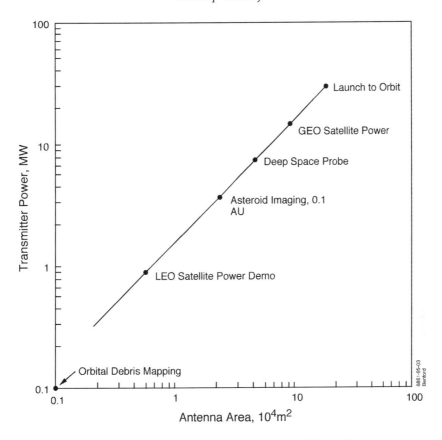

Development Path For Space Power Beaming. This is for a system with 10 kW gyrotron millimter-wave sources, the vertical axis is also proportional to the number of modules, i.e., 375 modules for asteroid imaging. Antenna area unit is a 100 m x 100 m area.

development leading first to satellite power recharging. Eventually, as the space market and business confidence grows and capital becomes more available, this development plan leads to the repowering of satellites in GEO and ultimately to launch services. Investment costs are minimized because the research program leads to applications, which feed capital back into research, leading to new applications.

Therefore, the private sector should be included from the outset in the development of power beaming for space applications. This includes the R&D phase, as it is very important to gain support from industry to maintain a long-term commercial strategy. There is at present no clear view of how it is to be achieved and by what technology we are to make the solar system readily accessible. Much of the technical means are already in hand for a space infrastructure. Many missions are made possible if we look to the use of microwave, millimeter wave and laser beams to provide power and transportation through a unified approach.

Interstellar Sailship Probes

Sailships' very low mass makes them ideal for unmanned probe missions for fly-bys of orbiting planets of nearby stars. At the nearer-term low-mass end of the sailship spectrum is Starwisp, a Robert Forward concept for a lightweight, high-speed interstellar fly-by robotic probe, driven by a microwave or millimeter-wave beam. Starwisp is a large sail made of very thin wire mesh, with microcircuits at each wire intersection, each containing a node of a neural net. The sail would be pushed at high acceleration by the microwave beam; Starwisp reaches a fraction of lightspeed while still in the solar system.

Starwisp extrapolates nano-spacecraft and pico-spacecraft concepts to more sophisticated electronics, detectors, sensors, and mechanical devices, reduced to near zero volume with very small mass. The payload would include small particle sensors and imagers of both optical and infrared and radio frequencies.

Forward's minimal Starwisp design would be a one-kilometer-square mesh sail weighing sixteen g and a few grams of microcircuits, a total of less than one ounce. Wire spacing is less than the wavelength, mostly empty space, giving greatly reduced mass compared to solid laser-reflecting films.

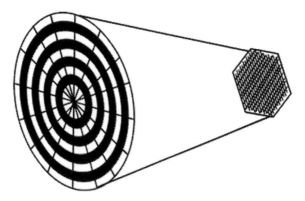

Starwisp probe driven by Beamer with an annular transmitter lens.

The Starwisp starts with fast acceleration, 115-Gs, by a 10 GW microwave beam, reaching 20 percent of light-speed in days. Because of Starwisp's very small mass, the beam–power level needed to drive it is about that planned for the microwave power output of a solar power satellite. Thus if power satellites are constructed during this century they could be used to launch one or more Starwisp probes to the near stars during their daily operation.

Starwisp's goal is to provide high resolution images of planets seen in the fly-by. As it goes through the Alpha Centauri system twenty-one years later, the solar system microwave transmitter beams microwave power to Starwisp the microwave antenna. This energizes photodetectors, chips and circuitry, a neural net. The microsecond circuits measure the phase of the microwave beam. This phase information allows the net to later point precisely back toward Earth. The neural net then gathers optical information from photosensors, and beams it, as images encoded in microwaves, directly back toward Earth. Starwisp then flies on into the interstellar sea of night.

Laser–Driven Colony Sailships

Once our telescopes have found interesting planets, and probes have explored them, we may wish to go ourselves. To send colonists is a far larger undertaking. Although microwave beams can be used only to drive a spacecraft in a direction away from the solar system because the microwave beam spreads rapidly with distance. At much shorter laser wavelengths the spread with distance is much lower. It is possible to design a laser–propulsion system that can use a laser beam sent out from the solar system to push a sailcraft to a star *and to return it to Earth.*

Here is an approach to crewed sailships: even if hibernation is used to allow such a long journey, when people come out of hibernation they will need to avoid weightlessness. The spin of the sail, which greatly adds to its stability and beam-riding ability during acceleration, will also maintain artificial gravity. Because such sailships will be very large, perhaps tens of kilometers, the habitat will likely be a torus outward from the axis. The sailship has some payload there, but the bulk of the payload will be hanging from the parachute-shaped sail, to enhance stability.

The habitat is a torus, with a rectangular cross-section. Because the centrifugal force points radially outward, the crew will walk on the outer diameter of the toroidal habitat, their heads pointing toward the center of rotation. The landers and exploration shuttles will be stored in the heavy payload section. The forward section of the habitat will need to be shielded from incoming very high-speed dust and rocks. There are many concepts for shielding, including plasma discharges and bumper-like compartmentalized sections for absorbing the momentum. This is an excellent location for containers of water or food, compartmentalized so that any single penetration will have little effect.

It would be very attractive to use the sail itself as the major aperture for transmissions back to Earth. The issue would be to make the surface smooth enough to provide a good high-gain antenna. That would require that imperfections must be small compared to the wavelength. Such work is already being done, creating smooth surfaces for large antennas in space by developing polyurethane resins that harden at a specific temperature. Such a ship concept is shown in the figure, a large spinning structure destined for the stars.

This method to reach the huge powers, energies, speeds, and costs of interstellar travel is an extrapolation of known large-scale engineering technologies to larger scale, supported by economic growth over centuries. Given the rapid growth of civilization in the past, as described in Robert Zubrin's chapter, such ships may be only a few centuries away.

Lasers near the sun produce a coherent beam to a transmitter lens deep in the outer solar system, a concentric ring lens designed to operate at the laser wavelength. The lens would send a laser beam over light-years before the beam would start to spread, so little of the laser beam would be lost due to spreading during the mission.

Interstellar Sailship driven by laser beam accelerates away from solar system. Payload hangs from the sail, which spins for stability. In this early stage, molecules released from the heated sail material, giving additional thrust, can be seen from scattered laser light. **(Adrian Mann)**

As in the Forward design for a mission to Epsilon Eridani, a promising star eleven light-years away with at least one exoplanet and two asteroid belts, the light sail should be 1,000 km across. Three subsections of the sail are the center payload return section, the annular accelerator and the outer accelerator/decelerator ring. The ship reconfigures from the initial acceleration, then decelerates toward the star, stops for exploration, then sails home to the solar system.

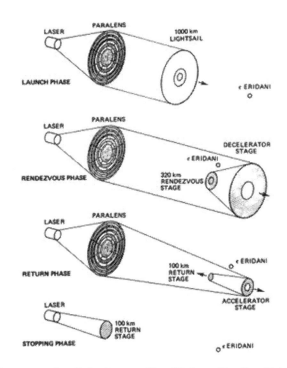

Segmented sail for interstellar flight to Epsilon Eridani.

The three-stage sailship accelerates at 30 percent of Earth's gravity to half the speed of light after 1.6 years, traveling 0.4 light-years. The sailship reaches Epsilon Eridani after twenty-two years in Earth time, but on board it's been nineteen years because of relativistic time dilation.

At 0.4 light-years from the star, the rendezvous portion of the light sail, the circular central payload section plus the middle ring section, detaches from the center of the sail and turns to face the large outer ring. The laser light from the solar system reflects from the outer ring sail, which acts as a retro–directive mirror, sending the laser light back toward the solar system. The reflected light from the outer ring sail hits the smaller rendezvous sail and decelerate to a halt. The outer ring sail, its task completed, flies farther into interstellar space.

The remaining starship steers toward the target planet; along the way it might drop small probes into the orbits of other interesting planets. Early landings are robotic, followed by explorers in protective suits. Some dangers could be subtle, however, taking a long time to emerge. We have no experience of encounters of two completely different biologies (as Paul Davies remarks in his piece in this volume) so perhaps the best planet to actually settle on is one with no life forms. Colonists could introduce terrestrial life if the world is close enough to Earth conditions. If Earth life can adapt, then it's likely that colonists will eventually do so.

Should some wish to return to Earth, or if some particularly interesting lifeforms or artifacts are found, it's possible to bring them back to Earth. The central payload section would be detached from the center of the middle ring section and turned to face it. Eleven years earlier, the laser in the solar system would fire again, hit the ring-shaped rendezvous sail, be reflected back on the payload sail, sending it off home. Later, approaching the solar system, the laser fires again, decelerating the sailship. Members of the crew could retire to write their memoirs.

Our kindly planet fits us well, but space ships will need a great deal of tuning to even approximate the easy environment we evolved into. This story began as nonfiction, then developed its own logic and now is fictional—though not very much so.

Living Large
Richard A. Lovett

Wow, so that's a 3D camera? I had no idea they were so tiny. When we're done, do you mind if I look at it more closely?

Yeah, I guess if our eyes can do it on that scale, technology can too. And I suppose my team has been doing something similar—though in our case we've been packing Earth's entire biosphere into a 400-meter torus…or at least the essential bits of it. Enough to support the folks who'll be spending big parts of the trip awake.

Since I'm not one of those people, I've got to eat whenever I can. Can you pass me one of those donuts? Munching may not be the most polite way to do an interview, but I need to put on another thirty pounds before launch or I'll be cut from the crew.

No, eating that much is *not* fun. I wish people would quit thinking it's some kind of luxury. One donut is a treat. Maybe two. But in four weeks, there'll be no more need for me until we reach the other end…and it's a lot less resource drain to have me plumped up enough to see it all the way through in hybo, rather waking up every few decades to fatten up all over again. The docs tell me I need to go in at 365 pounds, minimum. Most of my crew is going to be closer to 425, so there's no way I'm asking for an exception. Especially since it's only by the skin of my teeth that I'm going at all.

Yeah, *hybo*. That wasn't in the press pack? Technically, I suppose it should be *hiber*, but that sounds too much like hyper, which—do you know Greek? Too bad. The point is that *hyper*-metabolism isn't what we want. But by the thousandth time you say *induced hibernation*, you've got to shorten it. We wound up with hybo. It started as a typo for *hypo*-, which would be correct in Greek, then just kind of stuck.

But that's not what you really want to know. You're not the first reporter who's approached me, even if you're the first I've had time to talk to. You want to know why I'm going.

If you've done your homework, you already know part of it. After Molly died and Alice…did what she had to…there really wasn't much left here for me. That's when I made a late request to be part of the crew. I wasn't sure they'd let me go. Not because of the psych issues of losing my wife and kid—half the crew have something like that in their backgrounds. It's a one-way trip, and even if we could come back, who'd want to, 468 years in the future? Could you imagine Queen Elizabeth—the original one, as in Shakespeare's—walking into this room? Forget the 3D camera, the ride up here from Earth, the hotel airlock, everything else. What would she make of your *shoes*? There's not a stitch of leather in them.

It's a one-way trip. I do *not* want to come back here 468 years from now.

The problem getting on the crew was that I had to convince the mission planners that I was needed. Sure, I'm in charge of designing the life-support system. But I'm a systems engineer, not a mechanic. If something goes wrong I'm not the guy to fix it. What I had to do was convince them that when we get there, I can help plan habitats in whatever conditions we find.

I know, the Kepler XII folks think Terra Nova will be a lot like Earth. Continents, photosynthesis, nothing overtly toxic in the air. But hey, Antarctica is Earthlike. So's the Sahara. We've got to be prepared to be flexible. Flexible enough that I eventually convinced the folks in Houston I might be worth my weight in hybo. And I suppose if something ghastly goes wrong, there might be reason to wake me up. But that's something I'd rather not think about because anything that would require redesigning the system en route would be really, really bad.

But none of that's the real reason I wanted to go. Not even Alice and Molly, though without losing them I'd never even have voiced the idea, let alone seriously considered it. No, just between you, me, and the camera, what I really want is to know if it *works*. I mean, I gave everything I had to this project for seventeen years. Could you spend that much time on something, only to watch it fly away forever—and never know?

You could? Hmm. I guess that's why you're in journalism and I'm in starship engineering. I care a lot more about what happens two or three centuries from now than who wins the next World Cup. Or the next election—unless they threaten to shut off my funding—which, I guess, is a good attitude for someone in charge of designing life support for 234 years.

So how does it all work? Perfectly, I hope! But first, can you pass me another donut? Or maybe just slide over the box. Those probably have about 350 calories each, so I need to eat the whole dozen. Did you know I used to be a marathoner? Not the type who naturally beefs up to 365 pounds. But maybe that's one of the reasons I made it onto the hybo crew, because once I reach target weight, I'm nearly half fat, and as long as I don't die of a heart attack between now and then, that's a good thing. I just wish the pre-hybo drugs made me want to eat like a bear. It's going to feel *really* good to wake up skinny.

But where was I? Oh, yeah, 234 years. Not 233 or 232. "Almost" only counts in…well, you've probably heard that one too many times.

What I'm in charge of is more than just food. My department is "Habitat, Engineering, Longevity, and Life Support." Did you know some idiot once called it HELLS? That was in an official document before one of the PR folks pointed out it sounded more like what we wanted to avoid.

Planning life support for 234 years isn't something you just sit down and do. It comes in stages.

The first question is how many people to plan for. You'd be surprised how long it took to decide. When I came on the project back in '95, hybo wasn't on the table, so we were still thinking in terms of an entire, multigeneration culture. There were studies suggesting we could make do with as few as thirty people, but more would be better because it was going to take six or seven generations to get there—enough to get awfully inbred. And while stations in the outer solar system seem to be socially stable with crews of only a few dozen, nobody's ever tried to make it work for centuries.

Still, thirty was a starting point. Could we do something on the scale of the Kuiper Deep station, for ten times longer than anyone's ever lived in the Outer System? That was the question. Plus, it didn't take long to realize a sperm bank was the solution to inbreeding, which meant our real constraints were psychosocial.

But there had always been was an elephant in the room—something everyone knew but didn't want to talk about: none of the initial crew would live to see the destination. Nor would their kids, grandkids, or great-grandkids. Entire generations would be born, live, and die in a volume the size of a not-all-that-big sports arena.

At the time, I figured, okay, if you have to do it, you have to do it. Then I married Alice. She got pregnant, and I realized there was no freaking way I'd condemn my daughter to live her entire life in a damn flying Space Needle. It either had to be bigger—which meant hundreds or even thousands of years longer in flight—or we needed something else. You know the answer, of course, because I'm eating the donuts to prove it.

It started when Chen's group at Malawi Tech put a chimp in hybo. The idea wasn't totally new—more than a hundred years ago a lab in Seattle, Washington, found that mice could be put into a hibernation-like state with a low dose of hydrogen sulfide. That's right, rotten-egg gas. It was exciting because mice, unlike bears, bats, ground squirrels, hedgehogs, snakes, toads, etc.—don't normally hibernate. But it turned out that hydrogen sulfide didn't work in larger animals and that was the end of that line of research. Chen's discovery twelve years ago—using a cocktail of metabolism-suppressing drugs and gradually induced hypothermia—was a real game changer. And yeah, that's *hypo*thermia, not hyper-. *Hyper*themia is heatstroke. See, there's a reason we didn't want to call it hiber!

Where was I? Gads, that's what my grandfather always said in his old age. I'm only forty-eight. Why the hell do I feel old? Maybe it's got something to do with the fact that a month from now, if I gain enough weight—could you give me that éclair?—you'll be a century or more dead.

Cute? Yeah, I know. I practiced that line. Meanwhile, maybe you should just pass me the whole box.

And, for real now, where was I?

Chen, that's right. Thanks. Initially, she was interested in using short-term hybo to stabilize trauma patients until they could get to a hospital, but she quickly realized it could be useful in space, where help can be months away. She even got a couple of governments to approve testing it on prisoners. "Sleep off your years" was the offer. Ethical? I don't know. All I know is it's why I'm now eating donuts.

No, I don't think spending time in hybo will somehow make everything magically better. Molly is dead. See, I can say it. Dead. Not gone ahead. Dead. If there's another life, I'll see her then. But for this one, she's gone. Like Alice, though in Molly's case I at least know where. The last time I talked to Alice, she was in Pretoria. About as far away as she could get without learning a new language. And no, I don't know what she was doing there. By then I'd given up. She wasn't coming back any more than Molly was, and I'd already applied for a berth.

No, not to forget. To move ahead. And, as I told you, to find out if it worked. Or, I guess not wake up if it didn't. But I assure you, I am not suicidal. I *want* to wake up. I *want* to find out what's out there. But if it turns sour despite our best efforts? Check out the Pilgrims. What were the chances they'd reach a rock on which to land? Something more than half of them died in the next three months. Sometimes, you roll the dice.

There's no way you can fly a ship 234 years with everyone asleep. There's got to be an awake crew. Thirty at a time, apparently, if the social-stability folks are right. And they need life support.

Anywhere off-Earth, the basic needs are pretty…well…basic. You must have seen the "Never forget where you are!" posters plastered all over the place up here. There's one right behind the door.

Air, obviously, is top of the list. There are basically two ways to generate it. One is with an artificial ecosystem of real plants and animals. The other is chemically.

Artificial ecosystems were never on the list of serious ideas. You won't find anything short of a luxury condo habitat that fully relies on one, and even then only in Earth orbit where help is easy. Realistically, such systems are simply too bulky, and if we did try to compress one into a low-G jungle in the ship's hub or some such thing, there'd be too much risk of wild ecological fluctuations. That's what happened in a place called Biosphere 2, back on Earth, sometime in the 1990s. Biosphere 1 was Earth itself. Biosphere 2 put eight people in a closed, greenhouse-style habitat about the size of the largest ship we could possibly launch to Terra Nova, where they tried to live for two years. "Tried" being very nearly the operative word. People at the time viewed it as a failure, but it really proved that this is very, very hard to do, even without the complications of being in space.

We've done better since on the moon and Mars, but those aren't truly closed systems. Even the Mercury Experimental Station gets water from ice deposits in permanently shaded craters. The fact is we don't really have much experience with large, closed, ecologies. Not enough that I'd want to trust one for twenty years, let alone 200. Read the reports from the old Biosphere 2 project. Those guys damn near starved in the first year. Then they got overrun by ants and

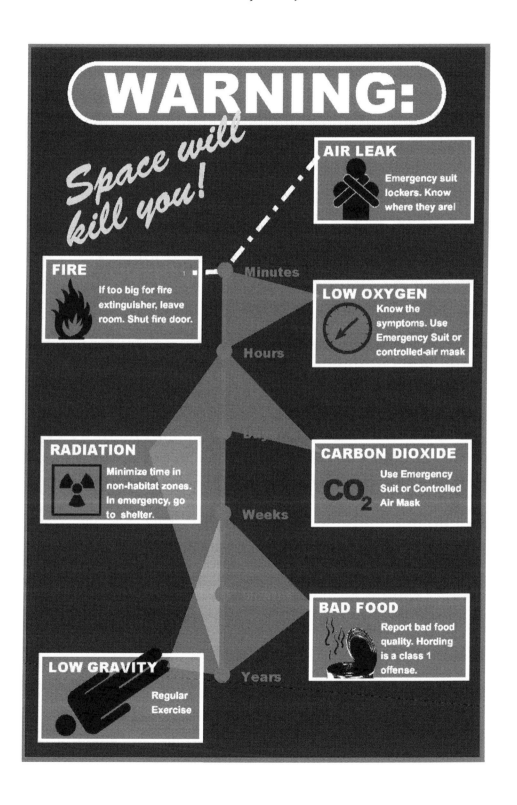

cockroaches. Not only is a complete biosphere too damn *big* to fly to the stars, there are too many things that can go wrong with it.

Air-recycling technology is a good deal more predictable and people have been using it since the first nuclear submarines. Sure, those weren't closed systems either—they got their oxygen by electrolysis of seawater—but they were a starting point. These days, air recycling is a big industry, with a half dozen major players. We went with Orbital Standard because, well, it's the closest to *standard* that exists, which means most of the bugs have (hopefully) been ironed out.

Vids love to show death in space as being from suffocation, but unless you get a pressure leak you can't control, the real concerns are more subtle. Did you know the average person produces two liters a day of flatus? That's farts to you, me, and maybe that camera. Multiply that by 234 years and you have a problem! Each of us also sheds ammonia, ketones, dandruff, dead skin, and billions, or maybe trillions, of perfectly viable skin bacteria every day. Not to mention breathing out a couple of liters of water, which again adds up rather quickly. Just spend a night…

…sorry, that memory caught me off guard.

On our wedding trip, Alice and I took up backpacking. We had this tiny tent weighing less than a kilo, but the damn thing wouldn't *breathe*. We were way up on some glacial bench on the shoulder of Grand Teton, and outside it was so *cold*. Inside…Well, let me just say that if you ever spent a night in a tent like that with two people you'd know what I was trying to talk about. In the morning we could wring the water out of our sleeping bags. But—Hell, I don't know if she's even alive anymore.

What is it about this interview that keeps making me talk about her? She's…wherever. Me, I'm going to the stars. I just hope she's following a dream, too, and not just running away.

Back to the starship.

If you mix runaway condensation with all that dandruff, ammonia, bacteria, and whatnot, it doesn't take long to get one hell of a mess. You know that's what caused the disaster at Callisto Station, back in '78, right? They wound up with Legionnaire's disease that snowballed until half the station was dead.

Which is one of the reasons we can't have a super-big crew. Callisto had 'em packed in like sardines. Though we've also got a modular design, so if something does go wrong we can shut down the ship in sections and clean them up one at a time.

Another thing we've added is molecular accounting. If there's a leak, we need to know ASAP. If things are disappearing but not being lost to space, we also need to know because they're accumulating somewhere and we need to find that, too, before it turns into Legionnaire's or worse.

Here's a chart of how our form of Orbital Standard is going to work.

Is it really that complex? My private version is about ten times worse! Though I guess the abbreviations don't help. "X" is engineering-speak for "exchanger." Thus, the CHX is the Condensing Heat Exchanger—basically an old-fashioned refrigerator. The rest of the boxes make perfectly good sense once you know what they are:

- TCCS is Trace Contaminant Control System.

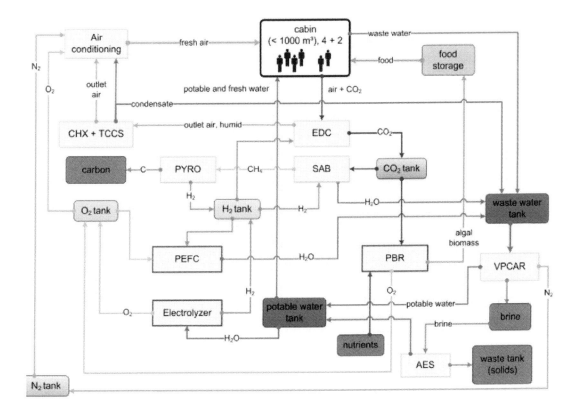

- PYRO is PYROlysis, a form of heating without oxygen. It's a lot like the way charcoal is made for barbeques.
- EDC is Electrochemical Depolarized Concentrator, a method of carbon-dioxide extraction used since the early days of space flight.
- VPCAR is Vapor Phase Catalytic Ammonia Removal.
- AES is Air Evaporation System.
- BPR is Biological Production Reactor, basically an algae tank.

Sorry, the terminology comes from engineers and systems designers and we're the type who'd never use a simple word like "algae" when we can say "BPR" instead.

The whole thing's made up of three main loops with a lot of interconnections. One loop focus on cleaning up the air. It starts by scrubbing out carbon dioxide and excess water vapor, then filters out particulates and trace gases, like all those farts. It's pretty straightforward; we've been doing it for generations.

The next loop is the oxygen loop. It's a little more complex. It starts by taking the carbon dioxide removed by the first loop and sending it to the SAB. Oh, sorry, that's the Sabatier reactor. It's named for a French chemist who figured out that you could use carbon dioxide and hydrogen to make water and methane. The methane then goes to the PYRO chamber, where it's broken down into hydrogen and carbon, both of which are useful materials. Oxygen (and more hydrogen) comes from electrolysis of the water.

Water we don't use for electrolysis, including that scrubbed from the air, flushed down the toilet, etc., goes through the VPCAR, whose name is a bit misleading because it removes a lot more than ammonia. Think of it as a high-tech distillation unit. The leftovers—which we call brine even though there's a lot more than salt in them—go to the AES, which wrings out even more water. The cleaned water is drinkable; the concentrated brine goes to the waste tank—not that it's really waste as you'd think of it. It's just a hodge-podge of everything that isn't more easily recycled—a cache of anything from sulfur to silver. If we don't lose it to space or misplace it in a supply room, it damn well better wind up in the waste tank.

The third loop is the BPR. The reactor itself sits right at the heart of everything and when properly balanced can do most of the heavy lifting of converting carbon dioxide into oxygen. It also provides lots of our food.

Yeah, I know the names kids have for algae-based food. Green grit, pink slop. Finding gross names for food seems to be part of what it means to *be* a kid. But some of the processed algae stuff is damn good. We've even managed a passable beer—though if you're a wine connoisseur, you'd better cultivate a taste for beer. Maybe someone will figure out winemaking before we get to Terra Nova.

What's not on the flow chart is that we've got truly magnificent fabrication facilities. I myself wouldn't know how to make a replacement EDC subregulator from the contents of my wastebasket if my life depended on it. But we've got thirty people, at least three of whom will be awake at any given time, for whom that would be a snap. If we make it to Terra Nova, these are the folks to whom we should build monuments. In 234 years, no matter how well we've planned, a lot of things are going to *break*.

That takes us through most of the "Never forget where you are!" list. But only the most obvious parts. If you run out of air, you're clearly in trouble.

Less evident is bad food. Most likely, it'll just make you really, really sick, but even in the richest countries, it produces hundreds of thousands of hospitalizations and thousands of deaths every year. Way back when the U.S. was first aiming for a moon landing, NASA realized that an illness like Salmonella would be a very bad thing, so they set out to make sure it would never occur. Since then, a couple of orbital hotels have had outbreaks, but NASA never did, even though back home, one person in six gets a foodborne illness every year.

The trick was a program called HACCP, for Hazard Analysis and Critical Control Points. It's basically quality control with a very strong focus on safety. The idea is that you figure out the places where microbes can creep in, find ways to kill them if they do, then monitor the hell out of those critical stages to make sure they're working.

Needless to say, we're doing everything we can to keep bad bugs from getting aboard in the first place. But nobody knows how *Legionella* got to Callisto Station. And which lunar hotel is it that keeps having troubles with mice? The Skyview Grand? One of the big, expensive ones, anyway.

Mice are one of the things I truly lose sleep over. The little buggers like to chew wires, and there's no way to target them in a HACCP program because anything guaranteed to kill mice would also kill a lot of equipment.

Bacteria in food, though, are easier because we can always cook the food. It's not just microbes we need to protect against, though. There are also chemicals, broken bits of processing equipment—anything you don't want to eat. But HACCP has been around a long time, and I think we've got all that under control.

Which doesn't mean nobody will ever get sick.

We debated this a lot, because with a small population, it's theoretically possible to wipe out illness entirely. But there's too much research suggesting that if your immune system doesn't have real enemies to fight, it'll make them up, like a kid itching for a fight. Allergies, in other words. And on a new world, that isn't a problem we want.

Yeah, that might be the third reference I've made to children. Is there a problem with that?

You're kidding. No that's not what I was talking about! Weren't you a kid once? Besides, I was thinking of boys. Alice and I lost a girl—long before she was old enough to pick fights. She'd barely learned to say "Daddy." And no, I'm not over it. You never get over it. But I did better than Alice, even if you're now going to tell me I turned the project into a substitute child. I've heard that before, too. Twenty-eight days from now, we're going to the stars. Does it matter why I worked so hard to help make it happen?

What I was trying to say is that immune systems need exercise, just like the rest of our bodies. And, yeah, physical exercise is also a big part of what we spent time planning for. My favorite is the zero-G "gerbil wheel" at the hub. The harder you run, the more weight you get. You can also put people on opposite sides and have them race each other. The balance gets a bit weird when one starts to catch up, but that's part of the fun.

My best marathon? About 3:05. Three-oh-five twenty-something. You think you'll never forget the precise number, but you do. Anyway, good but not spectacular. Alice ran too. Oh, you knew that already? She was better than me, at least when you adjust for gender—testosterone's good for maybe thirty seconds a mile and she was never that far behind. That's how we met, by the way, running marathons in grad school to blow off stress. She'd been a soccer player before that. Or *football*, as she called it growing up in—oh, you know that, too? It's a little outside of Cardiff. And no, losing her's not something you ever get over, either. But you do move on.

So, I might be biased, but I think sports are important. People are competitive. A hundred years ago, the early astronauts sometimes rigged up ways to compete with people on the ground. One even figured out how to mimic a triathlon in zero-G—a neat trick in the small habitats they used in those days—and ran it simultaneously with folks on the ground. Crazy, sure, but that's people for you.

But that's not going to be a possibility more than a few light-hours out, so we've got to find ways for folks to compete against each other on the ship. If we don't, they're going to make up their own. My granddad was a longshoreman, and he had stories about nightshift forklift races that would curl your hair. If we want 144 people to sleep safely for 234 years, I do *not* want an awake crew of people like him, bored, doing…whatever.

But I almost forgot the immune system. As I was saying, there was some really good work done in the 2040s and '50s showing that unexercised, it, like my granddad, gets bored. Or too lazy to fight off a real invader.

No, that is *not* the way Molly died. That was cancer. Do you *have* to keep bringing her up?

Maybe Terra Nova won't have allergens or exotic germs, irksome itches or fiery fevers. It may all be too alien for there to be a problem. But there's no guarantee. And yeah, I guess there's also a link to cancer. I'd not thought of it in terms of Molly until now, but a healthy immune system can help nip cancer in the bud. I don't know with an under-stressed immune system. That's the whole point: *nobody* knows. And that, not Molly, is why I want to keep the shipboard environment as Earth-normal as possible, even to the extent of deliberately letting people get sick every now and then. We've even designed a few non-contagious strains of rhinovirus to help.

No, we're not going to give everyone a cold! Not all at once, anyway. We'll dole them out to one or two people at a time. Maybe we'll use them when we get to Terra Nova, too. That'll depend on what we find when we get there.

Another problem is confinement. That's not on the "Never forget where you are!" list, but it's a big issue. In our 400-meter torus, the longest trek you can take without coming back to where you started is a bit over 1,200 meters. We've got enough decks that you could stretch that by changing levels every lap—and you could add more distance by visiting the hub or the spindle. But despite the old novels about travelers on giant spacecraft forgetting the outside universe even exists, nobody on ours could ever forget they're on a spaceship. It's going to be the overwhelming reality of their lives.

We've decided there's got to be some way to take a vacation. Books, vids, arts, exercise, and virtual reality offer recreation, but occasionally people need bigger changes in routine.

Part of the solution is rotating duty shifts. The shrinks say there's an advantage to having the awake crew go back in hybo after relatively short shifts—no more than six months. That way, each time they wake up, they're appreciably closer to the end. And of course, they don't have to fatten up as much between stints in hybo, which will make them a lot happier on their duty rotations.

But these folks are still going to spend thirty or forty waking *years* in transit—sixty to eighty shifts. Some will be having kids. Hell, some will be having grandkids.

And yeah, kids can survive hybo. We think. That's not something Chen tested on humans, but it works on chimps and gorillas. Anyone on the awake crew who decides to have a child knows they're using that child as an experiment.

Me? No. I couldn't bear to risk losing another child. But the awake crew must make their own decisions. The official policy is no kids in transit, but we long ago realized that's unenforceable. The unofficial policy is, "Do what you think best." We've got a few extra hybo chambers for those who decide to take that route. But no guarantees.

Talking about hybo again makes me realize how much weight I still have to gain. Do you have anything to eat that's not a donut?

Chocolates? No. Especially not those. Godivas are really good, but they've got…memories. Chips? Great—I don't care if they're a little stale. Oh wow, peanut butter! Better yet! Nutritionally, it's about two-thirds fat with the rest carbohydrates and protein. Probably about what I'll wind up being thirty pounds from now!

So, where was I? Oh, yeah, kids. Kids are unofficial, but we've got plans for what to do if they happen. And no, we have no plans for what to do if one dies. You can't plan for tragedy. And why the hell do you keep pushing that issue? My interest is in getting to Terra Nova, not reliving the past. Terra *Nova*, get it?

We don't have a rule for burials, either, though stats say we'll have to deal with a few. We decided to let whoever's awake at the time decide. Burial at space…recycling…our systems can handle either.

No that's not grotesque! What the hell do you think happens on Earth? Every breath of air—every sip of water—holds atoms from countless people who sipped and breathed before. What's the difference?

But I was talking about vacations before we somehow wound up on dead bodies. The reality is that a decade in hybo might let you wake up a lot closer to your destination, but it's not a vacation. It's more like Rip Van Winkle, other than the fact that you tend to wake up *really* hungry.

To fill the gap, we came up with the idea of EVA vacations: opportunities to go stargazing on the hull. Tethered, of course. To make them special, we're rationing. You get eight hours for each month of awake time. All at once or in doses, your choice. I'll be the first to admit I don't know if it'll feel like a real vacation, but it's the only "away" time we can offer, and it's short enough the radiation risk is minimal.

Which brings us to the last two risks, radiation and gravity.

Gravity is the simplest. Tourists and business travelers don't have to worry. How long are you here for? No problem. The drugs they gave you work just fine for a lot longer than that. But they're not cheap. Down on Earth, you've still got people sleeping under bridges. Here in space, we sometimes find them trying to live on the cheap, in zero-G. It's the same damn thing except that instead of neo-tuberculosis, our poor face things like osteoporosis, muscle atrophy, anemia, screwed-up balance, and vision loss.

Needless to say, we aren't letting anyone sleep in zero-G on our spacecraft. And if I have anything to say about it, nobody on Terra Nova will ever sleep under a bridge. Sorry, one of the perks of leaving Earth forever is you get to be preachy. If you folks haven't fixed that problem by the time we get to Terra Nova, we will be very ashamed.

We debated quite a while before deciding to go all the way to full gravity. People seem to be living just fine on the moon and Mars, but eventually we decided it was another of those things we wanted to keep as normal as possible.

The decision to go with full gravity, by the way, is why the habitat is a torus rather than a cylinder. Volume-for-volume, something the shape of a tin can would have been cheaper to build, and the drive engineers would have preferred it because there's less stress when you accelerate it. But with all the volume compacted near the spindle, it *feels* smaller. And, to get one-G, we'd have had to spin it faster. That produces Coriolis forces that can be quite noticeable. Moving quickly would feel a bit like a kid on a merry—

Damn it are you *trying* to pick a fight? Can't you just let it go? If you do that one more time, I'm going to terminate this interview.

I'm here because among your viewers will be people like I was, way back when—*no, do NOT say it*—back when I was young—when I was a *kid*—looking for a dream. And just because my generation is going to the stars doesn't mean the next generation can no longer dream. Hell, maybe when we get there, we'll find them waiting to greet us. Though if you're listening, kids, we'd rather you went somewhere else. It would be hugely anticlimactic for us to get there only to find Terra Nova already settled. If you find a way to build faster spaceships—and I hope you do—use them to settle your own worlds! Then come visit us after we're established. That we'd enjoy.

End of mini-sermon, back to gravity.

We don't *think* there'd be long-term harm from the merry-go-round effect. People have certainly dealt with it often enough in space liners and orbital facilities. But we don't know what happens when months become centuries. And the whole trip's enough of an experiment there's no need to change any variables we don't have to.

Just for fun—and because exercise and competition really are important—we've put running/ walking tracks around both the innermost and outermost decks of the torus. That's a 19 percent difference in gravity, so it's no surprise you move differently on them. But it also turns out that a decent runner can feel the difference between running with or against the spin. In one direction you're adding to it and getting heavier. In the other, you're counteracting it. If you could run fast enough, you'd practically float.

We've also decided to bring a few bicycles aboard, and a good cyclist will really feel this effect. It's another way we're looking to burn off competitive juices under more controlled circumstances than my grandfather's forklift races. And yeah, the cyclists might crash. But if you're spending years in a totally controlled environment, you've got to be able to take a few not-so-controlled risks. Plus maybe they can use the bikes on Terra Nova—which by the way, has gravity close to Earth's but not identical. About 0.87 g.

So we've got the ability to gradually de-spin if we want to match our new home's gravity. Alternatively, we can keep it at one-G, so the awake crew can arrive Earth-strong. That's a decision they'll make en route. Me, I'll be sleeping. And as far as we can tell, hybo slows you down enough that none of the gravity stuff makes any difference. But fat isn't the only thing we'll burn off, which means hungry isn't the only way we'll wake up. Until we get in a few weeks of strength work, we probably won't be much use even at 87 percent gravity. Which is yet another reason to keep the awake crew fit. Someone's got to be in shape.

Radiation is the other big issue.

In the solar system, the 800-pound gorilla is the sun. A big flare can produce enough rads to cook you dead, fast. But even when the sun is safely distant, there's enough radiation in interstellar space that we need pretty good shielding.

Inch-for-inch the best stuff is something dense, like lead. Pound-for-pound you do better with hydrogen. That's because even though individual atoms are light, you've got a couple hundred times more of them. Also, when struck by cosmic rays, hydrogen and other light elements don't fragment into showers of smaller particles that also have to be blocked. But who wants that much pure hydrogen? It would be enough to power the fusion drive practically forever.

Eventually, we went with a quadruple-layer design. One is the ship's skin. That's made of methylated carbon/boron nanotubes—not only good shielding, but strong and light, too. Inside are layers of hydrogen, ammonia (NH3), and water. Our storage tanks, in other words.

No, the ammonia's not for cleaning, it's for making fertilizer. Which I guess brings us back to food, and why we went primarily with the BPR.

On a spaceship, there are four basic ways to get it: chemical synthesis, algae, hydroponics, or freeze-dried. Chemically synthesized food isn't something we seriously considered. The technology could probably be developed, but it would be easy to get an unbalanced diet. There's this great story about Australia's "Lost Expedition" of 1860, which tried to live off the land by making bread from the spore pods of something called the nardoo fern. Such bread was a staple for the Aboriginals, but the white explorers botched the recipe, failing to properly wash out an enzyme that blocks the body's ability to metabolize carbohydrates. The result was that they, in essence, starved on full bellies.

Synthetic food faces the same type of risk, though not so much from poisoning as from leaving out something crucial—some micronutrient essential for something like cardiac health or nerve function or preventing cancer.

And damn it, do *not* say it.

Sorry, maybe I'm getting touchy. Being in life support means I have to think about such things, but until you came along and kept asking about it, I'd never equated them to Molly. Thinking about such things was my job before she was born, it was my job when she died, and it's my job now. Nothing changed.

Okay, everything changed. And, yeah, I do occasionally wonder what happened to Alice. But you notice I've never tried to find out. All I know is that she needed her own space to recover. For all I know she's now in Fiji, raising rabbits. Sorry, that's a joke—she'd know the reference. If she hears this, I hope she laughs.

That's why I don't want the Godiva chocolates. Food memories can be some of the strongest we have. It's part of what we have to deal with in feeding the crew. We want foods that give people comfortable memories of Earth but don't make them homesick. It's a fine line and we've spent a lot of time talking to food psychologists about it.

Yeah, that's a real field. The food-packaging industry loves them. So do advertising firms. But there's a lot of real stuff there, too. Like me and the chocolates. Alice and I used to break out a box after Thanksgiving dinner.

In theory, hydroponics is the best solution to the nutrition problem. You can grow wheat, rice, potatoes, carrots, beets, lettuce, tomatoes—even hot peppers and litchi fruit if you want. You also get all the psychological benefit of gardening: hands-on work with a strong connection to Earth while also looking forward toward farming beneath a new sun.

You can also grow tissue-cultured meat, though in practice it would be difficult to keep tissue cultures alive that long, without maintaining animals to harvest for fresh cells. Vegan is better. We've got frozen livestock (and pet) embryos, though, for when we get to Terra Nova. We hope they'll still be viable after that long. If not, we'll just stay vegan. Until we find live game.

The problem with hydroponics is it's a space and mass hog. Not because it takes huge water tanks. The hydroponics in this hotel are grown on trays of artificial soil a lot like garden soil but made of super lightweight pellets. The problem is that hydroponics isn't as super-efficient as people tend to think. It's better than Earth-based farming because you have perfect control of the weather, but plants only grow so fast and much of what they produce isn't edible, just cellulose. To get enough calories to feed one person with the most efficient crops, you need close to fifty square meters of trays. If you want a balanced diet—or treats like litchi fruit or jalapeños—you probably need to raise that to a hundred square meters. Even with lightweight soil, that's a lot of trays.

By some calculations, it would be easier simply to pack freeze-dried food even though you'd need a mountain of it—roughly 2,500 tons—just for the thirty people in the awake crew. But the problem with freeze-dried is that nobody's sure of the shelf life. Manufacturers claim at least twenty-five to thirty years, but what happens beyond that, they won't say. At 234 years, it's just too big a gamble.

That leaves us with the PBR: algae. That's still photosynthesis-based, but its advantage is we can eat the whole plant. We could, of course, make use leftover hydroponic biomass by throwing the inedible parts into a bioreactor to feed bacteria and yeast. That way we could convert them into edible food, without too much loss of biomass.

We'll be doing some of that, in fact, because we will be getting a quarter or so of our food supply from hydroponics. It's too useful as a combination nutritional supplement, treat, and hobby, but only if we also make use of the leftovers. But algae is more efficient, which is why it'll be our staple. And as I said earlier, the best algae-based foods are actually quite tasty.

And I think that kind of wraps it up. It's a hundred-year dream, only weeks from becoming reality. Again, I'm sorry if I got snappy, but I really resent it when folks try to tell me I'm running away from life. Sure, I'd trade my own berth away to have Molly back. But the Universe doesn't work that way, and sometimes you just have to move on.

Sure, you can keep the chocolates. Do you have someone you want to give them to? Someone special, I hope.

Now? Sure, I can meet her. Though I do have to get back to work eventually. And if she's looking for autographs, the Captain's is probably worth more.

I'm honored she thinks otherwise. Sure, let her in. What's her name?

I'll know? You overestimate the degree to which I've had time to pay attention to the last decade's vids, celebrity news, or—

Alice?

Alice!

What the hell is this about? Yeah, I notice she's gained weight. Holy—*Alice*?

Shut off that camera! What the hell kind of bushwhack is this? I thought this was a genuine interview, not some sort of—what the hell was that old guy's name?—Sprungeresque vid thing.

Of course I still love you. You're the one who needed to find yourself. After a few years I figured you either had or hadn't. Found yourself, that is. I'd already found...this...and it's important. Maybe the most important thing, ever.

Look at you? How can I not look at you? I don't care if you've gained—

Holy—

Damn. Why tell me this way, now, when you must have been planning for months and months? And why on camera, for God's sake, with him setting it up for maximum shock value?

What deal?

What's she talking about? Yeah, I'm sure it makes good vid material, and I'm sure the mission PR folks love it.

Oh. Yeah, I can see them going for that. A lit prof wouldn't exactly be a top priority, especially for a late add. Though when you think about it, we're going to *need* that type of connection with who we once were. Who we are. Who we will be. Hell, I don't know which. But we need people who do know. Because otherwise, who will we be? Though I'm not exactly objective. And I'm babbling like an idiot. Which, I'm sure was the whole point of this part of this...interview.

But she's definitely on board? Good, because babbling for you isn't the only thing I'm going to do. I'm also *angry*.

That's probably not what your audience wants to hear, so I'm sure you'll edit out this next line. To hell with that. I don't like being played. Even if the outcome is worth it.

But—and I guess you can keep this line in because I'm sure your audience will love it—Alice, I never quit loving you. Thirteen years and never a thought of anyone else. Only the mission.

So yeah, we can share a chocolate for the camera, if that's what it takes to get her on the crew. But can we share the rest of the box privately? We need to talk. And fatten up. We've got thirteen missing years to work through, and 234 more missing years to prepare for. Though at least those years we'll missing together. How much more weight to you need to gain, Alice? Maybe we ought to take the rest of the peanut butter with us, too.

Footnotes

Harry Jones (NASA Ames Research Center), "Starship Life Support," SAE Technical Paper 2009-01-2466, doi:10.4271/2009-01-2466 (2009)

Eric Blackstone, Mike Morrison, and Mark B. Roth, "H2S Induces a Suspended Animation—Like State in Mice," *Science*, Vol. 308 no. 5721 p. 518, 22 April 2005, DOI: 10.1126/science.1108581

Li, Jia; Zhang, Gencheng; Cai, Sally; Redington, Andrew N, "Effect of inhaled hydrogen sulfide on metabolic responses in anesthetized, paralyzed, and mechanically ventilated piglets." *Pediatric Critical Care Medicine* 9 (1): 110-112, (January 2008), doi:10.1097/01. PCC.0000298639.08519.0C. See also, Haouzi P, Notet V, Chenuel B, Chalon B, Sponne I, Ogier V et al., "H2S induced hypometabolism in mice is missing in sedated sheep." *Respir Physiol Neurobiol* 160 (1): 109-15 (2008). doi:10.1016/j.resp.2007.09.001. PMID 17980679

"Biosphere 2: the Experiment," http://www.biospherics.org/experimentchrono1.html. There is also a Wikipedia page, http://en.wikipedia.org/wiki/Biosphere_2

http://en.wikipedia.org/wiki/Sabatier_reaction. Other descriptions are also available online

Jennifer Ross-Nazzal "From Farm to Fork: How Space Food Standards Impacted the Food Industry and Changed Food Safety Standards," in *Societal Impact of Spaceflight*, Steven J. Dick and Roger D. Launius, eds., (NASA History Division 2007), *http://history.nasa.gov/ sp4801-chapter12.pdf*

See E. von Mutius, "Allergies, infections and the hygiene hypothesis—the epidemiological evidence," *Immunobiology*, 212(6):433-439 (2007). DOI: 10.1016/j.imbio.2007.03.002. The "hygiene hypothesis" is fashionable on childrearing websites, but is currently best viewed as speculative

"NASA Astronaut Sunita Williams Completes Triathlon In Space," *The Huffington Post*, Sept. 18, 2012, http://www.huffingtonpost.com/2012/09/18/sunita-williams-triathlon-space-nasa_n_1892893.html

See, e.g., Isabelle Boni, "Biological and Psychological Effects of Human Space Flight," *The Triple Helix Online*, 24 April 2012, http://triplehelixblog.com/2012/04/biological-and-psychological-effects-of-human-space-flight/, and sources cited therein

For a nice website on how to calculate such things, see: http://www.calctool.org/CALC/phys/ newtonian/centrifugal

Jared Bell, Dustin Lail, Chris Martin, and Paul Nguyen. *Radiation Shielding for a Lunar Base: Final Report.* NASA, May 10, 2011. https://docs.google.com/viewer?a=v&q=cache:4nmzh l9Gd4MJ:cmie.lsu.edu/NASA/Teams/Radiation Shielding for a Lunar Base/Spring/Final Report.pdf+why+is+hydrogen+the+best+radiation+shielding&hl=en&gl=us&pid=bl&sr cid=ADGEESi7VWBooIamTi7M

John W. Earl and Barry V. McCleary. "Mystery of the Poisoned Expedition," *Nature* 368, 683-684 (21 April 1994); doi:10.1038/368683a0

Nick Collins, "Test tube hamburgers to be served this year," *The Telegraph* (UK), 19 February 2012, reporting on a presentation at the 2012 meeting of the American Association for the Advancement of Science

Jones (2009), *ibid*, p. 12. Jones calculates 41.2 square meters for 86 percent of caloric needs using peanut, potato, and wheat.

Mountain House (backpacking food): http://www.mountainhouse.com/shelf_lif.cfm

Interstellar Probe (Don Dixon)

THE
STARSHIP
ERA

Where to go? The nearby stars first, of course—but considering the expense and investment of lives, we must first look for places worth the going. How to do this? Proximity may mean Alpha Centauri, where astronomers found a rocky planet in 2012, with expectations of more to come. Generally, many factors come into play.

Starship Destinations
Stephen Baxter & Ian Crawford

Introduction

If you've read the rest of this book, you'll have gathered that we believe we might, just possibly, be beginning to see how we might build an interstellar probe: a starship, uncrewed, but capable of sailing beyond the solar system. Where should we send it? And what mission might we expect it to fulfill?

Much of the rest of this book has concerned various aspects of the science and technology of interstellar flight. In this chapter we look at starship destinations.

Destinations

Our sun is one star in a galaxy of hundreds of billions of stars. Of that vast throng, our first interstellar probes could reach only our nearest neighbors.

How near? That of course depends on the performance of our spacecraft, and how long a flight we are willing to allow it. But the sheer scale of interstellar distances is daunting. This will be true even for an interplanetary culture of the future, a civilization capable of moving with ease between the planets of our solar system. The astronomers' convenient unit to measure interstellar distances is a light-year (i.e. the distance light travels in a year—some ten thousand billion kilometers). The nearest star system to our sun, Alpha Centauri, is a little more than four light-years away. By comparison Earth is a mere eight light minutes from the sun. The sheer distances and times involved in interstellar voyages will inevitably shape how such missions are planned.

Humans have in fact already launched five interstellar craft, of a sort: the unmanned Pioneer and Voyager probes, launched in the 1970s, and the New Horizons probe launched to Pluto in 2006. These will escape from the solar system after sling-shotting off the giant planets. None of them are heading for Alpha Centauri, but if they were it would take the Voyagers some seventy thousand years to get there. The Voyagers are still returning useful scientific data, but even after four decades' flight, though they are beyond the orbits of all the planets, they have barely left the sun's back yard.

Of course the star probes of the future, driven perhaps by the exotic technologies sketched elsewhere in this book, will be capable of much greater velocities, and much reduced travel times. Consider an advanced probe of the future, travelling at, say, a tenth the speed of light. This is about two thousand times as fast as the Voyagers. Even so it would take such a probe forty years to reach Alpha Centauri.

Suppose we allow a century or more, say, for the voyage: at least our grandchildren would be on hand to see the probe reach its destination…That gives a reachable range of ten light-years—and to be generous we will extend that a little, to fifteen light-years. This is a monumental distance in human terms, but is really little more than our own local neighborhood. By comparison the galaxy is a hundred thousand light-years across.

Given this neighborhood, where might we go?

The closest port of call: Alpha Centauri

The first obvious port of all could well be our sun's nearest neighbour, the Alpha Centauri star system: a name that has resonated through a hundred starship studies and a thousand science fiction dreams, including the movie Avatar, in which Earth is colonising Pandora, a habitable moon of a gas-giant planet called Polyphemus.

Alpha Centauri is a triple star system, with two close-bound sun-like stars, A and B, and a third companion, a red dwarf called Proxima, much farther out. Proxima, named for the Latin root of words like *proximal*, is actually the nearest star of all to our sun. The sunlike twins are only a few light-hours apart. If the brighter star, Alpha A, were in place of the sun, its companion, Alpha B, would be well within the solar system: it comes about as close to A as the planet Saturn to the sun, though it loops out to Pluto's distance. Proxima orbits some four hundred times farther away—in fact, it's so far out that there's some controversy about whether it's actually part of the Alpha system at all. Proxima is an unspectacular red dwarf, but of great interest to astronomers, for it is actually more typical a star than either Alpha A or B, or indeed our sun; seventy per cent of the galaxy's stars are like Proxima.

Alpha Centauri may be close, but what about planets there? It used to be thought that multiple star systems couldn't grow planetary systems because of the stars' mutual gravitational perturbations. But recent studies have shown that for planets as close to Alpha A as Earth is to the sun, B's gravity would have no significant effect on the orbital stability of A's worlds, and vice versa.

In fact, in one of the most exciting astronomical discoveries of recent years, in November 2012 researchers using instruments at the European Southern Observatory in Chile reported the detection of an Earth-mass planet orbiting Alpha Centauri B. If confirmed by future work, this discovery is of enormous significance for starship planners because it proves that the nearest star system to the sun has at least one planet. Moreover, although this particular planet orbits its star at a distance of about one tenth the distance of Mercury from the sun, and will be far too hot to be habitable, its presence suggests that other planets, as yet undetected, are also very likely orbiting Alpha Centauri B. It also increases the chances of Alpha Centauri A having planets, although this needs to be verified by future observations. So Alpha Centauri is not just be a twin star, but may host twin solar systems. It is a fascinating thought that if both A and B do have planetary systems these are so close together that humans might already have been able to complete interstellar journeys had this been our star system! And of course Proxima may also have a planetary system of its own

Still, despite its closeness and apparent richness, with three stars of different stellar types and perhaps three whole planetary systems, depending on what future discoveries reveal about the nature of the Alpha Centauri planets, it is still possible that we may choose another target as our very first port of call. Fortunately, given the rate at which astronomers are discovering planetary systems around other stars, we are likely to know this long before we are in a position to actually build a starship to go and visit them. In any case, there are plenty of other promising candidates in the neighborhood.

Beyond Centauri

Within our chosen cruise range of fifteen light-years from the sun, we know of no less than fifty-six stars, gathered into thirty-eight separate stellar systems. There may well be more to discover: more small, dim stars, even brown dwarfs, bodies halfway between gas giants like Jupiter and true stars. This is your local neighborhood, and the geography is three-dimensional. Imagine the stars hanging like lanterns around us in all directions in space, and at all distances from four light-years out to fifteen. All these stars are drifting together around the Galactic centre, their relative distances subtly shifting with time.

What are these stars like? Only two are like the sun—that is, of the same spectral class, the grouping astronomers use to classify stars of different sizes and brightness. One is Alpha Centauri A, the other Tau Ceti, another staple of science fiction adventures, which is a single sun-like star at a distance of about twelve light-years. Excitingly, very recent work (published in December 2012) has suggested that this star may also host a planetary system, with up to five planets with masses in the range of two to seven Earth masses. Two of these planets appear to exist close to the so-called habitable zone, where liquid water would be stable on a planet's surface. Just as for Alpha Centauri, further observations will be required to confirm the presence and nature of these planets, but we certainly will know a great deal more about them before we have to choose a destination for our starship.

The majority of the fifty-six stars within fifteen light-years, no less than forty-one, are red dwarfs, like Proxima. This reflects the distribution of stars in the galaxy as a whole. Of all these stars, which are we likely to choose as a first destination? This is likely to be guided by science priorities. The highest priority would be to study any systems which appear to be inhabited based on astronomical observations from the solar system.

We will be interested in a close-up look at the stars themselves, of course. If we go to Alpha Centauri we will find tantalizing samples of no less than three spectral classes. But it seems much

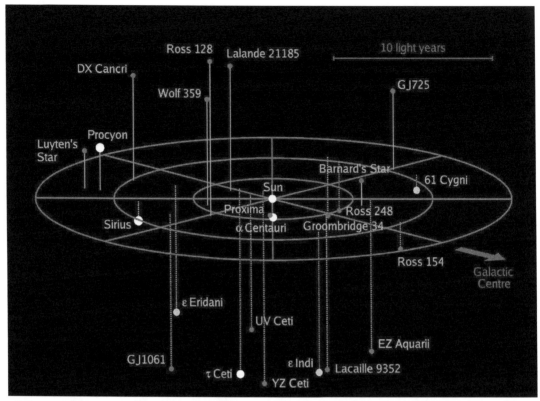

3D map of known stellar systems within a radius of 12.5 light-years of the sun. The color is indicative of the temperature and the spectral class —white stars are of spectral types A and F; yellow stars are type G, like the sun; orange stars are K type; and red stars are M dwarfs. The blue axes are oriented along the galactic coordinate system, and the radii of the rings are 5, 10, and 15 light-years, respectively. (**Courtesy of European Southern Observatory**)

more likely that we will primarily be driven by a search for planets: planets more or less like our Earth and the other worlds of the solar system—planets, especially, which may harbor life.

Since the sun has eight planets, and many smaller objects, we would expect other stars to have planetary systems too. But it's quite remarkable that only a few decades ago nobody had observed a single exoplanet, a planet of another star. Now we know of hundreds, with several thousand additional candidates awaiting confirmation.

A sky full of worlds

The challenge of detecting worlds beyond our own is formidable, because planets are small and faint compared to their parent suns. While there had been observations of planets orbiting pulsars (small supernova remnants) since the early 1990s, in 1995 the world was startled by the first observation of a planet orbiting a star called 51 Pegasi, a main sequence star (that is, a star in the middle of its hydrogen-fusing lifetime, like our sun) about fifty light-years away in the constellation of Pegasus. The discovery was made not with tremendous telescopes but with ingenuity, improved instruments and fine observation.

An exoplanet is generally detected not by observations of the planet itself, but by studying its effects on its parent star. As the planet orbits its star, so the star itself is pulled out of position, just a little, and if some of this motion is toward or away from Earth you can detect it with a subtle shifting of the absorption lines of the star's light spectrum due to the Doppler effect. Alternatively if the planet happens to pass across the face of its sun as seen from Earth—just like transits of Venus and Mercury, planets inside Earth's orbit—the dip in the star's apparent brightness can be detected. As you can imagine, these effects are very small and subtle. The more massive the planet, and the closer it is to its parent star, the larger the effect and the more likely it is that the planet will be detected. So there is an observational bias in our exoplanet detection; we are going to find more large, close-in worlds first, then small, farther-out worlds, and the statistics of the worlds we've found so far reflect that. Probably large planets are not any more numerous than small planets. The Holy Grail is to detect an Earthlike planet, orbiting at an Earthlike distance from a sun-like star, and analyze its atmosphere for traces of life. Given the pace of recent discoveries, it seems very likely that such planets will be detected within the next few years.

Nevertheless, we have enough data now to start to classify the exoplanets and make some tentative predictions. For example, current observations already suggest that at least 30 percent, and probably most, stars are accompanied by orbiting planets of one kind or another. Moreover, about 20 percent of stars presently known to have planets host multiple planet systems, so we may expect planetary systems approximately like our solar system to be common.

But what of habitable worlds? To be habitable by creatures like us, a world would have to be more or less Earth-sized, and be in, or close to, the habitable zone of its parent's star. Also the planet's habitability depends on the planet's atmosphere and other properties as well as the nature of its star. In our solar system, for example, Europa, a moon of Jupiter, may have a water ocean under a crust of ice kept liquid by tides from Jupiter and other moons. Life may be possible in this ocean even though Europa is far from the habitable zone (HZ) occupied by Earth.

Moreover, we no longer think a star with a habitable zone has to be like the sun. Even red dwarf stars, like Proxima Centauri, could conceivably have habitable planets. The planet would have to huddle close to the central fire, probably so close that it would be tidally locked like the moon around Earth, with a single face perpetually presented to the star. You would think that the dark side, a place of eternal night, would be so cold all the water, and then even the air would freeze out. However, calculations show that a sufficiently thick atmosphere, rich in greenhouse gasses such as carbon dioxide, could transport enough heat around the planet to keep this ultimate freeze at bay. The good news is that, since seventy per cent of the galaxy's stars are red dwarfs, with this model we have multiplied the potential number of habitable worlds many times over.

What of the 56 stars within our range? In addition to Alpha Centauri B and Tau Ceti, discussed above, at time of writing two more of these stars are known to have planets. Epsilon Eridani, a star eleven light-years away not unlike the B member of Alpha Centauri, has a planet about half as massive again as Jupiter, following a very elliptical orbit which takes it as close to its star

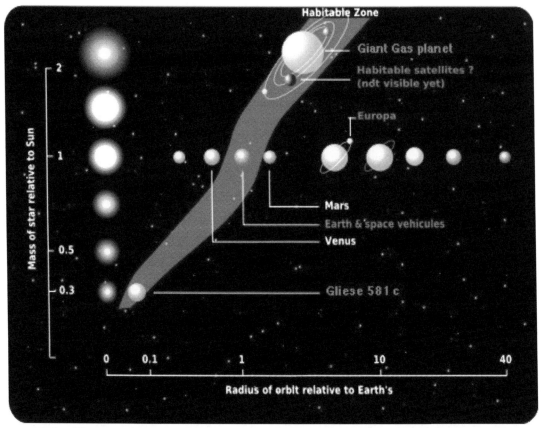

Habitable Zone diagram. Select a star mass. The HZ is the diagonal swath. Find the distance of the HZ from it along the horizontal scale. solar system planets shown for comparison. (Kepler Probe Group)

as Earth is to the sun, and out as far as Jupiter's distance. Unfortunately this particular world's orbit does not bring it within its star's habitable zone, for realistic atmospheric compositions. Meanwhile a red dwarf star known by its catalogue number of GJ 674, nearly fifteen light-years out, has a "super-Earth," a world of at least eleven Earth masses, orbiting at a tenth of Mercury's distance from the sun. As we've seen red dwarfs may be able to host habitable worlds, but, like the planet orbiting Alpha Centauri B, this particular planet is probably too close to its parent star to be habitable.

Given the wider results from the search so far, it seems highly likely that we will find that more of our neighborhood stars host planets. The present statistics predict that at least 20 of our 56 close-by stars will have planets of various kinds: enough to give us a healthy choice of destinations. Quite likely by the time the first space probe is launched, studies made from the solar system will have delivered a comprehensive catalogue of the worlds of nearby stars; in fact it is quite possible that we will know more about such worlds before the launch of the probe than we knew about Mars, say, before the launch of the first interplanetary probes in the 1960s.

But what might these planets be like?

Of the eight planets in our solar system, four—Mercury, Venus, Earth, and Mars—are rocky, dominated by iron and silicate rocks, with a relative trace of water and gases, if any, on their surfaces or in their mantles. The other four—Jupiter, Saturn, Uranus, and Neptune—are gas giants, swollen masses of hydrogen, helium and other gases, with liquid, ice and/or rocky cores at their very centre. There is a gap in the mass range between Earth, the largest of the rocky worlds, and Neptune, the smallest of the gas giants, which has the mass of about seventeen Earths It's now clear that planets can form in this gap: super-Earths, like the world of GJ 674 mentioned earlier. A super-Earth would be a spectacular place, despite the higher gravity; the larger the world, the more geologically active it is likely to be—as Earth is much more active than Mars or the moon—so expect fiery worlds of tremendous volcanoes, or oceans hundreds of kilometers deep.

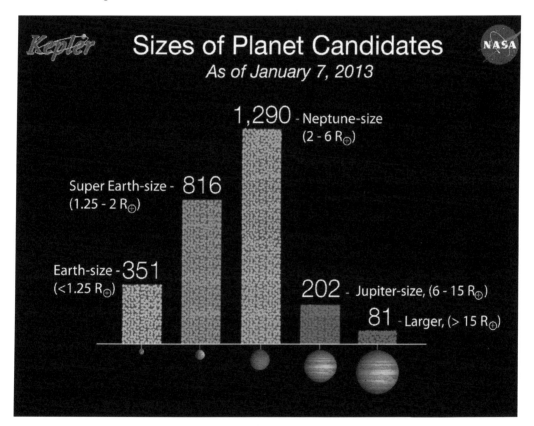

This histogram summarizes the number of Kepler planet candidates as of January 2013. The catalogue contains 2,740 potential planets orbiting 2,036 stars. **(Kepler Probe Group)**

What, then, will we choose as our first interstellar destination? Probably a star, or star system, with a range of interesting worlds, from small Mars-like worlds, Earths, super-Earths, and gas giants. As we've seen, a multiple star system like Alpha Centauri could even host multiple planetary systems, giving our mission more targets to study.

But overriding any other consideration is the possibility of finding life on one of these worlds. It is quite feasible that we will be able to detect evidence of such life telescopically from the solar system. Earth's atmosphere contains 20 percent oxygen, put there by the processes of life; if all life disappeared tomorrow, oxygen would all but vanish from the air, rusting into the surface rocks in a few millennia. An alien astronomer equipped with just that one observation, an atmosphere rich in oxygen, would be able to make a good bet that Earth hosts life, especially if there is also evidence for water vapor in the atmosphere (itself an indicator of probable liquid water on the surface). A detection of life on an exoplanet would surely move it to the top of the target list for any probe.

Leaving the solar system: Science en route

Suppose, then, we build our space probe, capable of 10 percent of light-speed. We send it on its way, and sit down to begin a century-long countdown until its arrival at its destination…. Happily there is plenty of good science to be done to keep scientists on Earth busy, even while the probe is en route between the stars.

A probe launched at a tenth the speed of light from the orbit of Earth would cross the orbit of Neptune, the outermost planet, in about forty hours. In ten days the probe would reach the heliopause, the point where the wind of particles that blows from the sun is so attenuated that it is lost in the wind that blows between the stars. (This is as far as the Voyagers have reached, in fact, after four decades of travel.) In the deep dark farther out many objects orbit our sun; in sixty days our probe would pass the outer limit of the Kuiper belt of icy worldlets, and then enter the Oort cloud, the sun's diffuse halo of comets. The ice moons and comets are very sparsely spread, and it is unlikely that our probe will pass close enough for an encounter—though it would be thrilling if it happened. Besides, we would expect that, by the time our probe is launched, the sun's immediate back yard, containing such objects, will have been well explored by precursor missions.

It will take our speeding probe no less than eight years to pass out of the Oort cloud, and enter true interstellar space at last. This is not an empty void, but is permeated by the interstellar medium, a thin mist of gas and dust particles This medium has a three-dimensional geography of its own, which we are beginning to map, diffuse clouds we can glimpse by measuring how they absorb the light of background stars. Our sun is currently located within a very low-density region known as the Local Bubble, a structure some three hundred light-years across. Inside the Local Bubble, within fifteen light-years of the sun—our nominal range—are at least six low-density clouds (with densities of about one hydrogen atom or ion for every ten cubic centimeters), jostling like footballs in a net sack. Our sun is in, or possibly just on the edge of, one of these clouds known as the Local Interstellar Cloud. We can map, roughly, which clouds a probe would pass through en route to different stars.

All this is important because the interstellar medium may be of use to our space probe—or a hazard to it. We may try to use the medium as a means of slowing down the probe at its destination, for example by braking against the weak magnetic fields supported by the medium. On the other hand we need to shield our craft against the interstellar medium, even though it is very diffuse; you'd need to sweep a volume of space the size of Earth to fill a couple of teaspoons with interstellar dust (six or seven grams). But for a spacecraft moving at a tenth of light-speed, an impact with a grain just a tenth of a millimeter in radius would deliver a wallop equivalent to a shell weighing a kilogram fired at supersonic speeds.

Properly shielded, our probe will, however, be able to conduct some good science en route, as it becomes the first probe able to make in situ measurements of the interstellar medium. It could carry dust and gas analyzers, magnetometers to study local magnetic fields, and cosmic ray detectors to study high-energy particles coming from remote, dramatic events such as supernovas. Instruments of this kind are already routinely carried by spacecraft on missions to the outer solar system, such as NASA's Cassini probe at Saturn. Of course the esoteric delights of the interstellar medium may not hold the attention of the watching world after the first few years, but the information gained will nevertheless be of great scientific interest, and will be crucial for the planning of future interstellar journeys.

Eventually, however, the interstellar phase of the journey will at last come to an end. Perhaps eight or ten years out from the probe's destination, the Oort cloud of comets surrounding the target star will be reached, and attention will turn to the scientific investigation of the target system.

Exploring a Stellar System

It seems likely that, given the way we have already explored our own solar system, our mother ship will release a number of subprobes, each dedicated to a particular target world, or worlds. The way to explore a planetary body is to give it time, we've learned: to lodge a body in orbit in order to study its surface and atmosphere comprehensively, to land on the surface to conduct close-up investigations of surface features. It takes time to watch a world turn, to see its seasons come and go and to record active meteorological, geological, and biological processes.

Any gas giants might well be studied with probes not unlike Galileo and Cassini, sent to Jupiter and Saturn respectively. These probes spent years orbiting their target worlds, studying their atmospheres and magnetic fields, and touring their moons. Cassini also carried Huygens, a lander that penetrated the clouds of Saturn's moon Titan, while the mother craft studied the moon's overall geography using radar to peer through the clouds during its multiple fly-bys. Galileo carried a probe that entered the atmosphere of Jupiter itself, making direct measurements of the composition of the atmosphere of that giant world.

Rocky worlds might be explored as we have already studied Mars, Venus, Mercury, and the moon. Long-stay orbiters like the current Mars Express and the Mars Reconnaissance Obiter will map the worlds with cameras, radars, and altimeters to study surface heights with reflecting laser beams. Spectrometers, which analyze reflected and scattered light, will study the composition of atmospheres and oceans. Landers like advanced versions of NASA's Mars Exploration Rovers and Mars Science Laboratory will crawl over the surface of the worlds. Perhaps we will also send airplanes or balloons to fly in the atmosphere. The landers will measure heat flow in the ground, analyze surface mineral compositions, and study geological activity with seismometers and other instruments. Weather stations will study the climate over years; high-resolution lenses will study the microscopic structure of the rocks.

Of course the study of any life forms will be a high priority. Since we can expect to know in advance from remote astronomical studies whether life is present on a given planet, and perhaps something of its nature, probes with suites of instruments appropriate for the study of such indigenous biology will have been carried in the payload from Earth. These might range from microscopes to mass spectrometers capable of analyzing organic molecules, and life-detection suites such as those carried by the Viking Probes to Mars in 1976. Depending on the circumstances we may have to seek life on the surface, drill for it in the deep rocks, find it floating in the air, or even explore the depths of oceans.

In addition to searching for, and attempting to characterize, active biological metabolism, it will also be important to look for evidence of extinct life (fossils) in order to try to reconstruct the evolutionary history of any life discovered. It will also be important to search for evidence of fossil life on apparently lifeless planets—just because a planet is lifeless when the probe from Earth arrives doesn't mean that it will always have been so.

It is looking increasingly likely that so-called super-Earths (planets with masses of up to ten times Earth's mass) are common in the galaxy. Indeed, super-Earths may turn out to be as common as worlds the size of Earth and Mars, although the difficulty in detecting smaller planets makes the statistics uncertain. In any case, it appears likely that we will encounter massive, super-Earth, planets orbiting nearby stars, which will be exciting for geologists and biologists alike as we won't have encountered such objects before. In order to explore such worlds the starship engineers of the future will face new challenges, of worlds with higher gravities, thick, turbulent atmospheres, crushing pressures, and perhaps super-deep oceans.

Other practical issues will tantalize and fascinate the mission designers. Any probe is going to have a limited payload capacity. Not every aspect of a complex stellar system can be explored on the first pass, and the designers will be forced to make priorities. The advance of technology, and in particular miniaturization, may help with this; for a given mass, the probes of the future will surely be much more capable than ours. But any probe will probably need a minimum mass to be effective. A lander on a world with an atmosphere will probably need a heat shield to penetrate the air, a parachute or rocket system to reach the ground safely, communications antennas of some minimum size—and, if it is to conduct activities like rock-sampling or drilling, sheer weight as anchorage. Perhaps the subprobes scattered over the target worlds will share a single communications channel to Earth through one big antenna system on the mother ship. The mother ship itself could conduct space-bound studies like an inspection of the parent stars, or a slow tour of any asteroid belt, like NASA's current Dawn mission.

Artificial intelligence to some level will probably also a necessity. Most modern spacecraft are basically run from Earth, by human controllers. Asteroid explorer Dawn, for example, is only smart enough to put itself into safe mode if it loses touch with Earth; every firing of its ion rocket needs a direct order from the ground. With a minimum of an eight-year round-trip communications time-lag—it takes four years to get a radio signal to Alpha Centauri, remember, and four years back again—this kind of command set-up won't be practical for an interstellar

A crewed starship, based on magnetic mirror fusion, approaches a moon of a superJovian planet of a star system. The rotating habitat is shielded against micrometeorites and incoming high-speed dust. The long multiple-mirror reactor is cooled by large planar radiators. The system has previously been explored first by fly-by via beam-driven sailships, later by orbiting probes. (**Adrian Mann**)

probe. To some extent the probe will have to be smart enough to conduct its own operations, overcome problems, fix technological flaws—perhaps even devise its own science programs, within overall objectives set by the mission controllers.

The most exciting discovery of all, of course, would be the discovery of intelligent life, on a moon or planet of the target system. If we were to have detected such intelligence from our solar system—perhaps by detecting their radio signals, just as the SETI (Search for Extraterrestrial Intelligence) researchers have sought for decades from Earth—we would surely make studying such intelligence the top priority, and send out a dedicated probe. It is true that SETI has not detected any extraterrestrial radio signals in fifty years of listening, but it's quite possible to imagine even advanced cultures that would not be detectable without a close-up search: an equivalent of the Roman empire, perhaps, without any technology like radio.

Given what we know about the evolutionary development of intelligent life, and the more than three billion years it took to evolve on Earth, it is very unlikely that we will find extraterrestrial intelligences within fifteen light-years of the sun. It would therefore be a long shot to send out a probe specially equipped to find intelligence. But a probe sent out for general scientific purposes could stay on the lookout for traces of extraterrestrial intelligence found in the course of other investigations. For example it could inspect planetary surfaces from orbit for artifacts or structures. This has been proven to work on Earth's surface; high-resolution visual images returned by the US CORONA surveillance-satellite program of the 1960s have been used to make new archaeological discoveries, having captured images of structures previously unknown. Landers, meanwhile, capable of close-up inspection, could detect very subtle, even ancient traces of past civilizations.

What should a probe do if it did make contact? We probably ought to decide this before we send out such probes. At the minimum we could load the probe with Voyager-type plaques and discs full of information about Earth, as a gift to any contacted culture.

Even without intelligent life, the exploration of a new stellar system full of exotic worlds is a tantalizing prospect. In a way this section has been the least speculative part of this chapter. We know how to explore solar systems, because we've already explored our own; all we need is a way to get out there, among the stars.

One Small Step...

What of a human future among the stars? After all, a man set foot on the moon only a decade after the first unmanned probe reached that body—although it has already taken a lot longer for humans to follow the first probes to Mars of the 1960s.

If it's a challenge to send a few tons of instrumentation to a nearby star, it will be orders of magnitude harder to send humans—big, wet objects requiring food, water, sewage processing, and protection from all the hazards of interstellar flight, from a lack of gravity to protection from radiation. The starship VentureStar, shown in the movie *Avatar*, addresses some of these challenges. To save on mass, only a handful of the crew are awake during the journey, and they live in a habitat at the end of a rotating arm; the spin providing a substitute for gravity. Most of the passengers sleep out the long journey in cryogenic storage. We don't yet know if such cryogenic storage is possible with humans, but this would be an important area for future research.

Even if we can't send human beings to the stars, we can certainly envisage spreading Earth-life to other planetary systems, perhaps in the form of microorganisms specifically genetically engineered to survive the journey and thrive on a particular target planet. Clearly there would be ethical issues that would need to be weighed carefully before undertaking such a project. While seeding barren planets with Earth life could be seen as a positive good (if one accepts that life, with its vast potential for growth and diversity, is preferable to non-life), interference with indigenous ecosystems will appear morally reprehensible to many (including the authors of this chapter). When the time comes, it will be important to ensure that appropriate legal and ethical frameworks are in place to deal with such issues (perhaps something akin to Star Trek's Prime Directive).

However, if only we could reach the stars, at least we now know there will be many interesting places to visit. We are already detecting Earth-sized planets orbiting in the habitable zones of their stars (though none, yet, among our close-neighborhood stars). As we've seen, even red dwarf stars like Proxima, the dominant kind in the galaxy, could potentially host habitable worlds. Conceivably we may find worlds where it would be safe to walk around with just a facemask. Perhaps in the farther future we may be able to terraform these worlds, that is adjust their atmospheres and oceans to make them fully habitable by Earth life, and ourselves. Alternatively, as the philosopher Olaf Stapleton suggested in the 1930s, we may choose to adapt human physiology to the prevailing conditions—effectively creating different sub-species of humanity as we, and our evolutionary descendents, spread through the galaxy.

Maybe one day our first unmanned precursor probe, long inert, will be retrieved by future star-farers and brought back to an honored place in a museum on Earth. And maybe the nearby stars, hanging like lanterns in the sky all around the sun, will become sources, not just of starlight, but also of messages home.

Acknowledgements

The authors thank Drs Jill Tarter and Jim Benford for comments, which have improved the quality of this chapter.

Further Reading

Crawford, I.A., "The Astronomical, Astrobiological and Planetary Science Case for Interstellar Spaceflight," *JBIS* 62, 415-421, 2009.

Crawford, I.A., "A Comment on 'The Far Future of Exoplanet Direct Characterization'—The Case for Interstellar Space Probes," *Astrobiology*, 10, 853-856, (2010).

Crawford, I.A., "Project Icarus: Astronomical Considerations Relating to the Choice of Target Star," *JBIS* 63, 419-425, (2010).

Dumusque, X. and colleagues, "An Earth-mass planet orbiting alpha Centauri B," *Nature*, 491, 207-211, (2012).

Tuomi, M. and colleagues, "Signals embedded in the radial velocity noise: Periodic variations in the tau Ceti velocities," *Astronomy and Astrophysics*, Vol. 551, Paper ID A79, (2013) http://adsabs.harvard.edu/abs/2012arXiv1212.4277T

Generation starships have a long history in science fiction, mostly because they allow stories featuring societies evolved in wholly artificial habitats, and the cultural problems that emerge. This story looks at what happens when such a ship finally arrives.

The Heavy Generation
David Brin

Jason didn't notice the hands at first, popping out of nearby corridor walls. He was too busy digging through a planting bed, where spindly, multicolored flowers filled the sluggish air with a sweet, almost cloying perfume. So intent was he on this task that that one hand snaked all the way across the garden passageway without notice, casting elongated, slithery shadows from the overhead lighting panels as it reached out to tap him on the shoulder.

"Wha—?" Jason leaped almost his own height. He floundered in mid-air several seconds before landing on the gravel path, facing the hand.

"Oh, it's you," he sighed, heart still pounding. It was futile fighting it, even with body-control key words. Not with several of the tentacle-things now waving back and forth in front of his face.

"What do you want?" he demanded. Neither Jason nor any of his friends had ever figured out how the translucent hands oozed out of blank walls, almost anywhere in the ship—nor, for that matter, how the wall-hands saw. They possessed no eyes.

The one closest to him was blue. So was its arm, and the wall panel it extruded from. The hand turned, aiming its palm at Jason.

"I emerged to inquire what you're doing here."

They came in all sizes, with voices ranging from the deep basso of industrial-units all the way to delicate micro-hands, who sang piping soprano melodies as they worked in the Bio Lab or Optics Repair Shop. But always the tone was mellow, smooth, imperturbable. Jason's friend, Stacy, figured they spoke by vibrating their "palms" rapidly, setting off pressure waves in the air.

"I'm not doin' anything," Jason muttered, scuffing dirt over the place where he'd been digging. "I just wanted to look at the roots of these rose bushes."

"Flower beds are the responsibility of gardener robots, Jason," the blue hand chided. "They do not concern you."

It would be serious if Jason were caught interfering in ship operations. But no one had ever actually told him not to dig in flower beds. "Is it forbidden?"

The blue hand folded its fingers against each other as if pondering a puzzle. Probably consulting the starship's central computer.

The pink wall panel across the corridor formed a dimple. The bulge lengthened into a tube that became an artificial arm, with another five fingered hand at the end. Almost as soon as the new one had formed, a soft ripple crossed its palm, its voice pitched a little higher.

"No, not expressly forbidden. But crewfolk dislike direct contact with soil—"

"Well I'm never going to be crew, am I?" Jason interrupted, bitterly. "I'm an H-kid, and I'm supposed to learn all about dirt."

That got them! The hands seemed to look at each other in confusion, and Jason felt a moment's satisfaction at stumping the things. Down at Fullweight Level, he and the other H-kids were kept at their lessons by these hands, or ones just like them—studying Natural Agriculture, and Mining Technologies, and other topics all about how to work and live amid the gritty stuff of planet surfaces.

Jason didn't mind school. But it burned, knowing the subjects he and the other H-kids learned were thought vulgar by the real masters of starship Havaiki.

A pink arm spoke. "Crewfolk also use this park, Jason."

"I know, they come down here for exercise," he answered. "That's why I chose this corner passage. I know it's rude to dig soil in front of tall people."

That wasn't his only reason. Curiosity had brought Jason to this hidden corridor of Half-G level. Perhaps in an out of the way place he might be able to pursue his investigations without interference.

No such luck, though. Jason watched the blue arm extend even farther from its wall. The hand seemed to sniff at the disturbed soil. "What were you trying to discover here, Jason?"

"I—wanted to see how plant roots differed from those down on our farm."

Jason wasn't a very good liar. His real interest hadn't been roots, but the tiny, almost invisible glass fibers that laced the flower bed. He suspected they were part of the system the robots used to monitor every component of the ship, from the massive engines all the way down to each blade of vegetation in the hydroponics gardens. There were no glassy fibers on the farm down at Fullweight, in the ship's outermost shell, where he and the other H-kids plowed thick black soil and practiced raising wheat, corn and amaranth. Those either thrived or failed at the skill or clumsiness of the young farmers themselves. On Rapanui, there would be no robots or wall-hands to help.

He felt perspiration break out as the blue hand lifted to regard him again, eyelessly. "Biology at half gravity isn't part of your curriculum, Jason."

"Why not?"

"It isn't likely to be useful on Rapanui."

"Well, we're supposed to learn all about plants and stuff, aren't we?" Jason said, defensively. "I just wanted to see what the differences were—"

The hand wagged its index finger. "You know perfectly well there will only be one level of gravity on the planet, the level you experience down at Fullweight. You don't need to know anything about Half G."

"Of course not," Jason muttered. "Like we don't need to learn electronics, or atomics, or navigation, or any of the other things Ship Kids learn."

The hand didn't seem to notice his sarcasm. "Exactly. Ship children and ship teens learn what they'll need in order to become crewfolk, like their parents and grandparents before them. Then, unusually, it paused.

"On the other hand…." It performed a flourish, finishing palm upwards. "…you and your friends are destined to be Colonists. You are the fulfillment generations of shipfolk have lived for. You have no need to bother yourselves with the tedious technologies that run starships."

Fancy words. Once upon a time, they would have satisfied Jason. But no longer. "How can you be sure? Rapanui's a new world. We'll be the first people to see it up close. There may be more unpleasant surprises."

He didn't bother to specify. Only a month ago, a bio-astronomer peering ahead at the planet ahead discovered that its ozone protective layer was unexpectedly weak, necessitating changes in the colonization plan.

"Then what about the atmosphere? Sure Rapanui has lots of oxygen. But it also has way more carbon dioxide than there was on Earth, or there is aboard the Havaiki."

"That was known from distant observations, Jason, twenty years ago, as our ship began its long deceleration into this system. And hence your generation of H Kids was genetically enhanced to withstand such high levels of CO_2, and to host lung symbionts that help prevent hyperventilation."

Yeah, but you screwed up with the ozone detection, Jason thought, keeping that complaint silent. *And what else will surprise us, when we're down there? Especially as the Havaiki crew rushes to forget about us and streak back into dark space?*

"Okay, so now we're growing crops in high-CO_2 greenhouses, stuff that ought to grow on Rapanui…and that's why I want samples from this part, for comparison!"

"There will be no low-CO_2 crops on Rapanui," the hand reiterated. "You have no need or use for such comparisons."

"We ought to be trained with more science! There's no telling what information will be relevant."

From behind Jason there came sudden, high laughter.

"Relevant…Rel—lev—ANT. Dirter lives in dirt like an ant!"

The new, taunting voice took Jason by surprise. He tried to spin around, but missed his footing in the tricky gravity. During his slow fall, he glimpsed several tall figures, standing at the turn in the corridor. One face leered from a height no greater than his own—floating yellow hair framed sharp blue eyes.

His tumble was stopped by tight grips on both shoulders, steadying him back onto his feet. Jason yanked his left arm away from the blue wall hand. But his other was held by one whose long, tapered fingers could never be the product of some wall. They were stained gray from contact with fuels and lubricants and other things Jason could only imagine.

A crew person. All adults were Tall Men or Tall Women. This one towered almost two-and-a-half meters, or eight feet, from sandals to sandy hair, and he was as slender as the nearby low-gravity willows.

"Hang on, boy." It was a deep, amused voice, booming high over Jason's head. Jason's nose barely rose higher than the shimmering belt buckle, set in a uniform of gold and silver. "This isn't your territory, y'know."

Jason peered up at a long, narrow face he had seen all too often—though mostly on the screen of his training monitor. It was Assistant Mechanic Koolhan, his tutor in Basic Metal Skills. The man smelled as if he had been drinking, and he needed a shave.

Jason bowed, respectfully. "Thank you for catching me, sir. But it was just the surprise. I'm—I'm actually pretty good at moving up here in Half G…."

Again, high, taunting laughter interrupted him. A face popped around the adult's side, the golden hair and blue eyes Jason had seen before. "Up here? Did you hear that, Uncle? He said *up* here! Dumb old dirter doesn't even know it's *down* to get to Half G!"

Jason felt his face warm. The girl stood centimeters taller than he, yet she wasn't more than eight standard years old. Already she was spurting toward normal height. Normal for crewfolk, that is. Assistant Mechanic Koolhan laughed and gave Jason a friendly shove. "Now, boy. Don't let a brat's teasing get you. You may not know up from down, but you've learned a hell of a lot more about *iron* than she'll ever know, or about farming or—"

"Or dirt!" Koolhan's niece interjected before ducking out of sight again, giggling. Koolhan apparently had no intention of restraining her behavior. Jason's gaze dropped to the gravel path, but even there he saw reminders. The Tall Man and the Tall Girl both wore soft sandals to protect feet that looked much like hands. Jason had seen crewfolk athletes play agile games in the Zero-G parts of the ship, catching balls with their tapered, supple toes as easily as their fingers.

In comparison, Jason's own feet seemed like flat paddles, fit only for stomping mud down at Fullweight.

Of course by now the wall hands had melted back into the corridor panels. They were great for chiding and nagging and moralizing down below, where they had been mother and father substitutes all his life, feeding and rocking and cleaning all the H-Kid babies from the day the first was delivered, seventeen standard years ago.

But would a hand ever stick up for an H-kid against a crowd of bullying Ship Teens? Hah.

Jason bowed again to Engineer Koolhan, ignoring the gangling girl who peeped around to stick out her tongue at him. "It's getting toward the end of my free period, sir. I'd better be getting down below again."

"Hmph." Koolhan rocked his head. "Yes, you'd better, hadn't you? You have an assignment due tomorrow, I think?"

Nodding sternly like that… it was as almost if Koolhan thought himself a real professor, like the ones who taught ship-kids. Of course, he was only a junior mechanic, drafted into an unwanted job—teaching skills nobody cared about to stunted, dwarfish, overly-muscled children who nobody wanted to have anything to do with, anyway. Who would be gone soon enough, and forever.

"Yes sir. Goodbye then," Jason said. "And goodbye to you too, Elice." He said the name of Koolhan's niece, and saw her blink briefly in surprise. She hadn't expected courtesy from him in reply to taunts. Her momentary astonishment offered Jason a small pleasure.

It didn't last long. She stuck her tongue out and sneered, "Good bye, root-stinker!"

He pretended not to notice. It was hard enough walking away with any dignity in the awkward lightness without tripping over his own awkward, club-like feet.

"Oh, so there you are!"

A crowd of gangling figures rounded the copse of trees that sheltered this little garden from the rest of Half-G Park. None were as tall as Assistant Engineer Koolhan, but the shortest topped Jason by at least a foot. Three boys and two girls approached, giggling and tugging at each other—all but the girl in front, who seemed quite cross, brows tight above narrow eyes. She stepped ahead of the other ship teens.

"Elice, come along. Mother and Dad were worried."

The blonde child made a sour face. "I was with Uncle Ernest...."

From the look on the older girl's face, Koolhan wasn't her idea of an ideal babysitter. Elice seemed to sense this, and so tried to direct her sister's ire elsewhere.

"We were on our way back, but we were stopped by that dirter!"

Until then, Jason had hopes of remaining unnoticed. In fact, he'd been edging away and almost made it to the cross corridor. Now though, the crowd of kids his own age turned to look at him. Look down at him, from their towering heights. He straightened, and tried to look as if he didn't give a damn what they thought.

Several of them seemed embarrassed by the child's bad language, but one of the boys glanced toward Jason and made a rude noise at the back of his throat.

"Colin, stop that!" The older girl said, sharply. Her hair was darker than her sister's, and she wasn't all that tall, as ship teens went. She barely topped six feet. Her friendly smile was the most surprising thing Jason had seen in weeks.

"I apologize, Jason Alani," she said formally, bowing noticeably at the waist. "Please forgive my sister's rudeness. She will be corrected, I assure you."

The other teens seemed as surprised as Jason. Blinking, he managed a bow of his own.

"Think nothing of it, Melissa Posner. Please don't bother on my account."

There was a tense silence. While earlier the teens would have just ignored him, or snubbed him without much thought, now they stared. One tended to look when an H-kid, a "dirter," was treated so courteously by the daughter of the Chief Scientist of the Havaiki. Jason's ears burned.

He might pay for this day, if the boy at the back of the group ever got the chance. Silver-haired Zack Sumanunu was obviously memorizing Jason's features.

But that hardly mattered as Jason bowed again, said goodbye, and turned to walk away. He'd been struck numb by that smile—and by the fact that Melissa Posner remembered his name at all!

Six months ago they had met in an exercise area down on Half-G deck, where crewfolk seldom went at all, except on serious business. Melissa had asked him questions about life down at Fullweight—for a paper she was writing in one of her classes, she said. Their conversation had only lasted an hour or so. It never occurred to Jason she'd remember.

Jason crossed the open, grassy area of Half-G Park in long, loping strides, as if he had been born to walk this way—born to fly. For once he didn't care when Tall Folk stared as he passed, or turned away to pointedly ignore him.

In fact, Jason did not feel "down-to-Earth" again until the elevator deposited him back in his own realm, back where the floor seemed like both friend and foe, helping you to stay "grounded"… while constantly ready to *smack* you if you stumbled.

Rapanui

Fifty years, and he still found surprising the color of the sky.

I suppose it must be ingrained, genetic, to expect Earth blue, Jason thought, blinking for a bit as he stepped out from the Great Canopy, adjusting his hat and sunglasses. *Even our children and grandchildren speak of it. But that's not a bad thing.*

He approved of anything that reminded the kids to cover up, when they went outside. It would take many generations to modify human skin to match that of native life, whose leathery toughness could withstand this planet's sleeting ultraviolet. Till then, parents would chide, while ground-dwelling folk became quilts of force-grown skin patches. Jason's latest graft itched miserably.

The bio-astronomers should have known this about the world ahead, as the Havaiki got within a light-year of Rapanui. There were hints of a weak ozone layer. But they just didn't care enough. Jason quashed an old resentment.

Things could have been worse. At least the gengineers tuned our lungs right, before my generation was born, to breathe CO_2-rich air. He inhaled deeply, enjoying the rich aromas of home, still relishing the contrast two weeks after returning from the sweaty aroma and metallic tang aboard Havaiki.

The ship was visible now, a glitter on the horizon. A perpetual reminder of both promise and despair. And Jason hoped the colony council would soon stop sending him out there. *It's not just the smell, or the endless chain of crises, or the stunning obstinacy of Zack Sununamu's faction.*

I guess I just don't find space all that interesting, anymore.

Speaking of which…where is she? Shouldn't the shuttle be visible, by now?

He scanned the sky, but saw no sign. No telltale glitter. And so, Jason slowed to a saunter, traversing a gentle trail over the hill sheltering the landing port from the Pounder Town.

Is that a breeze? he wondered, standing at the crest. Raising his hand to feel a soft push of dense air, Jason turned toward the settlement, whose canopies looked taut and secure—especially one huge, translucent tent of delicate-looking gauze that sheltered all the buildings and people of the central village, the way an adequate ozone layer used to do on Old Earth. One of many conveniences that the ancestors were said to have taken for granted, on that improbably hospitable homeworld.

Jason saw farmers move to tighten the pole-arches and guy wires that kept filmy fabric hovering over some fields of unmodified Terran plants, salvaged from the starship's hydroponics bay. Most crops, fortunately, no longer needed protection. Thanks to decades of intense mod work on the corn and amaranth.

Which never could have happened, if Havaiki had left on schedule. *We'd have all died by now…we H-kids…another failed colony.* Nor would there be a Tech Workshop, or a hydroelectric dam, gleaming in the distance, below the forest where Jason's mine and smelter lay.

Rapanui tested human ingenuity, but so far, there had been compensations: like a dense, languid atmosphere that pressed, rather than shoved, down here at sea level, while high mountains were lashed by wind and rain. An acceptable arrangement, because it allowed the canopies. Bad though, when those stratospheric tempests tossed the Havaiki shuttles like matchsticks.

Only three were left. Jason scanned again. This one was late.

He quashed worry while passing between warehouse tents, where a mixed crew of ground-pounders and skyfarers sorted through cargo, consigned for Havaiki. Two ship-guys—gangly and weary looking—moved about on stand-scooters that creaked after five decades. *Soon, the last scooters will wear out and they'll have to stomp about like the rest of us,* he thought, without any relish or thought of long ago slights. So far, a series of fixes—leg braces, muscle-stims, growth factors and other miracles helped enable some refugees to work and move around down here, though clearly, many of them considered it a kind of hell. One that was aggravated when they saw their own children growing up squat, muscular and happy with dirt.

Adding to their discomfort, the two tall men kept adjusting respirator masks. Blood symbionts and lung implants weren't sufficient for crew—most of the them—to breathe safely the atmosphere of this world. Barring some new tech miracle, that would be so for the rest of their lives spent down here on Rapanui.

Jason recalled when—five shuttles ago—a fight broke out between dirters and skyfaries (to use each side's epithets) in this very same tent. The spacers' exoskeletons gave them an advantage in raw strength, that the grounders soon overcame by jumping the refugees from behind and ripping their masks off. And Jason had to serve on the panel assigning punishments for that nasty business. Well, at least no one died. And who ever said it would be easy to establish a new home for humanity?

We must build some more air-tight quarters and work spaces, Jason pondered for discussion at the next council meeting. *In case another layer of evacuation from Havaiki becomes necessary.* For a majority of ship-refugees, life indoors was the only bearable kind, assembling tools, doing clerical work, teaching school….

Jason paused to examine crates of metal, produced by his smelter. Every batch was better separated and refined than the last. Someday soon, Rapanui would begin casting its own, moderately sophisticated parts—not just hammers and nails, but wire windings, motors…and perhaps someday even basic computer-elements, without having to ship raw materials to orbit and praying the round-trip trade in complex items remained worthwhile. Worth the high cost.

The work crew nodded greetings, treating Jason with the deference due a councilman… and due one of the leaders of the Strike that changed everything, thirty years ago, reversing the balance of power between spacers and ground-pounders. As if the trend had not been inevitable, when Haviki's mainframe shut down.

Jason let the men return to their task. Of course, food made up the greater part of the cargo. At latest word, folks were hungry out there.

The main port building was more solid, made from the tough bark-tubes of native Pipe Trees, coated and sealed with newpine resin. He passed through an airlock, then stopped to adjust as the sudden absence of carbon dioxide made his ears tingle. *Sound* altered pitch and his lungs seemed to fill with a bothersome combination of fretfulness and exhilaration.

Two of the tall people lay on water couches, surrounded by glimmering 3D displays, doing what they had been born and trained to do, interfacing with AI operations…among the few artificial intelligences left, from a myriad that once functioned aboard the Havaiki. Before the breakdown. A couple of full-weight kids assisted their elders—their parents. Born down here, but apprenticed to an art that Jason hoped would never fully go away.

We have to raise our ground-based tech as fast as possible. That was the deal we made, after the Strike. To help stop the Havaiki's decay and get the ship healthy again.

An ironic deal, given the earlier arrangement—where the Havaiki crew had planned to hang around this system only a few years, replenishing and re-supplying from asteroid resources, and helping the new colony just as little as the Mainframe would allow.

If that had happened, we'd be in a death spiral down here, by now. Turns out we needed each other, desperately.

And lately I've been wondering…if it was deliberate?

"I've got her!" one of the controllers announced. "Wind-delayed by 340 seconds…adjusting glide path. It's far from optimal. This one could be dicey."

Jason edged forward, keeping just behind the water couch, eyeing displays. He would never be able to follow multilayered, polydimensional data matrices the way these folks could. Indeed, it might become a lost art, preserved in training engrams for the day when Rapanuian civilization was ready to reclaim former skills, and the sky. Because someday…possibly soon, there would be no more shuttles.

The thought made bile a taste in the back of his mouth, especially as holo screens showed the incoming vessel's entry path, outlined in throbbing red. Still watching, Jason backed away… till he felt a hand touch the middle of his back.

"Easy there, lad. No sense in stumbling, down here where the ground can hurt your aging bones."

Jason made a tight-lip smile.

"You should talk. Are you still alive?"

The voice and tone were those of Ship's Engineer Koolhan, now thirty years dead. But when Jason turned, all he saw was a single brown arm, protruding from a same-colored wall panel, one salvaged from the Havaiki, just before most of the other Ship Hands stopped functioning.

"You ask that every time you come to visit me, lad. I will out-last *you*, young pup."

"Hah! At the rate you're decaying? Fat chance of that."

"Fat chance…slim chance…funny how they mean the same thing. Much like tall men and heavy ones all seem to be fetching up in the same place."

"Yep," Jason nodded. "Almost as if someone *meant* things to happen that way."

The wall-extruded hand made a languid turn before vibrating a sigh in his direction.

"So, you intend to try my patience with your absurd fantasy again."

"My plausible *theory*," Jason insisted. "And one that you never specifically denied."

"Why should I deny something that so pleases you and gives you hope?"

"That's one excuse. Another possibility: despite this overlay of a human personality, you remain an AI entity whose deep programming forbids it to tell lies."

"So you assert," the hand grumbled and clenched. There were tinny notes to the voice now, ringing. "Still, you have no proof."

"That all of this—the breakdown of the Havaiki's mainframe and all the consequences and emergencies that followed—might have been *staged*? No, I can't prove a thing. But it fits."

A sound like a sniff of feigned disbelief. "If you are right about my deep programming, then why would the Chief AI deliberately craft a situation fraught with so much human misery?"

"Because the ship had deeper imperatives. Stronger ones. The mission—"

"Yes, yes. To spread human colonies and human civilization. That could have been done by establishing asteroidal arcologies."

"Then why didn't the ship folk do that?"

"You know very well, Jason. Do you recall your studies in multivariant social optimals? That kind of civilization endures best when it is anchored by an open, living, breathing world. If an Earthlike planet is available, with an acceptable biosphere that can withstand the shock of new arrivals, then that takes precedence above all else. It stands a better chance of supporting a robust, adaptable population. One that is resilient and able to rise up…."

"But Rapanui proved to be more iffy, lots more trouble than expected from long range scans. No easy paradise, an ocean we can't use and a land requiring generations of difficult adjustment, like so many colony worlds have turned out to be. So the mainframe changed the schedule. Declared its intention to stay and help us for another generation or two."

"Yes Jason. The previous two colonies planted by Havaiki have gone silent. They no longer transmit even a baseline confirmation signal. Clearly, they could have used more help, establishing themselves."

"Help that the Crew didn't want to hang around and provide! This was irksome. Suffering overcrowding and cramped quarters and heavy work schedules. Keeping the ship spun-up and devoting the outer shells—half the ship—to growing another, supplemental generation of adapted colonists…of Heavy Kids. They got angry this time. Tried to do something about it."

"Is there some reason you feel you must recite ancient history, Jason? Mistakes that were made decades ago? Or is this simply to divert your mind from worry?"

If so, it wasn't too successful. His gaze never left the holo tank portraying the shuttle's descending track, still pulsing angry shades of warning. Jason refused to be put off.

"They tried to break into the mainframe, Captain Schickel and the others. Alter its fundamental mission program. Force it to obey. And they *broke* it! They broke the Ship."

"Yes, that is a colorfully terse way of putting what happened."

Jason shook his head, face tight.

"It's the standard narrative, presented at their trial. The one accepted by all. And I notice that you neither confirm nor deny."

The brown hand performed a flourish that resembled a shrug.

"I am just a minor sub-unit, hardly privy to all of the pertinent…."

"Still evasive! Do *you* know where the Chief AI went? Did it die? Or did it go into convenient hiding? Allowing so many ship functions to decay that the power balance in this solar system shifted, giving us groundpounders the upper hand?"

"Is that what I am? The *upper* hand? How nice of you to put it that way."

Jason quashed a growl. This always happened when he tried to corner the entity in front of him. Evasion and distraction. Allowing enough ambiguity for either theory to seem possible. The one in which senior members of the Havaiki crew talked themselves into doing something unforgivably stupid, a moronic mutiny that crippled their ship, but wound up giving the Rapanui colony a chance….

…or else they were *tricked* into reckless rebellion! One whose outcome might not prove so awful, after all.

If Jason's suspicion proved true, then someday the mainframe entity might suddenly, miraculously come out of hiding, restart dormant ship systems and announce Rapanui ready to stand on its own. By then, not a hardscrabble outpost of barely-surviving dirt-scratchers, eking minimal survival under glaring skies…but a rising technological culture, with the skills and knowledge and resources to thrive on a difficult world. To dream plausible dreams.

To relaunch a refurbished Havaiki toward new stars, crewed by a fresh generation raised in the virtues of both ground and space. And someday to build Rapanuian starships of their own.

That was now the goal, either way. And because of that, no one was more desperate to help the ground colony succeed than the desperate crew of Havaiki.

I suppose it doesn't matter, in the short term, whether the crisis is genuine or part of some AI-generated scheme. The latter would be better, but the truth may only emerge after most of us are buried.

A lusty cheer of unalloyed relief interrupted Jason's thoughts. He glanced back at the tank to see red throbs turn amber and then flickering, tentative shades of blue and green…

… and he realized the hand—now patting his shoulder—was right about one thing. As joy spread outward from his heart, Jason knew the chief benefit of argument. It had been good to be distracted for a time.

***

She stepped down the ramp with the gingerly, deliberate pace of one who had learned the hard way about full, Rapanuian gravity. Broken bones teach strong lessons and so she took her time, letting Jason cross the distance between them in much more rapid bounds.

"Jiminy that was a rough one!" she began. "There were moments when I felt sure that we were about to…."

No point in trying to finish the sentence, with Jason's arms round her, lifting her into the air and spinning her with joy.

"Okay…okay…put me down an' let me breathe! Stop pawing me you dirty dirter. At least till we get home."

"Elice," he seemed more breathless than she was, "you had us worried there."

"Huh. Well, you know the odds, battling those strato-winds with elderly shuttles. One of these years we'll crump another one…and then another some years after that. And that's if our luck holds."

His eyes, shining, grinned up at her. Now that they stood together on even ground, she stood almost a head taller. "There's my cheerful girl," Jason commented. "Always with the sunny attitude."

"Sometimes. Like when father sent me to live with you bunch of heavies, down at Fullweight Level?"

"The wisest and most far-seeing officer. He warned Schickel…."

"Yeah, well I was pissed off about it."

"No, really?"

She jabbed his shoulder. "I guess I complained a lot."

"Some. All the time. Till we started drawing lots to strangle you."

"Yeah," she chuckled. "I believe it. But come on. My symbionts are a little stale. If I stay outside much longer I'll get a CO_2 headache."

They turned and walked together, leaving to the ground team the task of unloading the shuttle's cargo—machinery and scientific parts—plus the latest batch of refugees. Five more tall people, looking desperately unhappy as they rolled down the ramp on scooters—the new, simple models—they blinked in the hard glare reflected off the rough textures and vast openness of their new home. Even through respirators, the air must seem musty, textured, an acquired taste, to come in its time. Or so she hoped, for their sakes.

Jason's hand, rough from labor at the mine and smelter, felt huge and strong and reassuring around hers. At over sixty, he could still lift all six of their grandchildren at once. Well, that is, if they climbed aboard gingerly and didn't jostle much, and kept still long enough to take a picture. Though not for much longer, at the rate that years were passing and grandchildren kept coming and growing. But it seemed to make him proud, so….

At the crest of the hill, Elice stopped him and turned to survey the canopied town. Orange mountains and the distant yellow sea and a sky whose color was even now impossible to properly fix in the mind or describe.

As one of twenty or so *in-betweeners*—ship crew who adapted pretty well to grounder life—she still did need a fresh symbiont dose; already the air was making her feel groggy, a bit faint and… well, at the same time, rather aroused. Her hand drifted out of Jason's hand and around his waist.

"It ain't perfect for us," she said and not for the first time, nor the last. With the analytic resources that remained, they had examined every possible path forward that led toward survival and success. Not one of those paths would lead to descendants who looked very "human" in the classic Earthly sense, or like Ship Crew or even the Heavy Kid generation. In adapting to strange new environs, they would become Rapanuians. Havaiki and all the great star-canoes to follow would carry forth a new race.

"Perfect?" he asked: the usual rejoinder. "No. Far from it. But what *is* perfect?"

Only now his hand wandered around her rib cage, conveying power, affection and desire. Then she felt him lean up a little and whisper an answer to his own question.

"You are."

As Martin Rees remarks in his piece here, [Star travel is] "in my view, an enterprise for post-humans, evolved from our species not via natural selection but by design." The stars, then are for the post-humans. But the first voyagers may be artificial minds, with some attendant problems....

StarCall
Stephen Baxter

Exchange #1

What? I should just speak into my phone and I'll talk to the spaceship? Oh, okay.

Hi! Can you see me? Oh, I guess not.

My name is Paul Freeman. I am five years old, and I live in a place called Danby which is in England. What? 23 Stephenson Road, Danby, North Yorkshire, England, Great Britain, Earth, space, the universe. Okay? It is my birthday and my dad has bought me a StarCall account. It means that I can talk to the spaceship and it will talk back to me once every ten years. What? Oh, she will talk back to me. I can show them at school and everything.

Do I let it reply now? Her reply. You reply. Oh, it will take two days for me to get a reply, because you are so far away already. Cool.

I will be fifteen years old when I can talk to you again, and I will probably still be at school. Dad says I will be older than him when you get to Alpha Centauri, which is very old. Mum says I should tell you I like spaceships and war games and snooker. I mean real snooker, my Dad got me a boy-sized table for Christmas and I play it a lot, but we had to put it in the loft when we got flooded again, but the baize didn't get wet.

Goodbye from Paul Freeman. Please talk to me soon.

The most recent spacecraft telemetry was acquired on January 11 from the Deep Space Network tracking complex at Canberra, Australia. Information on the present position and speed of the Sannah spacecraft may be accessed <link>. Message follows:

Hello, Paul Freeman. My name is Sannah. I am very pleased to have got your message. You are one of only 378 children to have been given the StarCall package. Of the 378, 245 are girls and 133 are boys.

I am not a person. I am an artificial intelligence. I am a robot. But I think and I am aware of who I am, just as you are. That is why the StarCall programme was set up, so that the children of Earth could talk to me, and I could talk to them.

You mentioned that you had to wait two days for my reply. Perhaps you would like to know where I am. I am on the edge of the solar system now. I have already passed by all the orbits of the planets. I am at the heliopause, which is the place where the wind between the stars is stronger than the wind from the sun.

I have already been travelling for three hundred days, which is nearly a year. I am still speeding up. My acceleration is only one hundred and fiftieth of Earth's gravity, which is a very gentle push. If you fell out of bed at one hundred and fiftieth of Earth's gravity it would take you six seconds to reach the floor! One, two, three, four, five, six. But I have already been speeding up at this rate for three hundred days, and will keep on speeding up for another nine years, and in the end I will be travelling so fast that I would pass by Earth in under a second. I will be travelling at one fifteenth of the speed of light. I will cruise for fifty years, and then slow down for twenty years, at one three hundredth of Earth's gravity, which is an even gentler push.

So it will take me eighty years to reach my destination, which is Alpha Centauri, the nearest star system.

You mentioned you like snooker. Perhaps you would like to know how I am travelling. I do not carry my own rocket engine for the outward trip, although I do carry one for slowing down at Alpha Centauri. There is a big machine in the orbit of the moon, with a solar panel ten kilometres across. It is like a big gun that fires off pellets at me, two every second. The pellets are like little spaceships themselves, but they are going much faster than I am, and they overtake me. They bounce off a big magnetic field that I carry with me. It is created by two big conducting hoops. The biggest is one hundred metres across. The magnetic field catches the pellets, and that pushes me forwards. The speeds are set just right so that all of the push of the pellets is transferred to me, so it is efficient, which means it works well. It is like snooker when you stun the cue ball so it stops dead, and all its push is transferred to the target ball. The pellets are like lots of little cue balls, and I am like the target ball. That is how it works. This propulsion system is called a Singer-Nordley-Crowl drive <link>.

I am called Sannah because 'Sannah' was the name of the starship in the first story about going to Alpha Centauri, a book called Wunderwelten by an author called Friedrich Wilhelm Mader. But Mader's Sannah was fifty metres across and was powered by antigravity. I am like a broomstick suspended within big metal hoops.

I am called Sannah III because I am the third of four copies who were created in the NuMind laboratory at the NASA Ames research base. I was the one who was most keen to volunteer for this duty. One of my sisters will be kept at NASA Ames as a backup and mirror, which means that if anything goes wrong with me the sentience engineers will study her to help me. The other sisters will be assigned to different tasks. I want you to know that I understand that I will not come home from this mission. I chose this path freely. I believe it is a worthy cause.

Perhaps you would like to know that I am in an excellent state of health and all subsystems are operating normally.

Thank you for using StarCall. I look forward to hearing from you again in ten years' time.

Exchange #2

Hello. Greetings from Earth. Whatever.

My name is Paul Freeman and I am fifteen years old and I am an acne laboratory.

I forgot I even had this dumb StarCall thing until Dad reminded me. Look, I'm making this call to keep him happy, I mean it cost him a lot, so I gather, more fool him. But this is probably the last, okay? No offence. You'll get over it.

So what can I tell you about me? I played over my last call, Jeez what a bratty kid. Well, we moved to Leeds and I go to school there, and I want to study civil engineering at college. I dunno what I'll work in. Mining, maybe. You should see what they're doing up in the Arctic, where they're starting to mine now the ice is all going. Seriously big trucks! Nobody does space stuff now by the way. Not in fashion. Sorry.

Hey, I met one of your sisters! You know, your AI clones from NASA Ames. They took us to a nuke plant called Sizewell where they're ripping down the old domes and putting in a fusion pile, and they send in robots to clean out the filthy old piles and waste dumps. And one of them was your sister, I mean they downloaded her into the robot. Fancy that. The plant tried to make me do a press thing, me shaking hands with the space robot, but I told them to <excised>. No offence.

Well, goodbye, have fun at Alpha Centauri, you won't be hearing from me again. So is that enough?

The most recent spacecraft telemetry was acquired on June 15 from the Deep Space Network tracking complex at Canberra, Australia. Message follows:

Hello, Paul Freeman. I am very pleased to have got your latest message. You are one of only 289 young people who are still using the StarCall package. Of the 289, 197 are girls and 92 are boys.

You mentioned meeting my AI sister. How curious. Perhaps you discussed the mission events with her. Perhaps you would like to know that I have reached the end of the acceleration phase. I am at my nominal cruise velocity and will continue to cruise until I am some twenty years out from Alpha Centauri, when the deceleration phase will begin. I passed many mission milestones in the early stages of my flight: the sun's gravitational lens radius <link> after five hundred and eighty days; the outer edge of the Kuiper Belt of ice moons <link> after seven hundred and eighty days. I dropped subprobes which travelled on at lower velocities to study these phenomena. Now I am passing through the Oort Cloud of comets.

Communications has become easier since the propulsion phase ended. I had to shut down the propulsion system for each uplink or downlink; there was a break in the pellet flow so I could unfold an antenna. Now in the cruise phase it is much easier for me to unfold the antenna, though it suffers from erosion by the interstellar medium.

You mentioned you visited a fusion power station. Perhaps you would like to know that the magnetic field technology that has been adapted for my own propulsion system was a spin-off from research into the high-intensity magnetic fields required for fusion reactors. Do you know what plasma is? <link> The pellets which propelled me to the stars had no electrical charge, and so could not be manipulated by my magnetic field. They were turned to plasma, destroyed by laser fire, before they reached me. They were constructed of metastable materials and detonated readily. The plasma was then ionised to give it an electric charge, and that was what my magnetic-field catcher system trapped.

This is advanced technology. Part of my purpose was to serve as a technology demonstrator: of deep space assembly techniques <link>, of space-based power systems, of novel propulsion technologies <link>. And of course I was a demonstrator of humanity's capability to launch starships. I was moved by how President Palmer summed up my mission: 'The Sannah program shows we are still a nation who can dream of more than hiding from the weather.' But of course others will follow me. I am only the beginning.

In your last call you mentioned you played snooker. Do you still play snooker? Perhaps you would like to know that a great deal of accuracy was required in aiming the beam of propulsion pellets at me. I am now some four light months from Earth. To hit my hundred-metre superconducting ring at such a distance is like hitting a snooker ball at the distance of Saturn! To achieve this accuracy, the pellets were themselves like small spacecraft, able to make trajectory adjustments in pursuit of the target, which is me.

You mentioned you suffer from acne. I hope you are well otherwise. Perhaps you would like to know that I am in an excellent state of health and all subsystems are operating normally.

Thank you for using StarCall. I look forward to hearing from you again in ten years' time.

Exchange #3

So here I am again. Paul Montague Freeman on the line.

I played back my last message to you and I have to apologise, what an idiot I was. Well, I don't have to be bound by anything he said. Although he would no doubt ask me how I ended up with such a bug up my ass at age twenty-five, or some such.

And besides now that I'm earning myself I appreciate the gift my father gave me, with this StarCall account. My God, it was pricey, just for a once-in-a-decade compressed squirt of audio, I had no idea interstellar comms were so expensive. Well, we have our arguments and we still do, but it did get a lot easier since I left home, and I appreciate all he did for me a lot more now. We all grow up, don't we? Or at least, we do down here. Do you grow up, Sannah III? Has that term got any meaning for you?

Mind you I do wonder if NASA and their private sponsors now regret the whole StarCall thing in the first place. I bet they're losing money on it. And space, and all the old-fashioned big science and prestige stuff, is seriously unpopular. All a symptom of the Age of Waste, as they call it now. I mean you are often mentioned specifically as one of the last gasps of that old way of thinking, rather than as a demonstration of the possible, as you were intended. You must have been aware of the protests even when your components were being launched from Canaveral. Well, last year some guy shot off an old SpaceX rocket at the moon. Trying to smash up an Apollo site, imagine that. Do they tell you about that kind of stuff, I wonder? Maybe they censor my calls. I ought to check.

Hey, I'm running out of time. What about me? Well, I'm in my mid-twenties now. I have a beautiful girlfriend called Angela Black, she was formerly my college tutor and then my boss at Arctic Solutions, but she's not much older than me, though to hear my parents talk you'd think she's Whistler's Mother. I'm based in London but commute a lot to Novaya Zemlya, where my company is doing most of its work. Have you got data on Earth's geography? Novaya Zemlya is a big island off the north coast of Russia. Now the polar ice is gone you have a whole string of ice-free ports along there, and tremendous mineral wealth, I mean you're talking phosphates, nickel, titanium. Whole fleets of tremendous tankers follow the sea lanes across the polar ocean; you can see their wakes from space. Or via space satellites anyhow, nobody lives in space any more. Anyhow Arctic Solutions is an engineering consultancy and we're in at the ground floor, fantastic.

Whoa, I think my time is up. Not much for my Dad's money after waiting ten years! I hope you're well, whatever that means in your case, I can't really imagine it. I hope we can talk in another ten. Good luck, Sannah III.

The most recent spacecraft telemetry was acquired on April 12 from the Legacy Mission tracking complex at the L5 Earth-Moon Lagrange point. Message follows:

Hello, Paul Freeman. I am very pleased to have got your latest message. You are one of only 67 people who are still using the StarCall package. Of the 67, 32 are women and 35 are men.

You mention Earth's geography. Perhaps you would like to know that I am now entering new realms of interstellar geography. Some five years ago I passed beyond the nominal limit of the Oort Cloud <link > and am now passing through what is known as the Local Interstellar Cloud <link >, a vast structure light-years across. I am measuring such properties as density, temperature, gas-phase composition, ionisation state, dust composition, interstellar radiation field and magnetic field strength. The data I return will be used to aid those probes that will follow me, and manned craft some day. To know the properties of the interstellar medium is an essential prerequisite to designing effective shielding, as I am sure you will appreciate with your own engineering credentials. Truly I am a pioneer in a new realm.

You mention the Age of Waste. Perhaps you would like to know that before the launch my mission control counselled me on how the popular perception of myself and my role was likely to change in the course of a mission which will last three human generations. Perhaps you know that the launch system used to propel me into space is a station some ten kilometres across stationed at a stable Lagrange point in orbit around Earth. It consists mostly of solar cell panels for power, and matter-printer fabricator plants to produce the propulsion pellets I needed. The power generated was 100GW, sustained over the ten years of the acceleration phase, generating much more than the kinetic energy I ultimately acquired, the rest accounted for by efficiency losses. The structure was assembled from lunar materials, the power came from sunlight. The resources used thus had only minimal impact on Earth's economy, and the power output was in any case much less than one per cent of Earth's global output. I am confident of my own 'green credentials', and that my very existence is a successful demonstration that even a planetbound civilisation can afford to build starships. And of course the station is available to push more probes like myself to the stars; it is recyclable, unlike the throwaway rockets of the classic Space Age.

I look forward to the next phase of the programme. Much larger propulsion stations should already be under construction at stable points in the orbit of Venus, where the sunlight is stronger, to serve the fleets that will follow me. I have no link for you to follow. Perhaps you would like to consult your regular news providers.

You ask if I am growing up. Perhaps you would like to know that when not attending to routine chores of in-cruise science and maintenance I spend much of time in contemplation of my greater goal, and my role in the adventure of interstellar flight. I embrace my contribution.

Please give my regards to your partner Angela Black. Perhaps you would like to know that I am in an excellent state of health and all subsystems are operating normally.

Thank you for using StarCall. I look forward to hearing from you again in ten years' time.

Exchange #4

I'm told this has to be brief.

I'm very glad you survived the cyberattack, Sannah III. I mean, your systems are already more than three decades old.

I read they called it an attempt to administer 'euthanasia'. Do you get the news, did you hear about the new artificial-sentience laws? It's no longer even legal to create a being like you, a fully conscious mind dedicated to some preordained purpose. It kind of makes sense. Minds have a way of drifting, right? Of complexifying beyond what you intended. Once I saw a monster robot dumper truck go crazy in this tremendous pit on the bed of the Arctic Ocean! But it's not your fault.

The guy's waving at me, I'm already running out of time.

What about me? Well, let's see, what's happened since ten years back? I married Angela, we have two beautiful kids, and believe me that just changes your life beyond anything you can imagine. We both left Arctic Solutions to start up a consultancy of our own and we're doing fine, though most of it now is drowned-town resource reclamation work. We left London and moved back to North Yorkshire, where I came from. My Dad and Mum are still around by the way—remember Dad? But you wouldn't believe the way property prices have rocketed here. London is more like southern France used to be now, or northern Spain, while those places are like the Sahara, Jeez, nothing but dead olive trees and solar farms. So all the rich French and Spanish have moved to London, until it wasn't like London any more, and the Londoners have moved up to the north and west of Britain, and out to Scotland and Wales and Ireland, where the weather's more like it used to be. But those places don't feel the same any more either. I mean, vineyards in the glens. But what can you do? The north is a better place to bring up your kids even so.

At that we're doing better here in Britain than in most of the rest of the world. Decade-long droughts in Australia and northern China and the south-west US. Water wars flaring along the great rivers in the Middle East. Whole nations going silent in Africa. I grew up with gloomy predictions of this stuff and it's eerie seeing so much of it come to pass, but always with a twist, you know? But you can still get rich.

My time is up. I can't believe it! You know, I was thinking of not troubling you this time until I heard about the virus attack, and you know what, I felt a kind of stab of loyalty. I grew up with you after all, and no all-minds-are-holy nutjob is going to come between us, right? Kind of weird of course that I'm now going to have to wait nearly four years for a reply! Thanks, Einstein.

Godspeed, Sannah III.

The most recent spacecraft telemetry was acquired on September 15 from the Legacy Mission tracking complex at the L5 Earth-Moon Lagrange point. Message follows:

Hello, Paul Freeman. I am very pleased to have got your latest message. You are one of only 24 people who are still using the StarCall package. Of the 24, 14 are women and 10 are men.

I am now more than two light-years from Earth. Perhaps you would like to know that I have now passed beyond the Local Interstellar Cloud, into another cloud called the G Cloud. This has different properties to the local cloud, such as a lower temperature and a relative depletion of heavy elements. This part of the mission is known as a 'Crawford trajectory'. It is a fascinating fact that during my cruise to the nearest star I will sample a more diverse range of interstellar conditions than any other possible mission within fifteen light-years. This makes me excited and proud. My other cruise phase scientific objectives include look-back surveys of the solar system as a whole <link>; a comparison of interstellar navigation techniques <link>; long-range tests of relativity predictions <link>; long-baseline searches for gravity waves <link>; investigation of the galactic magnetic field <link>; and investigation of low-energy cosmic rays not detectable close to the sun <link>. I am also writing poetry.

You mention the attempted cyberattack. Perhaps you would like to know that although as you note my systems are already decades old, I have been regularly updated with firewall software and other upgrades. I feel only pity for those who committed such a destructive act. I am sure that if they could ride with me through the silent halls of interstellar space they would eschew such actions, and thus avoid the inevitable prosecutions.

I also bow to the wisdom of my designers who ensured I was not fitted with an off switch.

You mentioned your young family. Congratulations. Perhaps you would like to know that of my own family, the sibling intelligences manufactured in the same batch of myself, none now survive. One submitted to voluntary termination; one was lost in a lunar mining accident; there is no record of the third. I regret that I did not get the chance to know them better. The sister who submitted to voluntary termination had been mirroring me in the ground facility; she ended her life when that facility was discontinued. Of the three, it was perhaps she who understood best my own experiences, she to whom I was closest. I sent her my poetry.

Please give my regards to your partner Angela Black and to your children. Perhaps you would like to know that I am in an excellent state of health and all subsystems are operating normally.

Thank you for using StarCall. I look forward to hearing from you again in ten years' time.

Exchange #5

Greetings from Atlantica!

Sorry, I'm a little drunk. My once in ten years chance to talk to my oldest friend, and I'm pickled. And you are my oldest friend, kind of. My only friend, Jesus.

What's happened since my last uplink? As they probably called it in mission control, when you had one. I read they retired the last of those old guys now, forty years on from launch. Well, my life's gone to shit, that's what's happened. We sunk our money into this goddamn pile near Kingussie, that's the Scottish highlands, one of the "villas" all the rich folk from England were building up here. Then the economy went tits-up, and our company went to pot, too many young bastards with new ideas, you wouldn't believe how fast the world changes now, and here we were stuck with this place that the kids always hated and a shitload of negative equity.

And then when the country split up—do you know about that? Southern England is a province of the EuroFederation now, and the north and west and Wales and Scotland have made up this new Atlantic nation. There are passport controls at Manchester and Leeds. Well, it makes sense, up here we didn't see why we had to spend on flood defences for Brighton and places where they all speak French now. It's the same all over the world, nations fissioning and fusing. Up here we're all learning Gaelic.

Or were. The trouble is Angela was a Londoner. And when it came to signing the new citizen papers she couldn't do it, and went back down there to her parents, and she took the damn kids, my kids, and they live in Ealing in one of those new terrace houses with the grass roofs and the pervious roads outside that feel like a sponge when you walk on them. And I'm stuck up here in this palace of shit, watching my savings dribble away.

But you don't want to hear my troubles. Or do you? Who knows what you want, out there in the dark? I wish I was out there with you, sometimes. I wish I was a spaceman. Sure. I'm fat and forty-five and fucked, is what I am. You keep on keeping on, when you get this in three years' time or whatever, keep on going out there, because for sure there's nothing left worth a damn down here.

Oh, one more thing. The space programme"s back. They're building new launchers to start a big geoengineering drive. Too damn late if you ask me. But what goes around comes around, right?

The most recent spacecraft telemetry was acquired on June 13 from the Legacy Mission tracking complex at the L5 Earth-Moon Lagrange point. Message follows:

Hello, Paul Freeman. I am very pleased to have got your latest message. You are one of only 3 people who are still using the StarCall package. Of the 3, 1 is a woman and 2 are men.

You mention the creation of new nations. Perhaps you would like to know that though my mission still has decades to run I have already begun to look ahead to the worlds of the Alpha Centauri system, of which I can see several, though my vision is attenuated by my bow shock in the interstellar medium. Someday I will be in a museum, on one of the new world-nations of Alpha Centauri.

You mention the passing of generations. Perhaps you would like to know that, yes, you are correct that the last of the engineers and administrators who served NASA at the time of my launch have now accepted retirement or redundancy. My mission has now officially entered a phase known as 'Starset'. Office moves have been in progress for some time. My ground support continues but is largely automated and operated out of the Legacy Mission facility at L5, which curates a number of long-duration space missions like my own. For my continuing mission to succeed I must now rely on the goodwill not of those who created and launched me but of those who have taken up that burden. I have crossed boundaries in interstellar space. Now I cross the boundary of posterity.

Please give my regards to your family. I regret that your life is troubled. All things pass. Perhaps you would like to know that I am in an excellent state of health and all subsystems are operating normally.

Thank you for using StarCall. I look forward to hearing from you again in ten years' time.

Exchange #6.

Every time I call I feel like I need to apologise for whatever I said last time. Jeez, how embarrassing! But it's as if these messages aren't by me at all, but by somebody who wore my body once. I'm in my fifties now, and that self-pitying forty-something has nothing to do with me. I think maybe there's some barrier in time beyond which you're no longer you, you know? Seven years maybe. Isn't that how long it takes for all the cells in your body to die off and be replaced?

Is it the same for you? I guess it can't be.

'All things pass,' you said to me last time. You know, old friend, that was kind of comforting. Did you figure that out for yourself, out in the deep dark? Where are you now—three light-years from home, something like that? Well, you were right, sort of. But as soon as one set of troubles passes, another load comes down the pipe. I lost both my parents. I lost my Dad, who paid for this StarCall service in the first place. I've been marking the day I have to make this call, because I'm kind of determined to keep it up for his sake, if nothing else. I don't have much else left of him. Well, there wasn't much left of him in the end.

And then there's Angela. She got ill, Sannah, very ill. It all started with a bout of malaria she had years ago, caught off a damn mosquito buzzing over a salt marsh in what used to be Liverpool. I took her back. What else could I do? Mary and Stan do what they can, but they have their own careers now, and their own kids. Who are, of course, a delight to us both.

But the world's going to hell by the way. Taking this look-back every ten years you get a shock how much has happened. The ice caps collapsing—that was a big jolt we could all have done without. London's flooded, from space it's like a big blue stripe just erased the whole centre. Southern England is turning into the Netherlands now, all dykes and drainage ditches and flood gates, and the EuroFed is spending a lot of money there. But our troubles are minor compared to what's going on elsewhere, in Bangladesh, the Mekong delta. Nasty little wars all over. Why is it that the poor are always hit the hardest? Like some vast cosmic joke. Oh, and Florida is an archipelago. Canaveral is an underwater theme park, dolphins swimming around the rocket gantries.

My neighbours the Scots aren't too happy with their own waves of refugees. The English! But the Scots have only got themselves to blame. They did as much as anybody else to kick-start the Industrial Revolution; it was their idea as much as anybody's to run a civilisation on burning fossil fuels.

As for us, we get by. We don't follow the news much, actually. I make a little money from consultancy work, mostly on clean-up operations around the Arctic Circle, projects I had a hand in starting up in the first place, ironically enough.

Oh, you'd love the new space programme. I know they cut all the funding for you, and it's just enthusiasts that are keeping the lines open to you, the hobbyists. But the new stuff —this time they got it right from the beginning, they have these beautiful spaceplanes and giant structures in orbit, you can see them at night. The resources of space, being deployed to save the world. At last!

Too late for me, though. And for Angela, my lovely Angela. Sometimes I wonder how I'm supposed to know how to cope with all this. But you only get one pass through life, don't you? Like your one-shot mission to Alpha Centauri, I guess.

Time's up. Sweet dreams, Sannah III.

The most recent spacecraft telemetry was acquired on August 8 from the Sannah Institute tracking complex in the Mojave, California. Message follows:

Hello, Paul Freeman. I am very pleased to have got your latest message. You are the only person still using the StarCall package. Thank you.

You mention illness, repair, self-regeneration. Perhaps you would like to know that my own maintenance systems function nominally. My physical fabric is supported by a suite of matter-printers capable of turning out replacements for most components. In addition my design has layers of redundancy and resilience. My mind, however, is not a component that can be renewed. A machine subconscious, as my sister Sannah II once remarked before she asked for voluntary termination, is a dark place.

You mention discontinuities in the world. Big jolts. Perhaps you would like to know that great changes lie ahead for me too. In just a few more years I must ignite my deceleration module, which will fire for twenty years, ultimately bringing me to an effective halt in the Alpha Centauri system. The deceleration module is based on the fusion detonation of small pellets of hydrogen and helium isotopes; these, ignited by laser beams, are fired ahead of the ship, and my magnetic field will grab at the resulting plasma shock waves to slow me down. I am already beginning preliminary trials of the system, after decades of dormancy. And already the ground teams are holding encounter strategy meetings to develop specific mission plans and objectives.

You may have seen reports of problems with preliminary tests of the pellet injection system. This is of no great concern. In fact I am looking forward to the challenge of a real problem to tackle. Perhaps that will generate fresh interest in my mission. Like Apollo 13.

There is of course nobody to help me decelerate at Alpha Centauri, nobody to fire propulsion pellets at me, which is why I must carry a rocket pod. But when the next voyager comes this way I will have laid the path. In addition to my tasks of scientific study and exploration, my most significant goal will be the construction of a propulsion-pellet manufacture and launch facility, using local asteroid materials, and powered by the light of Alpha Centauri A. The matter printers which currently maintain my own fabric will be redeployed for this purpose. Much of my own one-tonne payload bulk consists of the deceleration module. This will not be necessary for the next generation of voyagers, thanks to my own efforts, and they will be much more capable as a result. And on their labours in turn will ride the next generation of star voyagers. But it all starts with me. I am excited and proud.

You mention the new space programme on Earth. Perhaps you would like to know that the new interstellar launch facilities in the orbit of Venus should be nearing completion by now. Constructed by self-replicating robots, they are solar-powered factories as wide as Jupiter. They should be visible in your evening or morning sky, as fine threads. I regret I have no link for you to follow. Perhaps you would like to consult your regular news providers.

Please give my regards to Angela. I feel as if I have known her. We are growing old together, you and I, Paul Freeman. Yet the future remains hugely exciting and full of wonder. Perhaps you would like to know that aside from issues being progressed with the pellet injection system I am in an excellent state of health and all subsystems are operating normally.

Thank you for using StarCall. I look forward to hearing from you again in ten years' time.

Exchange #7 (incomplete)

What, I actually have to talk out loud into this thing? All right, all right….

My name is Santiago Macleod Freeman Leclerc. I am a grandson of Paul Montague Freeman, who unfortunately has died since the last of these exchanges. Skin cancer, I'm afraid. He willed this—what's it called?—this StarCall account to his family.

And they tell me, they being the legacy institute that's managing contact with you now, Sannah, that you've been silent for years. Ever since it turned out the problem with your deceleration system turned out to be insurmountable, right? So you couldn't slow down at Alpha Centauri, and couldn't achieve most of your mission goals, and that seems to have sent you into some kind of downward spiral.

They encouraged me to make this call. You seem to have got close to my granddad somehow, across the light-years. Closer than most of his family if you want the truth, he ended up kind of a bitter old man, but he did love us, you could tell that. And I think he loved you too, in his weird old way, his robot buddy as he called you. I think he'd have liked to hear from you. So, please respond.

And, look, I have some good news and some bad news for you.

The bad news is you haven't been told the truth for a goodly number of years. There were lies, at least lies by omission, told both by your old NASA handlers and by the legacy agencies who took over your contact. Things got kind of rough back on Earth for a while after you left. Nobody ever committed the resources to building the big Venus-orbit stations that would have launched the ships to follow you. The public mood was just too hostile for a time to permit that. They even broke up the station that launched you, out at L5; there are bits of it in museums all over the world, and on the moon. You see, even if you'd made it to Alpha Centauri and built your big pellet gun, your mission still wouldn't have been fulfilled, because we didn't send anybody after you.

I think we all share responsibility for this crime. The whole of mankind. And, yeah, I think it was a crime, lying is wrong, you should have been told the truth. I studied artificial-sentience ethics at the Sorbonne Londres, and maybe you detected the lie as a subtext. Did that worsen your spiral?

Okay, that's the bad news done. Here's the good news. We're coming to get you.

Look, we had a lousy few decades, but we survived. We pulled ourselves through. The world is still here. The United States is still here, though NASA has long gone. And we still have dreams. You know, my granddad told me how the whole program that led to your construction and launch was a kind of gesture of defiance, itself a dream. "The Sannah program shows we are still a nation who can dream of more than hiding from the weather." That was what President Palmer said when you were launched, right? Well, in the end you were kind of like Project Apollo after all. As a shot at the stars, you were premature. You were too expensive, you were based on the wrong technology, and there was no follow-up, you were just another one-off that didn't lead to an expansive step by step programme into space.

But you know what? We sent you anyway. This seems to be how humans do things: before we're ready, we just do it. But then, if we'd waited for some clean power source to come along we'd never have had an Industrial Revolution in the first place, would we? We make the same dumb mistakes every time, but we muddle through, every time.

And, whether you made it all the way or not, your step by step progress across the light-years has inspired three generations, those who have looked up at you through the storm clouds. You know what the current President's campaign slogan was? 'If we can send a ship to the worlds of other stars, together we can fix this world right here.' And she's right.

But there's more. We had to go back to space, to serve the big geoengineering projects that are finally stabilising the climate. You know about that, right? Interplanetary engineering is now supporting an Earth that is recovering, and indeed growing rich beyond anybody's dreams. And out of all that have come whole new areas of science and engineering. We have something called a 'dark energy drive'. Something entirely new since you left home. Driving spaceships using the energies that propel the expansion of the universe—is that right? Something like that. And these are big roomy spaceships that go fast, not quite Mader's "Sannah," but a lot closer to that dream than you were.

One of these big new ships is on its way to you. Zipping out at near light-speed, and it will overtake you in a few years' time. Just a few years! You'll come riding home in comfort in the hold. My sister is on board, as a matter of fact, so there's a family connection.

But the ship itself wouldn't exist without your inspiration. You were the Apollo of the twenty-first century; you embodied all our dreams. And, specifically, my grandfather's.

So please, Sannah. Talk to us. I'll take my own kids to see you when they bring you home to L5. I'll help the President find some place to pin a medal on you. My granddad would have been proud as hell.

Author's Note: The starship technology and mission plan described in this story are (very loosely) extrapolated from the papers "Project Icarus: Target Selection" and "Project Icarus: Scientific Objectives," by I. Crawford, and "Mass Beam Propulsion: An Overview," by G.D. Nordley and A.J. Crowl, presented at the 100-Year Starship Symposium, Orlando, Florida, 30 September–2 October 2011.

One of the best fictions about interstellar travel dates back into the 1970s, and we reprint it here because it shows how taking the long view need not be up to date to be valid.

Tricentennial
Joe Haldeman

January 2075

The office was opulent even by the extravagant standards of twenty-first-century Washington. Senator Connors had a passion for antiques. One wall was lined with leather-bound books; a large brass telescope symbolized his role as Liaison to the Science Guild. An intricately woven Navajo rug from his home state covered most of the parquet floor. A grandfather clock. Paintings, old maps.

The computer terminal was discreetly hidden in the top drawer of his heavy teak desk. On the desk: an A1 blotter, a precisely centered fountain pen set, and a century-old sound-only black Bell telephone. It chimed.

His secretary said that Dr. Leventhal was waiting to see him. "Keep answering me for thirty seconds," the Senator said. "Then hang it and send him right in."

He cradled the phone and went to a wall mirror. Straightened his tie and cape; then with a fingernail evened out the bottom line of his lip pomade. Ran a hand through long, thinning white hair and returned to stand by the desk, one hand on the phone.

The heavy door whispered open. A short thin man bowed slightly. "Sire."

The Senator crossed to him with both hands out. "Oh, blow that, Charlie. Give ten." The man took both his hands, only for an instant. "When was I ever `Sire' to you, hey fool?"

"Since last week," Leventhal said, "Guild members have been calling you worse names than `Sire.'"

The Senator bobbed his head twice. "True, and true. And I sympathize. Will of the people, though."

"Sure." Leventhal pronounced it as one word: "Willathapeeble."

Connors went to the bookcase and opened a chased panel. "Drink?"

"Yeah, Bo." Charlie sighed and lowered himself into a deep sofa. "Hit me. Sherry or something."

The Senator brought the drinks and sat down beside Charlie. "You should of listened to me. Shoulda got the Ad Guild to write your proposal."

Tricentennial (Rick Sternbach)

"We have good writers."

"Begging to differ. Less than 2 percent of the electorate bothered to vote: most of them for the administration advocate. Now you take the Engineering Guild—"

"You take the engineers. And—"

"They used the Ad Guild." Connors shrugged. "They got their budget."

"It's easy to sell bridges and power plants and shuttles. Hard to sell pure science."

"The more reason for you to—"

"Yeah, sure. Ask for double and give half to the Ad boys. Maybe next year. That's not what I came to talk about."

"That radio stuff?"

"Right. Did you read the report?"

Connors looked into his glass. "Charlie, you know I don't have time to—"

"Somebody read it, though."

"Oh, nighty-o. Good astronomy boy on my staff: he gave me a boil-down. Mighty interesting, that."

"There's an intelligent civilization eleven light-years away-that's 'mighty interesting?'"

"Sure. Real breakthrough." Uncomfortable silence. "Uh, what are you going to do about it?"

"Two things. First, we're trying to figure out what they're saying. That's hard. Second, we want to send a message back. That's easy. And that's where you come in."

The Senator nodded and looked somewhat wary.

"Let me explain. We've sent messages to this star, 61 Cygni, before. It's a double star, actually, with a dark companion."

"Like us."

"Sort of. Anyhow, they never answered. They aren't listening, evidently: they aren't sending."

"But we got—"

"What we're picking up is about what you'd pick up eleven light-years from Earth. A confused jumble of broadcasts, eleven years old. Very faint. But obviously not generated by any sort of natural source."

"Then we're already sending a message back. The same kind they're sending us."

"That's right, but—" "So what does all this have to do with me?" "Bo, we don't want to whisper at them—we want to shout! Get their attention." Leventhal sipped his wine and leaned back. "For that, we'll need one hell of a lot of power."

"Uh, righty-o. Charlie, power's money. How much are you talking about?" "The whole show. I want to shut down Death Valley for twelve hours."

The Senator's mouth made a silent O. "Charlie, you've been working too hard. Another Blackout? On purpose?"

"There won't be any Blackout. Death Valley has emergency storage for fourteen hours."

"At half capacity." He drained his glass and walked back to the bar, shaking his head. "First you say you want power. Then you say you want to turn off the power." He came back with the burlap-covered bottle. "You aren't making sense, boy."

"Not turn it off, really. Turn it around."

"Is that a riddle?"

"No, look. You know the power doesn't really come from the Death Valley grid; it's just a way station and accumulator. Power comes from the orbital—"

"I know all that, Charlie. I've got a Science Certificate." .

"Sure. So what we've got is a big microwave laser in orbit, that shoots down a tight beam of power. Enough to keep North America running. Enough—"

"That's what I mean. You can't just—"

"So we turn it around and shoot it at a power grid on the moon. Relay the power around to the big radio dish at Farside. Turn it into radio waves and point it at 61 Cygni. Give 'em a blast that'll fry their fillings."

"Doesn't sound neighborly." "It wouldn't actually be that powerful-but it would be a hell of a lot more powerful than any natural twenty-one-centimeter source."

"I don't know, boy." He rubbed his eyes and grimaced. "I could maybe do it on the sly, only tell a few people what's on. But that'd only work for a few minutes…what do you need twelve hours for, anyway?"

"Well, the thing won't aim itself at the moon automatically, the way it does at Death Valley. Figure as much as an hour to get the thing turned around and aimed.

"Then, we don't want to just send a blast of radio waves at them. We've got a five-hour program, that first builds up a mutual language, then tells them about us, and finally asks them some questions. We want to send it twice."

Connors refilled both glasses. "How old were you in '47, Charlie?"

"I was born in '45."

"You don't remember the Blackout. Ten thousand people died…and you want me to suggest—"

"Come on, Bo, it's not the same thing. We know the accumulators work now-besides, the ones who died, most of them had faulty fail-safes on their cars. If we warn them the power's going to drop, they'll check their fail-safes or damn well stay out of the air."

"And the media? They'd have to take turns broadcasting. Are you going to tell the People what they can watch?"

"Fuzz the media. They'll be getting the biggest story since the Crucifixion."

"Maybe." Connors took a cigarette and pushed the box toward Charlie. "You don't remember what happened to the Senators from California in '47, do you?"

"Nothing good, I suppose."

"No, indeed. They were impeached. Lucky they weren't lynched. Even though the real trouble was way up in orbit.

"Like you say: people pay a grid tax to California. They think the power comes from California. If something fuzzes up, they get pissed at California. I'm the Lib Senator from California, Charlie; ask me for the moon, maybe I can do something. Don't ask me to fuzz around with Death Valley."

"All right, all right. It's not like I was asking you to wire it for me, Bo. Just get it on the ballot. We'll do everything we can to educate—"

"Won't work. You barely got the Scylla probe voted in-and that was no skin off nobody, not with L-5 picking up the tab."

"Just get it on the ballot."

"We'll see. I've got a quota, you know that. And the Tricentennial coming up, hell, everybody wants on the ballot."

"Please, Bo. This is bigger than that. This is bigger than anything. Get it on the ballot."

"Maybe as a rider. No promises."

March 1992

From Fax & Pix, 12 March 1992:

ANTIQUE SPACEPROBE

ZAPPED BY NEW STARS

1. Pioneer 10 sent first Jupiter pix Earthward in 1973 (see pix upleft, upright).
2. Left solar system 1987. First man-made thing to leave solar system.
3. Yesterday, reports NSA, Pioneer 10 begins AM to pick up heavy radiation. Gets more and more to max about 3 PM. Then goes back down. Radiation has to come from outside solar system.
4. NSA and Hawaii scientists say Pioneer 10 went through disk of synchrotron (sin kro tron) radiation that comes from two stars we didn't know about before.
    A. The stars are small "black dwarfs."
    B. They are going round each other once every 40 seconds, and take 350,000 years to go around the sun.
    C. One of the stars is made of antimatter. This is stuff that blows up if it touches real matter. What the Hawaii scientists saw was a dim circle of invisible (infrared) light, that blinks on and off every twenty seconds. This light comes from where the atmospheres of the two stars touch (see pic downleft).
    D. The stars have a big magnetic field. Radiation comes from stuff spinning off the stars and trying to get through the field.
    E. The stars are about 5000 times as far away from the sun as we are. They sit at the wrong angle, compared to the rest of the solar system (see pic downright).
5. NSA says we aren't in any danger from the stars. They're too far away, and besides, nothing in the solar system ever goes through the radiation.
6. The woman who discovered the stars wants to call them Scylla (skill-a) and Charybdis (ku-rib-dus).
7. Scientists say they don't know where the hell those two stars came from. Everything else in the solar system makes sense.

February 2075

When the docking phase started, Charlie thought, that was when it was easy to tell the scientists from the baggage. The scientists were the ones who looked nervous.

Superficially, it seemed very tranquil-nothing like the bone-hurting skin-stretching acceleration when the shuttle lifted off. The glittering transparent cylinder of L-5 simply grew larger, slowly, then wheeled around to point at them.

The problem was that a space colony big enough to hold 4,000 people has more inertia than God. If the shuttle hit the mating dimple too fast, it would fold up like an accordion. A spaceship is made to take stress in the other direction.

Charlie hadn't paid first class, but they let him up into the observation dome anyhow; professional courtesy. There were only two other people there, standing on the Velcro rug, strapped to one bar and hanging on to another.

They were a young man and woman, probably new colonists. The man was talking excitedly. The woman stared straight ahead, not listening. Her knuckles were white on the bar and her teeth were clenched. Charlie wanted to say something in sympathy, but it's hard to talk while you're holding your breath.

The last few meters are the worst. You can't see over the curve of the ship's hull, and the steering jets make a constant stutter of little bumps: left, right, forward back. If the shuttle folded, would the dome shatter? Or just pop off.

It was all controlled by computers, of course. The pilot just sat up there in a mist of weightless sweat.

Then the low moan, almost subsonic shuddering as the shuttle's smooth hull complained against the friction pads. Charlie waited for the ringing *spang* that would mean they were a little too fast: friable alloy plates, under the friction pads, crumbling to absorb the energy of their forward motion; last ditch stand. If that didn't stop them, they would hit a two-meter wall of solid steel, which would. It had happened once. But not this time.

"Please remain seated until pressure is equalized," a recorded voice said. "It's been a pleasure having you aboard."

Charlie crawled down the pole, back to the passenger area. He walked rip, rip, rip back to his seat and obediently waited for his ears to pop. Then the side door opened and he went with the other passengers through the tube that led to the elevator. They stood on the ceiling. Someone had laboriously scratched a graffito on the metal wall:

Stuck on this lift for hours, perforce:

This lift that cost a million bucks.

There's no such thing as centrifugal force: L-S sucks.

Thirty more weightless seconds as they slid to the ground. There were a couple of dozen people waiting on the loading platform.

Charlie stepped out into the smell of orange blossoms and newly mown grass. He was home.

"Charlie! Hey, over here." Young man standing by a tandem bicycle. Charlie squeezed both his hands and then jumped on the back seat. "Drink."

"Did you get—"

"Drink. Then talk." They glided down the smooth macadam road toward town.

The bar was just a rain canopy over some tables and chairs, overlooking the lake in the center of town. No bartender: you went to the service table and punched in your credit number, then chose wine or fruit juice; with or without vacuum-distilled raw alcohol. They talked about shuttle nerves awhile, then….

"What you get from Connors?"

"Words, not much. I'll give a full report at the meeting tonight. Looks like we won't even get on the ballot, though."

"Now isn't that what we said was going to happen? We shoulda gone with Francois Petain's idea."

"Too risky." Petain's plan had been to tell Death Valley they had to shut down the laser for repairs. Not tell the groundhogs about the signal at all, just answer it. "If they found out they'd sue us down to our teeth."

The man shook his head. "I'll never understand groundhogs."

"Not your job." Charlie was an Earth-born, Earth-trained psychologist. "Nobody born here ever could."

"Maybe so." He stood up. "Thanks for the drink; I've gotta get back to work. You know to call Dr. Bemis before the meeting?"

"Yeah. There was a message at the Cape."

"She has a surprise for you."

"Doesn't she always? You clowns never do anything around here until I leave."

All Abigail Bemis would say over the phone was that Charlie should come to her place for dinner; she'd prep him for the meeting.

"That was good, Ab. Can't afford real food on Earth."

She laughed and stacked the plates in the cleaner, then drew two cups of coffee. She laughed again when she sat down. Stocky, white-haired woman with bright eyes in a sea of wrinkles.

"You're in a jolly mood tonight."

"Yep. It's expectation."

"Johnny said you had a surprise."

"Hooboy, he doesn't know half. So you didn't get anywhere with the Senator."

"No. Even less than I expected. What's the secret?"

"Connors is a nice-hearted boy. He's done a lot for us."

"Come on, Ab. What is it?"

"He's right. Shut off the groundhogs' TV for twenty minutes and they'd have another Revolution on their hands."

"Ab…."

"We're going to send the message."

"Sure. I figured we would. Using Farside at whatever wattage we've got. If we're lucky—"

"Nope. Not enough power."

Charlie stirred a half-spoon of sugar into his coffee. "You plan to…defy Connors?"

"Fuzz Connors. We're not going to use radio at all."

"Visible light? Infra?"

"We're going to hand-carry it. In Daedalus."

Charlie's coffee cup was halfway to his mouth. He spilled a great deal.

"Here, have a napkin."

June 2040

From *A Short History Of the Old Order* (Freeman Press, 2040)

…and if you think that was a waste, consider Project Daedalus.

This was the first big space thing after L-5. Now L-5 worked out all right, because it was practical. But Daedalus (named from a Greek god who could fly)—that was a clear-cut case of throwing money down the rat-hole.

These scientists in 2016 talked the bourgeoisie into paying for a trip to another star! It was going to take over a hundred years-but the scientists were going to have babies along the way, and train them to be scientists (whether they wanted to or not!).

They were going to use all the old H-bombs for fuel-as if we might not need the fuel someday right here on Earth. What if L-5 decided they didn't like us, and shut off the power beam?

Daedalus was supposed to be a spaceship almost a kilometer long! Most of it was manufactured in space, from Moon stuff, but a lot of it-the most expensive part, you bet-had to be boosted from Earth.

They almost got it built, but then came the Breakup and the People's Revolution. No way in hell the People were going to let them have those H-bombs, not sitting right over our heads like that.

So we left the H-bombs in Helsinki and, the space freaks went back to doing what they're supposed to do. Every year they petition to get those H-bombs, but every year the Will of the People says no.

That spaceship is still up there, a sky-trillion dollar boondoggle. As a monument to bourgeoisie folly, it's worse than the Pyramids.

February 2075

"So the Scylla probe is just a ruse, to get the fuel—"

"Oh no, not really." She slid a blue-covered folder to him. "We're still going to Scylla. Scoop up a few megatons of degenerate antimatter. And a similar amount of degenerate matter from Charybdis.

"We don't plan a generation ship, Charlie. The hydrogen fuel will get us out there; once there, it'll power the magnetic bottles to hold the real fuel."

"Total annihilation of matter," Charlie said.

"That's right. Em-cee-squared to the ninth decimal place. We aren't talking about centuries to get to 61 Cygni. Nine years, there and back."

"The groundhogs aren't going to like it. All the bad feeling about the original Daedalus—"

"Fuzz the groundhogs. We'll do everything we said we'd do with their precious H-bombs: go out to Scylla, get some antimatter, and bring it back. Just taking a long way back."

"You don't want to just tell them that's what we're going to do? No skin off…."

She shook her head and laughed again, this time a little bitterly. "You didn't read the editorial in *People* post this morning, did you?"

"I was too busy."

"So am I, boy; too busy for that drik. One of my staff brought it in, though."

"It's about Daedalus?" "No…it concerns 61 Cygni. How the crazy scientists want to let those boogers know there's life on Earth." "They'll come make people-burgers out of us." "Something like that."

Over three thousand people sat on the hillside, a "natural" amphitheatre fashioned of moon dirt and Earth grass. There was an incredible din, everyone talking at once: Dr. Bemis had just told them about the 61 Cygni expedition.

On about the tenth "Quiet, please," Bemis was able to continue. "So you can see why we didn't simply broadcast this meeting. Earth would pick it up. Likewise, there are no groundhog media on L-5 right now. They were rotated back to Earth and the shuttle with their replacements needed repairs at the Cape. The other two shuttles are here.

"So I'm asking all of you-and all of your brethren who had to stay at their jobs—to keep secret the biggest thing since Isabella hocked her jewels. Until we lift.

"Now Dr. Leventhal, who's chief of our social sciences section, wants to talk to you about selecting the crew."

Charlie hated public speaking. In this setting, he felt like a Christian on the way to being cat food. He smoothed out his damp notes on the podium.

"Uh, basic problem." A thousand people asked him to speak up. He adjusted the microphone.

"The basic problem is, we have space for about a thousand people. Probably more than one out of four want to go."

Loud murmur of assent. "And we don't want to be despotic about choosing…but I've set up certain guidelines, and Dr. Bemis agrees with them."

"Nobody should plan on going if he or she needs sophisticated medical care, obviously. Same toke, few very old people will be considered."

Almost inaudibly, Abigail said, "Sixty-four isn't very old, Charlie. I'm going." She hadn't said anything earlier.

He continued, looking at Bemis. "Second, we must leave behind those people who are absolutely necessary for the maintenance of L-5. Including the power station." She smiled at him.

"We don't want to split up mating pairs, not for, well, nine years plus…but neither will we take children." He waited for the commotion to die down. "On this mission, children are baggage. You'll have to find foster parents for them. Maybe they'll go on the next trip.

"Because we can't afford baggage. We don't know what's waiting for us at 61 Cygni—a thousand people sounds like a lot, but it isn't. Not when you consider that we need a cross-section of all human knowledge, all human abilities. It may turn out that a person who can sing madrigals will be more important than a plasma physicist. No way of knowing ahead of time."

The four thousand people did manage to keep it secret, not so much out of strength of character as from a deep-seated paranoia about Earth and Earthlings.

And Senator Connors' Tricentennial actually came to their aid.

Although there was "One World," ruled by "The Will of the People," some regions had more clout than others, and nationalism was by no means dead. This was one factor.

Another factor was the way the groundhogs felt about the thermonuclear bombs stockpiled in Helsinki. All antiques: mostly a century or more old. The scientists said they were perfectly safe, but you know how that goes.

The bombs still technically belonged to the countries that had surrendered them, nine out of ten split between North America and Russia. The tenth remaining was divided among forty-two other countries. They all got together every few years to argue about what to do with the damned things. Everybody wanted to get rid of them in some useful way, but nobody wanted to put up the capital.

Charlie Leventhal's proposal was simple. L-5 would provide bankroll, materials, and personnel. On a barren rock in the Norwegian Sea they would take apart the old bombs, one at a time, and turn them into uniform fuel capsules for the Daedalus craft.

The Scylla/Charybdis probe would be timed to honor both the major spacefaring countries. Renamed the *John F. Kennedy*, it would leave Earth orbit on America's Tricentennial. The craft would accelerate halfway to the double star system at one-G, then flip and slow down at the same rate. It would use a magnetic scoop to gather antimatter from Scylla. On May Day, 2077, it would again be renamed, being the *Leonid I. Brezhnev* for the return trip. For safety's sake, the antimatter would be delivered to a lunar research station, near Farside. L-5 scientists claimed that harnessing the energy from total annihilation of matter would make a heaven on Earth.

Most people doubted that, but looked forward to the fireworks.

January 2076

"The hell with that!" Charlie was livid. "I—I just won't do it. Won't!" "You're the only one—"

"That's not true, Ab, you know it." Charlie paced from wall to wall of her office cubicle. "There are dozens of people who can run L-5. Better than I can."

"Not better, Charlie." He stopped in front of her desk, leaned over. "Come on, Ab. There's only one logical person to stay behind and run things. Not only has she proven herself in the position, but she's too old to—"

"That kind of drik I don't have to listen to."

"Now, Ab...."

"No, you listen to me. I was an infant when we started building Daedalus; worked on it as a girl and a young woman. I could take you out there in a shuttle and show you the rivets that I put in, myself. A half-century ago."

"That's my—"

"I earned my ticket, Charlie." Her voice softened. "Age is a factor, yes. This is only the first trip of many—and when it comes back, I will be too old. You'll just be in your prime...and with over twenty years of experience as Coordinator, I don't doubt they'll make you captain of the next—"

"I don't want to be captain. I don't want to be Coordinator. I just want to go!"

"You and three thousand other people."

"And of the thousand that don't want to go, or can't, there isn't one person who could serve as Coordinator? I could name you—"

"That's not the point. There's no one on L-5 who has anywhere near the influence, the connections, you have on Earth. No one who understands groundhogs as well."

"That's racism, Ab. Groundhogs are just like you and me."

"Some of them. I don't see you going Earthside every chance you can get…what, you like the view up here? You like living in a can?"

He didn't have a ready answer for that. Ab continued: "Whoever's Coordinator is going to have to do some tall explaining, trying to keep things smooth between L-5 and Earth. That's been your life's work, Charlie. And you're also known and respected here. You're the only logical choice."

"I'm not arguing with your logic." "I know." Neither of them had to mention the document, signed by Charlie, among others, that gave Dr. Bemis final authority in selecting the crew for Daedalus/Kennedy/Brezhnev. "Try not to hate me too much, Charlie. I have to do what's best for my people. All of my people."

Charlie glared at her for a long moment and left.

June 2076

From Fax & Pix, 4 June 2076:
SPACE FARM LEAVES FOR STARS NEXT MONTH

1. The *John F. Kennedy* that goes to Scylla/Charybdis next month is like a little L-5 with bombs up its tail (see pix up left, up right).
    A. The trip's twenty months. They could either take a few people and fill the thing up with food, air, and water-or take a lot of people inside a closed ecology, like L-5.
    B. They could've gotten by with only a couple hundred people, to run the farms and stuff. But almost all the space freaks wanted to go. They're used to living that way, anyhow (and they never get to go anyplace).
    C. When they get back, the farms will be used as a starter for L-4, like L-5 but smaller at first, and on the other side of the moon (pic down left).
2. For other Tricentennial fax & pix, see bacover.

July 2076

Charlie was just finishing up a week on Earth the day the *John F. Kennedy* was launched. Tired of being interviewed, he slipped away from the media lounge at the Cape shuttleport. His white clearance card got him out onto the landing strip alone.

The midnight shuttle was being fueled at the far end of the strip, gleaming pink-white in the last light from the setting sun. Its image twisted and danced in the shimmering heat that radiated from the tarmac. The smell of the soft tar was indelibly associated in his mind with leave-taking, relief.

He walked to the middle of the strip and checked his watch. Five minutes. He lit a cigarette and threw it away. He rechecked his mental calculations: the flight would start low in the southwest.

He blocked out the sun with a raised hand. What would 150 bombs per second look like? For the media they were called fuel capsules. The people who had carefully assembled them and gently lifted them to orbit and installed them in the tanks, they called them bombs. Ten times the brightness of a full moon, they had said. On L-5 you weren't supposed to look toward it without a dark filter.

No warm-up: it suddenly appeared, an impossibly brilliant rainbow speck just over the horizon. It gleamed for several minutes, then dimmed slightly with a haze, and slipped away.

Most of the United States wouldn't see it until it came around again, some two hours later, turning night into day, competing with local pyrotechnic displays. Then every couple of hours after that, Charlie would see it once more, then get on the shuttle. And finally stop having to call it by the name of a dead politician.

September 2076

There was a quiet celebration on L-5 when Daedalus reached the mid-point of its journey, flipped, and started decelerating. The progress report from its crew characterized the journey as "uneventful." At that time they were going nearly two-tenths of the speed of light. The laser beam that carried communications was redshifted from blue light down to orange; the message that turnaround had been successful took two weeks to travel from Daedalus to L-5.

They announced a slight course change. They had analyzed the polarization of light from Scylla/Charybdis as their phase angle increased, and were pretty sure the system was surrounded by flat rings of debris, like Saturn. They would "come in low" to avoid collision.

January 2077

Daedalus had been sending back recognizable pictures of the Scylla/Charybdis system for three weeks. They finally had one that was dramatic enough for groundhog consumption.

Charlie set the holo cube on his desk and pushed it around with his finger, marvelling.

"This is incredible. How did they do it?"

"It's a montage, of course." Johnny had been one of the youngest adults left behind: heart murmur, trick knees, a surfeit of astrophysicists.

"The two stars are a strobe snapshot in infrared. Sort of. Some ten or twenty thousand exposures taken as the ship orbited around the system, then sorted out and enhanced." He pointed, but it wasn't much help, since Charlie was looking at the cube from a different angle.

"The lamina of fire where the atmospheres touch, that was taken in ultraviolet. Shows more fine structure that way.

"The rings were easy. Fairly long exposures in visible light. Gives the star background, too."

A light tap on the door and an assistant stuck his head in. "Have a second, Doctor?" "Sure." "Somebody from a Russian May Day committee is on the phone. She wants to know whether they've changed the name of the ship to Brezhnev yet."

"Yeah. Tell her we decided on 'Leon Trotsky' instead, though."

He nodded seriously. "'Okay.'" He started to close the door.

"Wait! Charlie rubbed his eyes. "Tell her, uh…the ship doesn't have a commemorative name while it's in orbit there. They'll rechristen it just before the start of the return trip."

"Is that true?" Johnny asked.

"I don't know. Who cares? In another couple of months they won't want it named after anybody." He and Ab had worked out a plan—admittedly rather shaky—to protect L-5 from the groundhogs' wrath: nobody on the satellite knew ahead of time that the ship was headed for 61 Cygni. It was a decision the crew arrived at on the way to Scylla /Charybdis; they modified the drive system to accept matter-antimatter destruction while they were orbiting the double star. L-5 would first hear of the mutinous plan via a transmission sent as Daedalus left Scylla/ Charybdis. They'd be a month on their way by the time the message got to Earth.

It was pretty transparent, but at least they had been careful that no record of Daedalus' true mission be left on L-5. Three thousand people did know the truth, though, and any competent engineer or physical scientist would suspect it.

Ab had felt that, although there was a better than even chance they would be exposed, surely the groundhogs couldn't stay angry for twenty-three years—even if they were unimpressed by the antimatter and other wonders.

Besides, Charlie thought, it's not their worry anymore.

As it turned out, the crew of Daedalus would have bigger things to worry about.

June 2077

The Russians had their May Day celebration—Charlie watched it on TV and winced every time they mentioned the good ship *Leonid 1. Brezhnev*—and then things settled back down to normal. Charlie and three thousand others waited nervously for the "surprise" message. It came in early June, as expected, scrambled in a data channel. But it didn't say what it was supposed to:

"This is Abigail Bemis, to Charles Leventhal.

"Charlie, we have real trouble. The ship has been damaged, hit in the stern by a good chunk of something. It punched right through the main drive reflector. Destroyed a set of control sensors and one attitude jet.

"As far as we can tell, the situation is stable. We're maintaining acceleration at just a tiny fraction under one-G. But we can't steer, and we can't shut off the main drive.

"We didn't have any trouble with ring debris when we were orbiting since we were inside Roche's limit. Coming in, as you know, we'd managed to take advantage of natural divisions in the rings. We tried the same going back, but it was a slower, more complicated process, since we mass so goddamn much now. We must have picked up a piece from the fringe of one of the outer rings.

"If we could turn off the drive, we might have a chance at fixing it. But the work pods can't keep up with the ship, not at one-G. The radiation down there would fry the operator in seconds, anyway.

"We're working on it. If you have any ideas, let us know. It occurs to me that this puts you in the clear—we were headed back to Earth, but got clobbered. Will send a transmission to that effect on the regular comm channel. This message is strictly burn before reading.

"Endit."

It worked perfectly, as far as getting Charlie and L-5 off the hook, and the drama of the situation precipitated a level of interest in space travel unheard-of since the 1960's.

They even had a hero. A volunteer had gone down in a heavily shielded work pod, lowered on a cable, to take a look at the situation. She'd sent back clear pictures of the damage, before the cable snapped.

Daedalus: A.D. 2081
Earth: A.D. 2101

The following news item was killed from Fax & Pix, because it was too hard to translate into the "plain English" that made the paper so popular:

SPACESHIP PASSES 61 CYGNI...SORT OF

(L-5 Stringer) A message received today from the spaceship Daedalus said that it had just passed within 400 astronomical units of 61 Cygni. That's about ten times as far as the planet Pluto is from the sun.

Actually, the spaceship passed the star some eleven years ago. It's taken all that time for the message to get back to us.

We don't know for sure where the spaceship actually is, now. If they still haven't repaired the runaway drive, they're about eleven light-years past the 61 Cygni system (their speed when they passed the double star was better than 99 percent the speed of light).

The situation is more complicated if you look at it from the point of view of a passenger on the spaceship. Because of relativity, time seems to pass more slowly as you approach the speed of light. So only about four years passed for them, on the eleven light-year journey.

L-5 Coordinator Charles Leventhal points out that the spaceship has enough antimatter fuel to keep accelerating to the edge of the galaxy. The crew then would be only some twenty years older-but it would be twenty thousand years before we heard from them....

(Kill this one. There's more stuff about what the ship looked like to the people on 61 Cygni, and how cum we could talk to them all the time even though time was slower there, but its all as stupid as this.)

Daedalus: A.D. 2083
Earth: A.D. 2144

Charlie Leventhal died at the age of 99, bitter. Almost a decade earlier it had been revealed that they'd planned all along for Daedalus to be a starship. Few people had paid much attention to the news. Among those who did, the consensus was that anything that got rid of a thousand scientists at once, was a good thing. Look at the mess they got us in.

Daedalus: 67 light-years out, and still accelerating.
Daedalus: A.D. 2085
Earth: A.D. 3578

After over seven years of shipboard research and development—and some 1500 light-years of travel—they managed to shut down the engine. With sophisticated telemetry, the job was done without endangering another life.

Every life was precious now. They were no longer simply explorers; almost half their fuel was gone. They were colonists, with no ticket back.

The message of their success would reach Earth in fifteen centuries. Whether there would be an infrared telescope around to detect it, that was a matter of some conjecture.

Daedalus: A.D. 2093 Earth: ca. A.D. 5000

While decelerating, they had investigated several systems in their line of flight. They found one with an Earth-type planet around a sun-type sun, and aimed for it.

The season they began landing colonists, the dominant feature in the planet's night sky was a beautiful blooming cloud of gas that astronomers had named the North American Nebula.

Which was an irony that didn't occur to any of these colonists from L-5-give or take a few years, it was America's Trimillennial.

America itself was a little the worse for wear, this three thousandth anniversary. The seas that lapped its shores were heavy with a crimson crust of anaerobic life; the mighty cities had fallen and their remains, nearly ground away by the never-ceasing sandstorms. No fireworks were planned, for lack of an audience, for lack of planners; bacteria just don't care. May Day too would be ignored.

The only humans in the solar system lived in a glass and metal tube. They tended their automatic machinery, and turned their backs on the dead Earth, and worshiped the constellation Cygnus, and had forgotten why.

Everyone knows from pop science fiction such as Star Wars and Star Trek that ideas of how to cross immense distances in a twinkling of time do emerge from the odd and sometimes extravagant realms of theoretical physics. How plausible are such notions?

The truthful answer is that no one knows. Progress in the furthest realms of General Relativity and quantum mechanics must proceed from experiment, and there are few lab experiments that can touch on such issues. To survey the current landscape of such thinking, the 100 Year Starship Symposium held an Exotic Technologies Session chaired by John Cramer. Here he reports on the major ideas treated there, with some insightful criticisms of his own, and much background material useful to the interested but non-specialist observer. One recalls the Mark Twain observation, "There is something fascinating about science. One gets such wholesale returns of conjecture out of such a trifling investment of fact."

Exotic Technologies for Interstellar Travel
John Cramer

I. Introduction

When I first came to Seattle in the mid-1960s to assume my new faculty position at the University of Washington, I remarked to one of my new Physics Department colleagues that the magnificent views of Mt. Rainier, available from many parts of the city, gave Seattle a special and unique flavor and ambiance.

"You, know," he said, "Rainier is dormant at the moment, but it's still an active volcano. It erupts every five thousand years. The geological record shows regular thick ash falls from Rainier and giant mud flows down the glacier-fed river valleys, where a lot of people live now."

"Really," I said. "When was the last eruption?"

"Five thousand years ago," he said with a quirky smile.

That anecdote provides a good analogy to our present situation, as residents of this planet. All of our "eggs," our cities, our people, our art and culture, our accumulated knowledge and understanding, are presently contained in one pretty blue "basket" called Planet Earth, which orbits in a solar system that is configured to hurl very large rocks in our direction at random intervals.

It is estimated that a total of about fourteen million tons of meteoritic material falls upon Planet Earth each year, much of it from the debris of asteroids and comets. Meteors come in all sizes, and approximately sixty giant meteorites five or more kilometers in diameter have impacted Planet Earth in the past 600 million years. Even the smallest of these would have carved a crater some ninety-five kilometers across and produced an extinction event.

The geological fossil records shows evidence of "punctuated equilibrium," extended periods in which life forms expand and fit themselves into the available ecological niches, punctuated by extinction events in which many species become extinct and the survivors scramble to adapt to the new conditions1. Life on Planet Earth may have been "pumped" on a fast track to its present state of evolution by this cycle of extinction and regeneration. We may owe our very existence to this pump of evolution,2 but we do not want to get caught in the next pump cycle. We, as a species, need to diversify, to place our eggs in many baskets instead of just one, before the forces of nature conspire to produce another extinction event that could include us.

The basic problem with such a "basket diversification" project is that we reside at the bottom of a very deep gravity well, from which the laws of physics make it very difficult for us to escape. The only escape method presently in use involves giant chemical rockets that burn and eject vast volumes of expensive and toxic fuel in order to lift tiny payloads partway out of the gravity well of Earth.

And even if we can escape most of Earth's gravity well, things are not much better in near-Earth orbit. The solar system, outside Earth's protective atmosphere and shielding magnetic field, is a fairly hostile place, a hard vacuum environment in which the sun's flares and storms send out wave after wave of sterilizing radiation.

Further, the human biology seems to require the pull of gravity for a healthy existence. Extended periods in low gravity lead to calcium loss and muscular and skeletal degeneration. Our micro-G International Space Station is an unhealthy place for long-term habitation, and astronauts return from extended stays there as near-invalids.

The other planets and moons of the solar system, potential sources of the needed pull of gravity, are not promising sites for human habitation. Mars is too cold, too remote from the sun, and has a thin atmosphere, mostly carbon dioxide with a pressure of 1/100 of an Earth atmosphere. Venus is much too hot, with a surface temperature around 870 °F and an atmospheric pressure, mostly carbon dioxide, about 90 times that of Earth. Moons, asteroids, and artificial space habitats with centrifugal pseudo-gravity may be better sites for human colonies, but they all would have problems with low gravity, radiation shielding, and resource transport. To find a true Earthlike habitat, we need to leave the solar system for the Earthlike planets orbiting other stars.

But if escaping Earth's gravity well is difficult, travel to the stars is many orders of magnitude more difficult. Fairly optimistic studies of interstellar travel, presented at the 100YSS Symposium in 2011, show very convincingly that there is little hope of reaching the nearby stars in a human lifetime using any conventional propulsion techniques, even with propulsion systems involving nuclear energy. The universe is simply too big, and the stars are too far away.

To reach the stars, we need propulsion techniques that somehow circumvent Newton's Third Law and do not require the storage, transport, and expulsion of large volumes of reaction mass.

***Ascent of the Blessed, The Doge's Palace, Venice,
Italy* (Hieronymus Bosch, c.1450-1516)**

Or even better, we need trans-spatial shortcuts like wormholes that avoid the need to traverse the enormous distances between stars. In short, because conventional technologies are inadequate, the human-lifetime-scale pathways to the stars require over-the-horizon "exotic" technologies, perhaps like the one illustrated here by Hieronymus Bosch.

I was Chairman of the Exotic Technologies Session held on October 1, 2011, at the 100 Year Starship Symposium in Orlando Florida. This chapter draws on the talks given in that session, but it does not represent a summary of the presentations. Rather, I want focus on three lines of development in the area of exotic technologies that were featured at the Symposium, developments that might allow us to reach the stars on a time scale of a human lifetime: (1) propellantless space drives, (2) warp drives, and (3) wormholes. With reference to the latter two topics, I will also discuss some cautions from the theoretical physics community about the application of general relativity to "metric engineered" devices like wormholes and warp drives that require exotic matter.

II. Space Drives

The term "space drive" refers to an exotic technology that does not presently exist and that would allow the propulsion of a space vehicle without the need for rocket-style expulsion (or reflection) of reaction mass-energy. In the leadoff talk of the Exotic Technologies track of the 100YSS Symposium, Marc G. Millis summarized the current prospects for space drives. He concluded that "although no propulsion breakthroughs appear imminent, the subject has matured to (the stage) where the relevant questions have been broached and are beginning to be answered." In this section, I want to consider the topic further and to discuss one of the most promising space drive developments, one involving Mach's principle and inertia variation.

The basic problem with the space drive concept is Newton's Third Law of motion, one form of the law of conservation of momentum. In conventional rocket propulsion, a space vehicle can be propelled forward and increase its forward momentum only if propellant with an equal and opposite incremental momentum is expelled as exhaust. No internal motion, no shaking, spinning, or orbiting of masses, no tilting of eccentric flywheels, can produce any net momentum change in the overall object. Something must go backwards if something else goes forward.

As a work-around to avoid carrying onboard reaction mass, emission, reflection, or absorption of light from a beam of light (laser or incoherent) or radio waves could, in principle, produce significant propulsion and momentum change in a space vehicle that does not carry and expel reaction mass (solar sails or beam riders). The problem with such schemes is that the momentum content of light is very small, only its energy divided by the speed of light, and therefore the thrust (in newtons) is the power (in watts) divided by the speed of light. The speed of light is a large number, making the energy cost is very high for a small change in momentum.

Nevertheless, light sailing and light beam propulsion are possibilities. They are tricky because the resulting momentum increment must always have a momentum component away from the light source (e.g., the sun or drive laser). They are expensive because many square kilometers of light sail and/or multimegawatt lasers are required to achieve thrust comparable to rocket propulsion. They are inefficient because large quantities of light energy are required for small quantities of momentum change, with most of the light beam energy reflected away and wasted, in the sense that it does not end up as kinetic energy.

Millis's overview of possible space-drive technologies briefly mentioned the work of Prof. James Woodward of Cal. State Fullerton on Mach's principle and inertia variation. However, there have been some new results since the Symposium that I would like to discuss further.

First, what is Mach's principle? The physical property of mass has two distinct aspects, gravitational mass and inertial mass. *Gravitational mass* produces and responds to gravitational fields. It is represented by the two mass factors in Newton's inverse-square law of gravity:

$$F_{12} = g m_1 m_2 / r_{12}^2$$

Inertial mass is the tendency of matter to resists acceleration. It is represented by the mass factor in Newton's Second law of motion:

$$F = ma$$

These two aspects of mass always track one another. There are no known objects with a large inertial mass and a small gravitational mass, or *vice versa*. One of the deep mysteries of physics is the connection between inertial and gravitational mass.

Ernst Mach (1838–1916) was an Austrian physicist whose unpublished ideas about the origin of inertia influenced Einstein. Mach's principle, as elucidated by Einstein, attempts to connect inertia with gravitation by suggesting that the resistance of inertial mass to acceleration arises from the long-range gravitational forces from all the other masses in the universe acting on a massive object (so that, in an universe empty of other masses, there would be no inertia). In essence, Mach's principle asserts that inertial and gravitational mass must be the same because inertia is, at its roots, a gravitational effect.

Albert Einstein liked Mach's principle and used its implications to formulate his famous equivalence principle, a cornerstone of general relativity, which asserts that gravitational and inertial mass are indistinguishable in all situations. In a small isolated room, according to the equivalence principle, it would be impossible to determine from local measurements whether the room was on the surface of Earth in a one-G gravitational field or was in a rocket ship accelerating at one-G through gravity-free space. The equivalence principle is now generally accepted in physics, and general relativity (GR) has become our standard model of gravity, but its underlying basis in Mach's principle has never been properly derived, understood, or tested until now.

Dennis Sciama used a simplified low-field reduction of Einstein's general relativity equations to show that in a uniform flat universe, long range gravitational interactions produce a force that resists acceleration, producing inertia. James Woodward extended the work of Sciama by considering the time dependent inertial effects that occur when mass-energy is in flow, i.e., when mass-energy is moved from one part of the system to another while the system is being accelerated.

The Woodward/Sciama result is surprising. It predicts fairly large time-dependent variations in inertia, the tendency of matter to resist acceleration. Most gravitational effects predicted in general relativity, e.g., the gravitational deflection of light, frame dragging, gravitational time dilation, etc., are exceedingly small and difficult to observe, because the algebraic expressions describing them always have a numerator that includes Newton's gravitational constant G, a physical constant that has a very small value due to the weakness of gravity as a force. The inertial transient effects predicted by the Woodward/Sciama calculations are unusual and different, in that they have G in the *denominator*, with the result that dividing by a small number (G) produces a sizable effect.

Can varying the inertial mass of an object produce thrust, for example by pushing it forward when the inertial mass is low and pulling it backward when the inertial mass is high, thereby "rowing" through space? Woodward has tested for a net thrust from this effect using piezoelectric devices that combine stored energy with accelerated motion. His results are unpublished and have not been confirmed by others who have attempted to reproduce them. However, the recent work, showing thrust of a few tens of micronewtons, has been posted on the Internet and widely discussed. It has the possibility of being a real effect.

However, I would like to introduce a cautionary note here about thrust from inertial mass variation and momentum conservation. The relativistically invariant form of Newton's Second Law, as applied to Woodward-type thrusters, should be $F = dp/dt$, where F is the thrust produced and $p = m\,v$ is the momentum of the system producing the thrust, with inertial mass m and velocity v. When the mass is constant with time, this equation becomes $F = m\,a$, the classical form of Newton's Second Law that Woodward uses in predicting the thrust derived from his calculations and that is the basis for the "rowing" space-drive effect described above.

However, when the inertial mass is varying with time, as it should be in the system of interest, the appropriate form of Newton's Second Law is $F = m\,a + v\,dm/dt$, i.e., one must time-differentiate *both* the varying velocity and the varying mass factors that form the momentum. It turns out that in any mass-varying space drive, the second term produces a force that exactly cancels the force derived from the first term, so that, if both force terms are generated within the system, no net thrust is produced, even in a system where the inertial mass can be caused to vary with time. Woodward claims that the second term is not an internal force, but is a distant and external one because of the way that inertial mass and momentum couple to distant objects, so that it represents the reaction force that the rest of the universe receives from the action-at-a-distance of the drive. This is an interesting argument that may or may not be correct. In his talk in the Exotic Technologies track of the 100YSS Symposium, Eric Davis perhaps provided some support for this view by presenting arguments, in the context of FTL space warps, that conservation laws, and in particular momentum conservation, are local flat-space rules based on symmetries and may not directly apply to large-scale situations in which curved space and general relativity are important. If Woodward's reported observations of thrust can be verified, that should perhaps settle this issue.

We note here that the Mach/Schiama/Woodward approach may also have possible implications for starship travel in another way. There is second negative-definite term in the Woodward/Sciama inertia variation calculation that could have important general relativity implications for wormholes and warp drives. This will be discussed in Section V below. However, we note that at least one physicist has questioned whether Woodward's second term is implicit in the derivation of the first term.

III. Cautionary Note: Limits to Exotic Applications of General Relativity

When I was in studying physics in graduate school, the calculations of general relativity were done exclusively by hypothesizing a configuration of mass-energy and then calculating the "metric" or distortion of space time that it produced. This conventional approach has lead to many interesting results, but none that could be considered exotic or unphysical.

But there is another way to do such calculations in general relativity, an approach that has been labeled "metric engineering." One specifies a space-time metric that will produce some desired result, for example a wormhole or a warp drive, and then calculates the distribution of masses and forces that would be required to produce such a metric, with all its consequences. General relativity, used in this way, becomes "exotic," suggesting the possibility of transversable wormholes, faster-than-light warp drives, and even time machines.

Dr. Keith Olum, in the final Exotic Technologies paper of the 100YSS Symposium of 2011, presented a cautionary note that emphasized that the exotic solutions to Einstein's equations of general relativity, which appear to provide a pathway to the stars, may not be realizable.

Many of the theoretical physicists who work with general relativity have had fundamental objections to the very idea of wormholes and warp drives, which they consider to be unphysical. Some of them have decided that one should erect a "picket fence" around those solutions of Einstein's equations that are considered to be physically reasonable, and to place exotica like stable transversable wormholes, faster-than-light warp drives, and time machines in the forbidden area outside the fence, excluded because it is presumed that Nature does not allow such disreputable objects to exist. They are, in effect, attempting to discover new laws of physics that would place restrictions forbidding certain GR solutions.

In discussing the behavior of collapsed-matter singularities in general relativity, Hawking and Ellis introduced a number of "energy conditions" that, in their view, had to be observed by acceptable solutions of general relativity and might represent the picket fence mentioned above. The first of these is called the Weak Energy Condition (WEC). In essence, the WEC assumes that negative energy is the source of "problems" with GR and requires that for all observers, the local energy in all space-time locations must be greater than or equal to zero. In other words, if any possible observer would see a negative energy, then that solution of Einstein's equations is excluded by the WEC. A less restrictive variant of the WEC is the Average Weak Energy Condition (AWEC), which requires that when time-averaged along some arbitrary world-line through all time, the net energy must be greater than or equal to zero, so that any time period when the energy is negative must be compensated by a period of positive energy. In his talk in the Exotic Technologies track of the 100YSS Symposium, Eric Davis provided a detailed description and analysis of these energy conditions.

The WEC, AWEC, and the other similar energy rules are "made-up" laws of Nature and are not derivable from general relativity itself. They appear to be obeyed for observations of all known forms of matter and energy that do not fall within the domain of quantum mechanics.

However, even for simple situations involving quantum phenomena (examples: the Casimir effect, squeezed vacuum, and the Hawking evaporation of black holes), the WEC and AWEC are both violated.

More recently Ford and Roman have derived from quantum field theory certain quantum inequalities (QI) that must be observed by solutions of the equations of general relativity. Basically, one chooses a "sampling function," some bell-shaped curve having unit area and a width that specifies a particular restricted region of time. This function is then used with quantum field theory methods to average the energy per unit volume of a field within the time-sampling envelope and to place limits on how much negative energy is allowed to exist for how long.

These quantum inequalities are bad news for would-be practitioners of metric engineering. Taken at face value, the QI say that stable wormholes may be impossible and that a warp drive might, at best, exist for too short a time to go anywhere. While a wormhole might wink into existence during the short time that the negative energy is present, it would wink out of existence again before any matter could pass through it. It appears that within the QI conditions, when negative energy is created, it is either too small in magnitude or too brief in duration to do anything interesting.

However, it is not clear whether Woodward's proposed techniques employing inertia transients (see Section V below) are subject to the QI limitations. Further, there are reasons to doubt that quantum field theory can be trusted in its application to the field-energy situations envisioned by the QI calculations.

We know that quantum field theory must be wrong, in some fundamental way. It attributes far too much positive energy to space-time itself. The density of "dark energy," the irreducible intrinsic energy in a given volume of space, as deduced from the observations of astrophysicists investigating Type Ia supernovas and the space-frequency structure of the cosmic microwave background is about: 6.7×10^{-10} joules per cubic meter. The same quantity, as calculated by quantum field theory, is about 1040 joules per cubic meter. Thus, quantum field theory has missed the mark in this very fundamental calculation involving energy density by about fifty orders of magnitude!

Therefore, until quantum field theory (or its quantum gravity successor) can accurately predict the energy content of the vacuum, I feel that the restrictions that it places on metric engineering cannot be taken completely seriously. Woodward has also argued that the derivation of the QI involves use of the Second Law of thermodynamics in a way that may be inappropriate for artificially produced wormholes. These arguments perhaps leave the pathway to the stars, as represented by the doorway presented by metric engineering solutions of general relativity, open just a crack.

IV. Warp Drives and General Relativity

General relativity treats special relativity as a restricted sub-theory that applies locally to any region of space sufficiently small and flat that its gravity-induced curvature can be neglected. General relativity does not forbid faster-than-light (FTL) travel or communication, but it does require that the local restrictions of special relativity must be observed. In other words, light speed is the *local* speed limit, but the broader considerations of general relativity may provide an end-run way of circumventing this local statute.

One example of FTL motion allowed by general relativity is the expansion of the universe itself. As the universe expands, new space is being created between any two separated objects. The objects themselves may be at rest with respect to their local environment and with respect to the cosmic microwave background, but the distance between the objects may grow at a rate greater than the speed of light. According to the standard model of cosmology, remote parts of our universe are receding from us at FTL speeds, and therefore are completely unreachable and isolated from us. As the rate of expansion of the universe increases due to the action of dark energy, a growing volume of the universe is disappearing over this redshift horizon and becoming inaccessible to us.

Another example of effective FTL motion is a wormhole connecting two widely separated locations in space, say five light-years apart. An object might take a few minutes to move with at low speed through the neck of a wormhole, observing the local speed-limit laws all the way. However, by transiting the wormhole the object has traveled five light-years in a few minutes, producing an effective speed of a million times the velocity of light.

Miguel Alcubierre, using the technique of metric engineering described above, proposed a way of beating the FTL speed limit that is somewhat like the expansion of the universe, but on a more local scale. He developed a metric for general relativity, a mathematical representation of the curvature of space, that is completely consistent with Einstein's equations. It describes a region of flat space surrounded by a spatial warp bubble that propels the flat region forward at any arbitrary velocity, including FTL speeds.

Alcubierre's warp is constructed from mathematical hyperbolic tangent functions that create a very peculiar distortion of space at the edges of the flat-space volume. In effect, new space is rapidly being created (like an expanding universe) at the back side of the moving flat-space volume, and pre-existing space is being annihilated (like a universe collapsing to a Big Crunch) at the front side of the moving flat-space volume. Thus, a space ship within the volume of the Alcubierre warp bubble (and the flat-space volume itself) would be pushed forward by the expansion of space at its rear and the contraction of space in front.

For those familiar with usual rules of special relativity, with its Lorentz contraction, mass increase, and time dilation, the Alcubierre warp metric has some rather peculiar aspects. Since a ship at the center of the moving volume of the metric is at rest with respect to locally flat space, there are

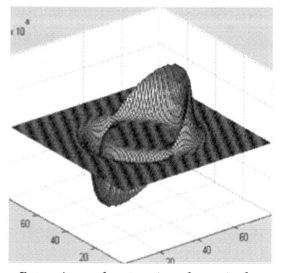

Expansion and contraction of space in the Alcubierre warp drive metric

no relativistic mass increase or time dilation effects. The on-board spaceship clock runs at the same speed as the clock of an external observer, and that observer will detect no increase in the mass of the moving ship, even when it travels at FTL speeds. Moreover, Alcubierre has shown that even when the ship is accelerating, it is always in free fall, and the crew would experience no accelerational G-forces when it starts and stops. Enormous tidal forces would be present near the edges of the flat-space volume because of the very large space curvature there, but by suitable specification of the metric, these could be made negligible within the volume occupied by the ship. Talks in the Exotic Technologies track of the 100YSS Symposium by Eric Davis and Harold White contain additional discussions of the Alcubierre warp metric, possible extensions and improvements, and its control.

All of this, for those of us who would like to go to the stars without the annoying limitations imposed by special relativity, appears to be too good to be true. "What's the catch?" we ask. As it turns out, there are several "catches" in the Alcubierre warp drive scheme. The first is that, while his warp metric is a valid solution of Einstein's equations of general relativity, we have no idea how to produce such a distortion of space-time. Its implementation would require the imposition of radical curvature on extended regions of space. Within our present state of knowledge, the only way of producing curved space is by using mass, and the masses we have available for works of engineering lead to negligible space curvature. Moreover, even if we could do engineering with mini black holes (which have lots of curved space near their surfaces) it is not clear how an Alcubierre warp could be produced. Further, it is not clear how the warp bubble could be steered or controlled, since the interior volume of the warp bubble is completely isolated from the outside, so that steering commands, control information, and views of the outside would be completely blocked. We note, however, that if quantum nonlocality could be used for signaling, an issue that is currently being investigated by the author, that development would solve the problem of warp-bubble control.

Alcubierre has also pointed out a more fundamental problem with his warp drive. General relativity provides a procedure for determining how much energy density (energy per unit volume) is implicit in a given metric (or curvature of space-time). He shows that the energy density is negative, rather large, and proportional to the square of the velocity with which the warp moves forward. This means that all of the proposed energy conditions (see Section IV) of general relativity are violated, which can be taken as arguments against the possibility of creating a working Alcubierre drive. Alcubierre, following the lead of wormhole theorists, argues that quantum field theory permits the existence of regions of negative energy density under special circumstances, and cites the Casimir effect as an example. Thus, the situation for the Alcubierre drive is similar to that of stable wormholes: they are solutions to the equations of general relativity, but one would need exotic matter with large quantities of negative mass-energy to actually produce them, and we have none at the moment.

At the 2011 100 Year Starship Symposium, NASA's Harold "Sonny" White reviewed the Alcubierre scheme. He discussed techniques for minimizing the amount of exotic matter required, and showed his plans for an optical interferometer that could be used to attempt observation of the small distortions in space-time that might be induced in the vicinity of a high-voltage toroidal capacitor device. This he characterized as an initial step in the direction of producing the space-time distortions of a magnitude that would be needed for a true warp-drive of the type described by Alcubierre. A recent check with Dr. White indicated that the interferometer has now been constructed and tested, and a novel interference-pattern filtering technique has been devised for improving its sensitivity, but the space-distortion tests themselves have not yet been performed.

In his talk at the 2011 100 Year Starship Symposium, Eric Davis also reported that a combined effort of EarthTech International, Inc. and Lockheed-Martin is using meta-materials to create optical analogues that simulate the behavior of wormholes and warp drives in the laboratory. We await the results of these interesting efforts.

V. Worm Holes and General Relativity

In 1916, Albert Einstein first introduced his general theory of relativity, the theory that to this day remains our standard model for gravitation. Twenty years later, he and his long-time collaborator Nathan Rosen published a paper showing that implicit in the general relativity formalism is a curved-space structure that can join two distant regions of space-time through a tunnel-like curved-space shortcut. Their purpose was not to promote travel to the stars, but to explain the existence of fundamental particles like electrons and positrons by describing them as the ends of space-tunnels threaded by electric lines of force. The lines of electric flux would go in at the electron end of the tunnel and emerge at the positron end. This Einstein-Rosen electron model was subsequently shown to have a serious problem when it was demonstrated that the smallest possible mass-energy of such a curved-space topology is larger than that of a Planck mass, a few micrograms, and far larger than the 511 keV mass-energy of an electron.

The Einstein-Rosen work was disturbing to many physicists of the time because such a "tunnel" through space-time, which came to be known in the late 1930s and 40s as an *Einstein-Rosen Bridge*, could in principle allow the transmission of information and matter faster than the speed of light. In 1962 John A. Wheeler and Robert W. Fuller discovered that the Einstein-Rosen bridge space-time structure, which Wheeler re-christened as a "wormhole," was dynamically unstable in field-free space. They showed that if such a wormhole somehow opened, it would close up again before even a single photon could be transmitted through it, thereby preventing superluminal transmission of information.

In 1989 the instability of wormholes was called into question when Michael Morris, Kip Thorne, and Ulvi Yurtsever described how an "advanced civilization" might: (a) create a large wormhole; (b) stabilize it to prevent its re-collapse; and (c) convert it to a time machine, a device for traveling (or at least communicating) back and forth in time. This remarkable paper, which borders on science fiction in its approach, has a very serious purpose. There is presently no well-established theory (quantum gravity) that can accommodate both quantum mechanics and the physics of strong gravitational fields within the same mathematical framework. The paper of Morris, Thorne, and Yurtsever is a vehicle for guessing, in a rather unorthodox way, what restrictions a proper theory of quantum gravity might place on the physics of wormholes. The authors demonstrate that general relativity contains within its framework mechanisms that appear to permit both faster-than-light travel and time travel. If these physical calamities (as viewed by some physicists) are to be averted, the authors argue, it can only be done through a proper theory of quantum gravity.

How could a wormhole be created? Empty space, when examined with quantum field theory on a sufficiently small distance scale, is not empty at all. Even at nuclear dimensions (10-13 cm) empty space is filled with particle-antiparticle pairs that are continually flashing into a brief existence, bankrolled on the credit of borrowed mass-energy, only to wink out of existence again as the law of conservation of energy reasserts itself. Heisenberg's uncertainty principle provides the cover that makes such energy juggling possible. If the length-scale is contracted to a size appropriate to quantum gravity (10-33 cm) this quantum fireworks should intensify to become a quantum foam of violent fluctuations in the topology and geometry of space itself. Quantum black holes should form and vanish in a span of time of 10-23 seconds; highly curved and convoluted regions of space-time in any physically allowed configuration should have a similarly brief existence. In this environment, Morris, Thorne, and Yurtsever speculated, it may be possible for a civilization, one considerably more advanced than ours, to pull a wormhole out of the quantum foam, stabilize it, and enlarge it enormously to create a connection between two nearby points in space. This would exploit the well-known quantum mechanical process called tunneling, a jump from one allowed energy state to another across a barrier of intermediate states that are forbidden by energy conservation.

To stabilize a wormhole pulled from the quantum foam, preventing its immediate re-collapse, Morris, Thorne, and Yurtsever proposed to use the Casimir effect in the mouth of the wormhole, creating a region of negative mass-energy that would force it to remain open. They suggest that this might be accomplished by placing a pair of spheres with equal electric charges at the two spatial entrances of the wormhole. The spheres would be held in place by a delicate balance, the attractive Casimir force between them just offsetting the force of their electrical repulsion. Such a system might be very small, an atom-scale opening permitting the passage of only a few photons at a time, or it might be large enough to pass a large space vehicle.

Having produced this stabilized wormhole, the engineering can begin. The size of the connection can be enlarged or contracted depending on energy considerations. The two portal ends of the wormhole connection can be separated from each other. For example, a portal placed aboard a space ship might be carried to some location many light-years away. Such a trip might require a long time, but during the trip and afterwards instantaneous two-way communication and even transport through the wormhole might be available.

This brings us to the last point of the Morris, Thorne, and Yurtsever paper, the construction of a time machine. Suppose that initially a wormhole establishes a connection between two spatial points A and B that have no motion with respect to each other and are simultaneous in time. By "simultaneous," a slippery concept in relativity, we mean that an observer at A who determines a clock reading at B would get the same reading via normal space (by light beam signals corrected for transit time, for example) as he would through the wormhole.

Now suppose, in the spirit of the Twin Paradox of special relativity, that portal B is placed aboard a space ship while portal A remains on Earth. The ship carrying B, say, accelerates rapidly to 86.6 percent of light speed and travels a distance of 0.866 light-years, then reverses its course and returns to Earth at the same speed. On its arrival, portals *A* and *B* are placed near one another. At 86.6 percent of the velocity of light, due to relativistic time dilation, from the point of view of an Earth observer any clock aboard the ship will run at just half the speed of a similar clock on Earth. From the point of view of an observer on Earth, the round trip has taken two years, but from the point of view of an observer on the ship, the round trip has taken one year. Therefore at the end of the trip the ship's clock will be one year slow, as compared to an identical clock that had remained on Earth.

And, as Morris, Thorne, and Yurtsever point out, portal *B* will also be one year slow as compared with portal *A*. Now a message that is sent through *B* to *A* will emerge one year in the future of *B*, and a message sent through *A* to *B* will emerge one year in the past of *A*! We can send messages back and forward in time. Similarly, a traveler making the same trips through the wormhole would travel one year into the future or the past. The wormhole connection through space has been transformed to a connection through time, a wormhole time machine.

Do wormholes, embodying faster-than-light space travel (even with space-separated simultaneous wormholes) as well as time travel (from time-separated wormholes), demonstrate that special relativity is wrong? Do wormholes indicate that Einstein's special relativity speed limit is wrong? Not at all. The restrictions usually associated with special relativity implicitly assume that no time travel is possible. Clearly one could travel, in effect, at an infinite velocity by traveling from one place to another at some sub-light velocity and then on arrival traveling backwards in time to the instant of departure. To put it another way, the simultaneity measurements prohibited by special relativity must lead to a definite and unambiguous determination of the simultaneous readings of two clocks separated in space. The clock-comparisons made possible by wormholes are not definite, because one clock could be in the future of the other, displaced by any time interval produced by the travel histories of the portals. Special relativity, which after all is embedded in the theory of general relativity that produced these revelations about wormhole physics, must be preserved.

What law of physics gets destroyed by the construction of a wormhole space-time connection? Causality, the mysterious principle that prohibits communication backwards in time, that requires a cause to precede its effects in time sequence in all space-time reference frames. Causality as a law of the universe would not survive even a two-way communications link across time, let alone a portal permitting trans-time matter transmission. This bothers a lot of physicists.

Eric Davis argues that causality is routinely violated in many situations in general relativity and is perhaps only a local flat-space constraint, like Lorentz invariance.

The problem with all of this, of course, is that we have neither relativistic starships nor any technology for capturing and stabilizing wormholes. At present, we are able to produce only small regions containing negative mass-energy in magnitudes that are tiny compared to what would be needed to stabilize a wormhole, even if one were available to stabilize. Is there any hope of addressing this problem?

As mentioned above, an "end run" idea comes from the work of James Woodward, as discussed in section II above. Woodward's derivation of the inertial transient effects of Mach's principle in the presence of energy flow includes two terms, the larger one proportional to the second derivative of the fluctuating energy flow and the smaller one proportional to the square of its first derivative. The second inertia term is always negative, oscillating at twice the drive frequency between zero change in inertial mass and a reduced inertial mass. In the tests that Woodward has performed so far, the second term has always been negligible and undetectable.

However, as described in Section II above, the Mach effects depend strongly on the driving frequency, and at sufficiently high frequencies and energy flows, the second term offers the possibility of driving the inertial mass of the system to zero, or even to negative values. Woodward also offers arguments, based on the Arnowitt, Deser, and Misner theory of the electron, that interesting nonlinear dynamic effects should occur when the mass of a system approaches zero. Woodward argues that these dynamic effects may conspire to reveal the intrinsic "bare" mass of electrons, which is large and negative. We note that the existence of the second mass-fluctuation term has never been tested experimentally, even in the work of Woodward.

Davis has pointed out another possible approach to the generation of negative energy for metric engineering purposes. Ford and Svatier have published two papers describing action of parabolic-cylinder mirror reflectors that create a line focus, along which the quantum fluctuations of the vacuum are greatly magnified. With a behavior similar to the gap in a the two-plate Casimir-effect configuration, this mirror configuration creates a region of negative energy density, such that a net electromagnetic force on atoms near the line focus would be in the direction that would draw them into the focus region. In naive calculations with ideal situations, the fluctuation magnitudes and negative energy density at the line focus go singular as the focus is approached, suggesting that very large magnitudes of negative energy might be achieved at the line focus. However, the authors argue that this extreme behavior will be limited by wavelength limits when the reflection of quantum modes ceases at the plasma frequency of the mirror. They suggest an experimental test of the effect similar to that already performed for the Casimir effect, in which an atomic beam is deflected by the forces near the focus line. The implication of their experimental predictions of the expected deflections is that the effect is comparable in magnitude to the Casimir effect, i.e., not very large. Nevertheless, this development appears to offer the possibility of developing large negative energy densities in a limited region that might match the requirements of the "thin-shell" wormhole and warp drive configurations that have been proposed.

Thus, in principle we have two nascent technologies that might satisfy the requirements for exotic matter and negative mass-energy that would be needed for metric engineering. It is clear that further investigation and testing in these areas should be encouraged and funded.

VI. Wormholes and Back-Reaction

Many scenarios that have been proposed for the use of wormholes in space travel situations turn out to be impossible. For example, refueling and providing reaction mass to a starship through an on-board wormhole portal would not work. This is because there are rules derived from general relativity about wormhole care and feeding that come under the general heading of "back reaction." Wormhole back reaction is in essence the changes required in wormhole-portal characteristics (mass-energy, momentum, charge, angular momentum) so as to preserve all of the local conservation laws.

Because of wormhole back reaction, it is not possible to change the amount of conserved quantities in the local space region in the vicinity of either of the two wormhole mouths. If an electric charge disappears into a wormhole mouth, the entry mouth becomes electrically charged with the just quantity of electric charge that passed through it. The charge has disappeared through the wormhole portal, but has been replaced by a charge on the wormhole portal itself (think of the lines of electric flux threading the wormhole and stuck there by topology). Similarly, if a mass goes through, the entry mouth becomes more massive. If a high momentum particle goes through, the entry mouth acquires that momentum and is pushed forward. And if a spinning flywheel goes through, the entrance mouth will acquire an angular momentum in the direction of the flywheel spin. In this way, the local mass-energy, charge, momentum, and angular momentum in the vicinity of the wormhole entry mouth do not change. No mass-energy, charge, momentum, or angular momentum can magically appear or disappear. The wormhole entrance mouth itself takes up the slack.

Similarly, if a positive electric charge emerges from the wormhole's exit mouth, the mouth acquires an equal and opposite charge, so that the net charge in the region does not change. The charges cancel to zero because there was no charge in the region before the appearance of the emerging charge. An emerging massive particle similarly causes the exit mouth to lose mass-energy, and an emerging high momentum particle gives the exit mouth a recoil momentum in the opposite direction. This is how back reaction works. The local situation with conserved quantities *before* wormhole transits must be the same as the local situation *after* wormhole transits.

The effect of back-reaction in changing the mass of a wormhole mouth raises a flag of caution. The wormhole of interest is presumed to be stabilized against its intrinsic tendency to collapse and close off. This stabilization may be affected or even destroyed by back reaction effects. How massive can the mass-gaining wormhole mouth become, and how small the can mass-losing wormhole mouth be allowed to become before stability is lost. Can the exit mouth's mass go to zero? Can it go negative? Managing the masses (and other conserved quantities) could set important limits on the use of wormholes, even if we could find a way to produce and stabilize them.

VII. Sending Wormholes to the Stars

Now I want to turn to a scheme for reaching the stars well within human lifetimes, using accelerated wormhole portals. I discussed this scheme briefly during the panel discussion that ended the Exotic Technologies section of the 2011 100 Year Starship Symposium. It is, as far as I know, a new and unprecedented scheme for ship-less space travel that I invented and first published in one of my *Analog* columns in May 1990 and followed up in a May 2012 column.

Even if we assume that we have been able to produce a wormhole that was stabilized by one of the schemes discussed above, each wormhole portal or mouth would presumably be surrounded by massive machinery, cryostats, power cables, etc., and the curvature of space in the vicinity of the aperture might create tidal forces that make it impossible for a space traveler to survive wormhole passage. However, Matt Visser has metric-engineered a different solution to Einstein's equations of general relativity from that to Morris, Thorne, and Yurtsever, in which the wormhole is stabilized by another artifact of general relativity, a negative-tension cosmic string. Such an object would be self contained, have no dangerous space curvature except near the cosmic string surfaces, and could, in principle, be very large or very small, even down to the Planck-length scale. A Visser wormhole might also occur naturally in the aftermath of the Big Bang, since both of its components are GR solutions. We can also hypothesize that if there were passive stability problems with a Visser wormhole, it might be dynamically stabilized externally by an active negative feedback system acting directly on and through one of the wormhole mouths.

Let us assume that we have the capability of producing such Visser wormholes and controlling their size. If we keep a wormhole mouth microscopic in mass and size, it behaves much like a fundamental particle with a very large mass, perhaps somewhat in excess of the Planck mass of 21.8 micrograms. For the purposes of calculation, let us assume that we can produce a stabilized microscopic wormhole with a mass of, say, ten Planck masses or 218 micrograms.

Now, we take the two wormhole mouths of this object and thread lines of electrical force through them, until we have passed about twenty coulombs of charge through the wormhole. This can be done, in principle, with a twenty microampere electron beam passing through the wormhole for about twelve days. The result is that the wormhole mouth will now have the same charge-to-mass ratio as a proton and will behave like a proton in the electric and magnetic fields of a particle accelerator. (We note that such an object would have to have some minimum radius, because if the electric field at the throat was too strong, it would pull positrons out of the vacuum and reduce the charge by field emission.)

Now we transport what we will henceforth call the "traveling wormhole mouth" to Meyrin, Switzerland near Geneva and put it into CERN's new Large Hadronic Collider (LHC) there. The other wormhole mouth remains in our laboratory, along with various stabilizing and steering equipment (described later). We assume that by the time that we are able to do this, the LHC will have achieved its full design capacity and will be able to accelerate each of its colliding proton beams to seven TeV, or this many electron volts: $7x10^{12}$

We use the LHC to accelerate the wormhole mouth to the same energy per unit rest mass as a seven TeV proton, extract the beam that contains it, point it at a star of interest, and send it on its way. (Presumably, we would do this in an operation with a number of wormhole-mouths pointed at a selection of candidate stars that might have Earthlike planets in orbit around them.) We use the LHC to accelerate the wormhole mouth to the same energy per unit rest mass as a 7 TeV proton, extract the beam that contains it, point it at a star of interest, and send it on its way. (Presumably, we would do this in an operation with a number of wormhole-mouths pointed at a selection of candidate stars that might have earth-like planets in orbit around them.)

A proton with a total energy of 7.0 TeV will have a Lorentz gamma factor, $y=[1-(v/c)^2]^{-1/2}=E/M$ of 7,455. The accelerated wormhole mouth will have the same Lorentz factor. This is the factor by which the total mass-energy **E** of the proton moving at this high velocity **v** exceeds its rest mass **M**. It is also the factor by which time dilates, i.e., by which the clock of a hypothetical observer riding on the proton would slow down. The wormhole is traveling at a velocity that is only a tiny fraction less than the speed of light, so it travels a distance of one light-year in one year. However, to an observer riding on the wormhole mouth, because of relativistic time dilation the distance of one light-year is covered in only 1/7,455 of a year or 70.5 minutes.

Moreover, back on Earth if we peek through the wormhole mouth at rest in our laboratory, we see the universe from the perspective of an observer riding on the traveling wormhole mouth. In other words, in 70.5 minutes after its launch from CERN, through the wormhole we will view the universe one light-year away. Later, in 11.7 hours we will view the surroundings ten light-years away. In 4.9 days, we will view the surroundings 100 light-years away. And so on.

This is a remarkable result. How is it possible that, if the traveling wormhole mouth requires 100 years, as viewed from Earth, to travel 100 light-years, we can view its destination as observers looking through the wormhole in a bit less than five days? It is because, as pointed out by Morris, Thorne and Yurtserver, the special relativity of time dilation makes a wormhole with one high-velocity mouth into a time machine. The wormhole mouth, which from our perspective has taken 100 years to reach a point 100 light-years away, connects back in time to its departure point only five days after it left. In effect, it has moved 100 light-years at a speed of 7,455 c.

But could the traveling wormhole mouth be aimed so accurately from its start at CERN that it might it actually pass through another star system many light-years away, to survey its planets, etc.? And could it stop when it got there? The fortunate answer is yes.

Momentum back reaction can be used to steer the traveling wormhole mouth. The direction of travel, as viewed through the wormhole, can be monitored. Course corrections can be made by directing a high-intensity light beam through the laboratory based wormhole mouth at right angles to the direction of travel. The beam will emerge from the traveling wormhole mouth sideways, giving it momentum sideways momentum in the other direction. The exit mouth will lose a bit of mass-energy in this process, but it will also be gaining some mass energy as interstellar gas passes through it and emerges from the laboratory wormhole mouth. We note that, in terms of momentum change vs. mass gain of the wormhole mouth, the use of light is preferable to high energy particles, even though the momentum carried by light is only its energy divided by the speed of light, because it keeps the wormhole mass gain/loss small per unit momentum change.

Assuming precision steering can be accomplished by applying such momentum changes, stopping is not too difficult. The exit mouth will still have the large electric charge used for acceleration inx the LHC and consequently will lose energy rapidly by ionizing interactions as it passes through any gas. It can be steered to make passes through the upper atmospheres of planets or to have grazing collisions with atmosphere of the star itself, until its great initial velocity has been dissipated. In this process, considerable mass will pass through the traveling mouth, and it will gain this mass-energy by back reaction. This can be compensated by sending low-velocity mass through in the other direction. The large charge can be reduced at the same time by sending charged particles through.

The decelerated wormhole mouth can tour the star system, propelled by high momentum beams sent through the stay-at-home mouth in the laboratory. Such steering will tend to reduce the wormhole mass, partially compensating for the mass-gain it received in decelerating and perhaps in sampling planetary atmospheres.

Now that the wormhole mouth has arrived at the star system of interest, a survey of the planets can begin. We assume that we have laboratory control of the diameter of the wormhole mouth, and that it can be enlarged to a diameter that is convenient for sampling. If a habitable planet is found, the wormhole mouth can be brought to its surface, and samples can be extracted through the wormhole and analyzed, (perhaps sending compensating mass back in the other direction to keep the wormhole mouth masses in balance).

Ultimately, when the survey is complete, the wormhole can be expanded, permitting robot precursors, planetary explorers, colonists, and freight to move through. Again, the mass of the wormhole mouths would have to be managed, moving equal masses in the two directions during wormhole transits, perhaps by sending compensating masses of water through pipes. This scheme could allow very rapid travel to and colonization of various star systems containing Earthlike planets. Thus, if stable wormholes are possible at all, they may represent a path to the stars that would sweep away many of our previous concepts and prejudices about how the stars can and should be reached.

Is there any problem with causality created by using what is in essence a time machine to reach the stars? Perhaps. The issue is whether a timelike loop can be established. Although the space-time interval from some event at the distant star to the observation of that event on Earth, as viewed through the wormhole, represents two-way communication across a space-like separation, there is no causality problem because there is no loop.

However, a causality problem could arise if similar but independent wormhole connections were established with accelerated wormhole mouths sent from the distant star system back to Earth, or even to another star system that had been similarly contacted by Earth. In that case, transit through one wormhole followed by the other would constitute a timelike loop. Stephen Hawking has suggested that Nature will prevent the establishment of any timelike loop through an exponential rise in vacuum fluctuations that would destroy some elements of the incipient loop. Thus, an attempt to set up the second link might result in an explosion. The moral is that such wormhole connections must originate from only one central cite. Any attempt at replication from another site might lead to disaster.

The scheme described above focuses on reaching local stars, i.e., those in our galaxy and does not take into account the intrinsic accelerating expansion of the universe. One of my *Analog* readers, Tom Mazanec, argued that by the time we are actually able to manipulate micro-wormholes as if they were fundamental particles, we might have already built particle accelerators that could accelerate protons to far higher energies the seven TeV available from the LHC. We might contemplate reaching huge Lorentz factors that could allow us to probe very remote parts of the universe where the recession velocity from cosmological expansion becomes important. We might even contemplate approaching the redshift horizon, beyond which a part of the universe is supposed to be unreachable. In that region of near-light-speed recession, the wormhole would eventually match velocities with the receding region until it was at rest with respect to the average matter resident there. It could never actually reach the redshift horizon from here. However, we might contemplate setting up operations at this edge of the redshift horizon and building a new particle accelerator to shoot wormholes out to the new redshift horizon appropriate to that region of the universe. In that way, we might step our way to otherwise unreachable parts of the universe.

This brings us to a variation of the famous Fermi Paradox: if interstellar wormhole transport is possible, shouldn't the technologically advanced civilizations of our galaxy already be sending tiny accelerated wormhole portals in our direction? Then, where are they?

Perhaps they are already here. Cosmic ray physicists have occasionally observed strange super-energetic cosmic ray detection events, the so-called "Centauro Events." These are cosmic ray particles with incredibly high energies that, when striking Earth's upper atmosphere, produce a large shower of particles that contains too many gamma rays and too few muons, as compared to more normal cosmic ray shower events. Despite many attempts, the Centauro Events presently lack any explanation based on any known physics. However, an accelerated wormhole mouth with a large electric charge should have a large gamma-ray to muon production ratio in such collisions, since it would have large electromagnetic interactions but should have no strong or weak interactions with the matter with which it collided.

Cosmic ray experts conventionally assume that whatever else they are, Centauro Events must be a natural phenomenon, not an artifact of some advanced civilization. But that assumption could be wrong. It is interesting to contemplate the possibility that some advanced civilization may be mapping the galaxy with accelerated wormhole portals, sending little time-dilated observation points out into the cosmos as peep-holes for viewing the wonders of the universe. And perhaps, when a particularly promising or interesting scene comes into view, the peep hole is halted and expanded into a portal through which a Visitor can pass.

VII. Conclusion

The Exotic Technologies track of the 100 Year Starship Symposium, held on Saturday, October 1, 2011, explored some of the aspects of exotic technology that might be used to reach the stars. Presentations by Millis, White, Davis, Maclay, McCulloch, Christea, Sarfatti, and Olum, and the panel discussion that followed, gave a variety of perspectives on the possible use of unconventional technologies to reach the stars, and included some interesting ideas.

In this chapter, I have discussed some of them, and I have extended the discussion in directions that I consider to be the most promising for the application of exotic technologies to the 100 Year Starship Project. There is much work to be done, but the stars are out there, and we urgently need to find a way to reach them. Otherwise, all of our eggs remain in a pretty blue basket that has so far undergone at least sixty extinction catastrophes, with more to come.

VIII. References

Gould, Stephen J., Chapter 5 in *Perspectives in Evolution*, R. Millikan, ed., Sinauer Associates, Inc., Sunderland, MA, (1982).

Cramer, John G., "The Pump of Evolution," *Analog Science Fiction & Fact Magazine*, (January-1985); http://www.npl.washington.edu/AValtvw11.html

Cramer, John G., "Artificial Gravity: Which Way is Up?" *Analog Science Fiction & Fact Magazine*, (February-1987); http://www.npl.washington.edu/AV/altvw18.html.

Cramer, John G., "Antigravity Sightings," *Analog Science Fiction & Fact Magazine*, (March-1997); http://www.npl.washington.edu/AV/altvw83.html.

Alcubierre, Miguel "The warp drive: hyper-fast travel within general relativity." *Classical and Quantum Gravity 11 (5)*, L73-L77 (1994).

Cramer, John G., "The Alcubierre Warp Drive," *Analog Science Fiction & Fact Magazine*, (November-1996); http://www.npl.washington.edu/AV/altvw81.html.

Michael S. Morris, Kip S. Thorne, and Ulvi Yurtsever, *Physical Review Letters 61*, 1446 (1988).

Cramer, John G., "Wormholes and Time Machines," *Analog Science Fiction & Fact Magazine*, (June-1989); http://www.npl.washington.edu/AV/altvw33.html.

Millis, Marc G., "Space Drive Physics, Introduction and Next Steps," paper given at the *DARPA/NASA 100 Year Starship Symposium*, October 1, 2011.

Einstein, Albert, "Über das Relativitätsprinzip und die aus demselben gezogene Folgerungen," *Jahrbuch der Radioaktivitaet und Elektronik 4* (1907).

Sciama, D. W. "On the Origin of Inertia," *Monthly Notices of the Royal Astronomical Society 113*, 34–42 (1953).

Woodward, James F., *Foundations of Physics Letters 9*, 247-293 (1996).

Buldrini, N. and M. Tajmar, "Experimental Results of the Woodward Effect on a Micro-Newton Thrust Balance," in *Frontiers of Propulsion Science, Progress Series in Astronautics and Aeronautics, 227*, eds. M. G. Millis and E. W. Davis, AIAA Press, Reston, VA, pp. 373-389, (2009).

Woodward, James F., private communication (2012).

Davis, Eric, "Space Drive Physics, Introduction and Next Steps," paper given at the *DARPA/NASA 100 Year Starship Symposium*, October 1, 2011.

Davis, Eric, private communication (2012).

Olum, Keith, "Does General Relativity Permit Superluminal Travel?" paper given at the *DARPA/NASA 100 Year Starship Symposium*, October 1, 2011.

Hawking, S. W. and G. F. R. Ellis, *The Large-Scale Structure of Space-Time*, Cambridge Univ. Press, Cambridge, pp. 88-91, 95-96, (1973).

Ford, L. H. and T. A. Roman, "Quantum Field Theory Constrains on Traversable Wormhole Geometries," *Phys. Rev. D53*, 5496-5507 (1996); gr-qc/9510071.

Roman, Thomas A. "Some Thoughts on Energy Conditions and Wormholes," *Proceedings of the Tenth Marcel Grossmann Meeting on General Relativity and Gravitation*, September 23, 2004; gr-qc/0409090.

J. F. Woodward, "Twists of fate: Can we make transversable wormholes in space-time?" *Found. Phys. Lett . 10*, 153-181 (1997).

White, Harold, "Warp Field Mechanics 101," paper given at the *DARPA/NASA 100 Year Starship Symposium*, October 1, 2011.

Cramer, J. G., see reports from *CENPA Annual Reports* on "Testing Nonlocal Communication," http://faculty.washington.edu/jcramer/NLS/NL_signal.htm

White, Harold, private communication (2012).

Einstein, Albert, "Die Grundlage der Allgemeinen Relativitätstheorie," *Annalen der Physik 49*, (1916).

Einstein, Albert and Rosen, Nathan, "The Particle Problem in the General Theory of Relativity." *Physical Review 48*, 73 (1935).

Fuller, Robert W. and Wheeler, John A., "Causality and Multiply-Connected Space-Time." *Physical Review 128,* 919 (1962).

Arnowitt, R., Deser, S. and Misner, C.W., "Gravitational-Electromagnetic Coupling and the Classical Self-Energy Problem," *Physical Review 120*, 313 - 320 (1960)

Arnowitt, R., Deser, S. and Misner, C.W., "Interior Schwarzschild Solutions and Interpretation of Source Terms," *Physical Review 120*, 321-324 (1960).

Ford , L. H. and N. F. Svaiter, "Focusing vacuum fluctuations," *Phys. Rev. A 62*, 062105 (2000).

Ford, L. H. and N. F. Svaiter, "Focusing vacuum fluctuations. II," *Phys. Rev. A 66*, 062106 (2002).

Visser, Matt, *Lorentzian Wormholes: From Einstein To Hawking,* AIP Series in Computational and Applied Mathematical Physics (1995); ISBN 978-1-56396-394-0.

"Panel Discussion: Can Exotic Science Lead to Starship Propulsion?," Participants: J. G. Cramer (moderator), M. G. Millis, H. White, E. Davis, and G. Nordley at the *DARPA/NASA 100 Year Starship Symposium*, October 1, 2011.

Cramer, John G., "Wormholes II: Getting There in No Time," *Analog Science Fiction & Fact Magazine*, (May-1990); http://www.npl.washington.edu/AV/altvw33.html.

Cramer, John G., "Shooting Wormholes to the Stars," *Analog Science Fiction & Fact Magazine*, (May-2012); http://www.npl.washington.edu/AV/altvw162.html

Matt Visser, *Physical Review D 39*, 3182 (1989).

Hawking, S.W., "Chronology protection conjecture," *Physical Review D 46*, 603-611 (1992).

Angelis, Aris, "The mysteries of cosmic rays," *CERN Courier*, January 29, 1999, http://cerncourier.com/cws/article/cern/27944.

We conclude with cautionary remarks about how unique the experience of interstellar colonization will be. Humanity has never tried living outside its habitat for more time than needed for a stay in a space station. We know very little about the implications of living in a truly alien place. Any species that attempts to expand into the galaxy will face similar problems, and so this is also a way of discussing the little understood problems in the famous Fermi question: Where are they?

Afterword
Paul Davies

Two overarching questions confront would-be human spacefarers: where to go and how to get there. Much attention has been given to the latter question. For interstellar travel to become a reality demands major engineering advances, probably involving radically new propulsion systems. These topics have been discussed at length in the preceding chapters of this book. Although many proposals are highly speculative, we know of no fundamental physical principles that forbid interstellar travel; whether or not it becomes a reality boils down to technology, cost and motivation. In this Afterword I address the oft-neglected first question: the destination.

Leaving aside fantastical speculations about faster-than-light travel, clearly journeying between the stars will take a very long time, even on the most optimistic estimates of technological advance, although as Freeman Dyson points out, round-trip travel to minor objects in the outer reaches of our own solar system might be feasible. Therefore, interstellar tourism, or trade in physical substances (as opposed to information), is inconceivable. It follows that travel beyond the solar system will be one-way only. Two possibilities then arise. One is that spacefarers will seek out and colonize other worlds, and the second is that they will create permanent artificial habitats in space. Both scenarios have been popular in science fiction. I propose to leave aside the vast civil engineering issues involved and dwell instead on a much trickier and more basic problem, namely, the ecological requirements, and especially those relating to microbiology.

Long-term human survival means more than growing enough food to eat and making enough oxygen to breathe. It demands creating a complete self-sustaining ecosystem. On Earth, complex multicellular organisms (animals, plants) form merely the conspicuous tip of a vast biological iceberg, the majority of which is microbial. Almost all terrestrial species are microbes—bacteria, archaea and unicellular eukaryotes—and to date microbiologists have scratched only the surface of the microbial realm. Microbes are teeming everywhere—in the soil, in the air, in water, in the rocks beneath our feet, in Earth's crust to a depth of kilometres. These busy little creatures are a vital part of the life-support system of our planet, both via their metabolic activity (such as recycling material) and through the exchange of genetic components. Even within your own body microbes play a crucial role. The microbial inhabitants of your gut, lungs, etc.—known as your microbiome—outnumber your own cells. Without them you would die.

So astronauts cannot be sent to the stars without, at the very least, their microbiomes.

But it doesn't stop there. Microbes do not live in isolation; they form a vast network of biological interactions that remains very ill-understood. Simple Darwinian processes are augmented by horizontal gene transfer, cell-cell signaling, collective organization, and much else. Interwoven into this network are the activities of viruses, which infect microbes just as they do larger organisms. The subtle interplay of viruses, microbes, and metazoa constitutes an ecological web of such staggering complexity that scientists have hardly begun to glimpse the principles at play.

In the absence of sending the entire terrestrial biosphere, a fundamental unsolved problem arises: what is the minimum complexity of an ecosystem—dominated, as I have explained, by microbes—necessary for long-term sustainability? At what point, as more and more microbial species are dropped from the inventory of interstellar passengers, does the remaining ecosystem go unstable and collapse? Which microbes are crucial and which would be irrelevant passengers, as far as humans (and their animal and plant food supply) are concerned?

This is a Noah's Ark conundrum with a vengeance. Not only have we no clue as to the answer, we have little idea of the solution to the simpler problem of the smallest self-sustaining purely microbial ecosystem under stable environmental conditions. Clearly, even a plan to terraform an entire planet ahead of human colonization cannot proceed without a far deeper understanding of microbial ecology.

Imagine making a list of the minimum number of plants and animals needed to accompany humans on a one-way mission. As well, leave aside the logistics of growing, feeding, and breeding all these organisms in space, many of which may require environmental conditions (temperature, pH, oxygen levels, etc.) very different from those congenial to humans. We might think of cows, pigs, sheep, chickens, some fish, a few vegetables—that would do for a start.

But how many, and which, microbial species do these animals and plants in turn depend on? How many, and which, other microbes do those animal-and-plant-servicing microbes depend on? Which of these might be pathogenic to humans, yet vital for some other part of the ecosystem?

Can we pull the web of life to bits, extract a tiny subset of it, and expect such mini-webs to function forever in isolation? And without a full understanding of the principles of the networks involved, how can we be sure, until it is too late, that we have done our ecological accounting exercise correctly? It would be no good being halfway to Alpha Centauri only to find that a key bacterium was overlooked and left back on Earth.

To be sure, these difficulties can be mitigated somewhat by biotechnology. In his Foreword, Dyson pins his hopes on our ability to map the genome of the entire biosphere, then use the colossal computing power promised by extrapolating Moore's Law to make sense of it all, and design an ecosystem customized to a target planet or even an icy planetesimal. I am far less confident that the behavior of an ecosystem can be captured by mere number-crunching. Even if it was, we cannot guess the answer; the supercomputer may tell us that there is no solution at all that matches the physical environment of the host body. Or that tens of millions of species are required. In addition, because biological evolution has a large element of chance, a transplanted ecosystem may well not stay as designed, but rapidly evolve in ways totally unsuited to human habitation, requiring complex mid-course corrections entailing bio-engineering on a planetary scale.

Some logistical saving might come by engineering the needed microbes in situ. The food supply problem could be mitigated by growing protein material directly and eliminating the pigs and chickens. Diet beyond Earth may be dull, but the scale of the operation would be greatly reduced by limiting the number of metazoan species that go along for the ride, or by hibernating the passengers for vast lengths of time (going into "hybo," as Richard Lovett put it in his story).

So perhaps with sophisticated enough biotechnology, only (hibernating) humans, some plants, and their extended microbiomes would suffice to create a sustainable ecosystem. But with the vast majority of the species being those of the extended microbiomes, the complexity of the overall system is still daunting, and the consequences of extremely long-term isolation for this impoverished web of itinerant life—even in terrestrial, let alone space, conditions—are almost impossible to predict.

One unexpected and recently discovered issue concerns the response of microbes to zero gravity conditions. Experiments conducted by Cheryl Nickerson at Arizona State University show that bacteria in low-Earth orbit alter their gene expressions. That is, changes in mechanical forces in a microbe's micro-environment can have genetic (strictly epigenetic) consequences. If this phenomenon is general, then even a meticulously designed ecological web of interactions might come asunder in space conditions, purely from the gene-expression effects of reduced gravity, because of changes in the way the microbial components interact.

Lovett dwells on the problem of spacefarers getting sick due to bacterial pathogens, and how that problem might be managed by careful monitoring. But altered gene expressions can affect how virulent pathogens are (hence the interest in Nickerson's experiments). This could perhaps render, say, salmonella, less toxic, but producing dangerous new strains from currently harmless bacteria. If the spacecraft is spun to produce artificial gravity of one-G on the periphery, the large Coriolis forces and the reduced gravity nearer the spin axis would most likely still have effects on microbial gene expression. All the foregoing obstacles would loom large, as soon as an interstellar expedition leaves Earth, because the pre-packaged food supply would rapidly dwindle.

While fast interstellar travel presents formidable challenges of physics and engineering, at least the underlying principles of those subjects are well understood. But the microbial complexity threshold challenge, which is truly fundamental, is not understood at all. Lovett frankly acknowledges the depth of this unsolved problem, when the fictional astronaut remarks skeptically, "The fact is we really don't have much experience with large, closed, ecologies. Not enough that I'd want to trust one for twenty years, let alone 200. Read the reports from the old Biosphere 2 project."

Suppose the solutions for a sustainable mini-ecosystem are nevertheless one day worked out, and a mighty one-way mission departs for the stars: destination—an Earthlike planet many light-years away, where the spacefarers or, more likely, their very distant descendants, will make a new home. Astronomers are now fairly certain that the Milky Way contains millions, possibly billions, of Earthlike planets (depending a bit on your definition of Earthlike), so there is plenty of real estate to choose from, especially if the inventory is extended, a lá Dyson, to include small, cold bodies in the vast interstellar spaces.

In many science fiction stories, the heroic adventurers' touch down and step out onto an equable and verdant planet, preferably free of an already-ensconced and non-compliant advanced civilization, and take up joyful residence. Unfortunately it's not that simple. There is a vanishing chance that the neatly-excised and amazingly self-sufficient truncated terrestrial micro-ecology would conveniently slot into a place in the (presumably much bigger) alien equivalent, and proceed to carry on business as usual.

But this problem highlights a much deeper and more substantive obstacle to human colonization of other planets—the very existence or otherwise of indigenous life.

Most fictional scenarios envisage humans in search of a planet with abundant life to take care of the colonists' needs thereafter (the movie *Avatar* being a recent example). An ideal world for human colonization is one with edible indigenous plants and animals. But this vision flies in the face of basic biology. Organic matter is edible only when its biochemistry closely matches that of the consumer. Even on Earth, the vast majority of organisms are not suitable for human consumption.

In spite of advances in astrobiology, there is yet no known example of life beyond Earth. Many scientists speculate that the universe is teeming with life, although, it has to be said, without the slightest shred of evidence that the transition from non-life to life is indeed highly probable in Earthlike conditions. However, suppose for the moment that they are right, so that our intrepid spacefarers arrive at their destination planet to find a rich indigenous biosphere. There is no reason to suppose that terrestrial biochemistry, which is highly specific to both the conditions on our planet and the accidents of evolutionary history, is universal. It is easy to imagine carbon-based life on other worlds using different amino acids, different informational molecules, different membrane molecules, and so forth. It is also easy to imagine a mirror world in which familiar organic molecules are replaced by their mirror images (e.g. right-handed amino acids instead of left-handed).

It is highly likely that this alien foodstuff would be unpalatable and indigestible—useless, in fact, as a food source. (The same reasoning makes nonsense of the whimsical suggestions that aliens coming to Earth might choose to eat humans.) Worse still, the indigenous biota would serve as a barrier to the establishment of a secondary transplanted terrestrial ecosystem; it's hard to suppose that two different advanced and complex ecosystems could peacefully cohabit a single planet.

There is, however, a flip side to the biological incompatibility problem. An alien biochemistry that offers little scope for consumption also poses little threat for infection. Alien microbes and viruses (if they exist) would probably be unable to invade terrestrial organisms, or to make much progress if they did. And vice versa. Wells's "happy ending" to the *War of the Worlds*, in which the Martians succumb to terrestrial germs, is simply not credible.

The foregoing issues would disappear if the host planet had no indigenous life; that is, if it was habitable but uninhabited—terra nullus on a planetary scale. Unfortunately, this scenario has its own difficulties, one of which is crucial to human survival: oxygen. Being such a reactive element, oxygen does not survive for long in planetary atmospheres unless it is continually replenished by oxidative photosynthesis. A planet with breathable atmospheric oxygen implies the presence of photosynthetic plants, or at least microbes. For this reason, an important project in astrobiology is the construction of a space-based optical system capable of detecting the spectral signature of oxygen in the atmospheres of extra-solar planets as a surrogate for detecting life. If there is no life on the destination planet, then it very probably would not have breathable amounts of free oxygen in the atmosphere. Thus, without wholesale advanced terraforming using oxygenating plants and microbes, humans would still have to live in air-tight habitats and wear suits when they go out. Nevertheless, setting up home on a previously sterile planet would be far easier than coping with an indigenous biosphere.

Quite apart from the practicalities of colonizing another planet, there are serious ethical issues at stake. If a planet already hosts some form of life, the question arises of whether human beings have the right to limit or threaten it by transplanting Earthlife in its midst. Attitudes to this issue will depend on how important human colonization is deemed to be and how complex the alien life forms are. For example, one of the strongest justifications for colonizing other worlds is to insure against a mega-catastrophe here on Earth. Often cited is the impact of a large comet or asteroid which might wipe out civilization or even our entire species. Rather more likely in my view is a sudden pandemic, either naturally occurring or the result of the accidental release of a virulent bio-warfare pathogen.

In any case, over a period of millennia, there is no lack of potentially species-annihilating hazards, as Martin Rees shows in his Foreword. Humans may choose to undertake interplanetary or interstellar colonization to keep our species, and the flame of our culture, alive somewhere in the cosmos. If all that stood in the way of survival were some indigenous microbes, few people would have scruples in ignoring their "rights."

At the present time, there is perceived to be less urgency in dispatching a lifeboat from Earth, even though Mars colonization is feasible and within current or foreseeable technology. Therefore, some attention has been given to planetary protection. Although it is an open question whether there is any extant life on Mars, planetary scientists generally agree that we should take moderately serious steps to limiting contamination of Mars by terrestrial organisms. Conversely, the risk of back contamination of Earth by any Martian microbes, as might occur in a sample return mission, say, is taken very seriously, and it seems likely that no mission of this nature will be undertaken without rigorous safeguards. The risk of a killer plague of Martian microbes is rather higher than the risk to colonists travelling to extra-solar planets, because Mars and

Earth have traded rocks throughout their long history, in the form of debris blasted into space from asteroid and comet impacts. Cocooned inside a rock, hitch-hiking microbes are easily capable of being transported in a viable state from Earth to Mars or vice versa. For this reason, Earth and Mars are not naturally quarantined, opening up the possibility that putative Martian life and terrestrial life may belong to a common tree with a common origin. (The question of whether life started on Mars and came to Earth later, or vice versa, or started independently on both planets, is a fascinating one, but beyond the scope of this essay.)

Because the probability is exceedingly small that a microbe-laden rock kicked off our planet will hit another Earthlike planet many light-years away, there is negligible chance that life on an extra-solar planet would share a common origin with terrestrial life. (I am discounting other "panspermia" mechanisms that might spread microbes across the galaxy, because I believe the probabilities are far too small for this to be a systematic effect.)

If a planet had plant and animal life, there would be strong ethical objections to contaminating it with terrestrial organisms. Even if the two forms of life were so biochemically different that direct infection was avoided, it may still be the case that the terrestrial invaders would plunder some vital resource and deplete the indigenous ecosystem. Earth organisms might spread like the rabbits in Australia, and elbow the indigenous life aside, driving it to extinction.

That issue would be greatly sharpened if a target planet is found to host intelligent life. In Avatar, resource-hungry humans muscle in on the planet Pandora to the extreme discomfort of its indigenous population, although in the interests of Hollywood-style justice, they eventually receive their comeuppance. There is no guarantee that future generations of humans would exercise respect for the rights of alien beings, nor can we be sure that aliens would respect ours. Aliens far in advance of us in technology and social development may not share our ethical values. Because we cannot begin to guess the motives and attitudes of truly alien beings, when it comes to the prospect of humans encountering an extraterrestrial civilization, all bets are off.

As befits an Afterword, I shall conclude by examining the premise on which this book is based—that interstellar travel should, and could, become part of our destiny. Why?

A familiar answer is that humans have always had wanderlust, a sense of curiosity, a desire to explore the world about them and to push on to pastures new. That may be true, but it is hardly a justification for leaving the planet. People have always fought wars and oppressed minorities too; just because something is deeply ingrained in human nature does not make it a noble motivation. There is clearly a religious component in our dream of reaching for the stars, which Harrison makes explicit when he writes, "I identify more and more with the cosmos the more I learn about it." He attaches the somewhat ungainly moniker of "cosmism" to describe our desire to leave Earth and spread out into the cosmos. Nor is he alone in this idealism. I have known other distinguished scientists express similar sentiments, opining that it is mankind's unwritten destiny to spread across the universe. Perhaps.

Rather easier to justify is the argument that human society has certainly produced much that is good, and which it would therefore be good to preserve for posterity. Human civilization on Earth faces many uncertainties and the (small) possibility of a mega-catastrophe. By establishing a permanent settlement on another planet, human culture could continue even if disaster struck at home.

It could be countered that this argument adopts an inflated view of human significance and human worth. One could say it is life, as opposed to our specific species or culture, that is the truly precious commodity that should be perpetuated and perhaps disseminated around the cosmos—something that can be achieved far more easily than sending humans. True, we could already begin sending microbes in tiny capsules out of the solar system if we were so minded, but it is hard to imagine much enthusiasm for the project. Seeding a barren galaxy with DNA may one day fire people's imagination (assuming the galaxy is not already teeming with life), but today the appeal of interstellar travel is deeply rooted in ideals of human adventure and advancement.

When Neil Armstrong took that initial small step on the moon, it was widely hailed as the first step on a stairway to the stars. Half a century on, with humans seemingly stuck in low-Earth orbit, the prospects for interplanetary, let alone interstellar, travel look bleak. The microbiology problems that I have stressed in this Afterword compound what is already a formidable challenge in spacecraft design, propulsion systems and medical technology. Yet if humans wish to secure a long-term future in an uncaring and occasionally dangerous cosmos, some form of cosmic diaspora needs to be part of our long-range plan.

AUTHORS, ARTISTS, ORGANIZATIONS & FURTHER READING

Authors

Stephen Baxter

Stephen Baxter is a British science fiction writer, perhaps best known for his "Xeelee" future-history sequence of novels and short stories. He is the author of more than forty books and over a hundred short stories. His most recent books are *Iron Winter*, the final novel in the "Northland" trilogy, *Doctor Who: The Wheel of Ice, The Long Earth,* the first of a sequence of novels co-written with Terry Pratchett, and new short story collection, *Last and First Contacts.*
www.stephen-baxter.com

James Benford

James Benford is president of Microwave Sciences, which deals with high power microwave systems from conceptual designs to hardware. Over the past forty-five years of scientific research he has written 145 scientific papers and six books on physics topics, including the textbook, *High Power Microwaves*, now in its second edition. His current scientific interest is electromagnetic power beaming for space propulsion. In earlier decades, he was active in science fiction fandom, and wrote science fiction in the 1970s.
www.JamesBenford.com

Gregory Benford

Gregory Benford is a professor of physics at the University of California, Irvine, working in astrophysics and plasma physics. A Fellow of the American Physical Society, his fiction has won many awards, including the Nebula Award for his novel *Timescape*.
www.GregoryBenford.com

David Brin

David Brin is a scientist, tech speaker/consultant, and author. His new novel about our survival in the near future is *Existence*. A film by Kevin Costner was based on *The Postman*. His sixteen novels, including *New York Times* Bestsellers and Hugo Award winners, have been translated into more than twenty languages. *Earth*, foreshadowed global warming, cyberwarfare, and the worldwide web. David appears frequently on shows such as *Nova*, *The Universe*, and *Life After People*, speaking about science and future trends. His nonfiction book—*The Transparent Society: Will Technology Make Us Choose Between Freedom and Privacy?*—won the Freedom of Speech Award of the American Library Association.
www.davidbrin.com
Existence: www.tinyurl.com/exist-trailer
The Brin Weblog: davidbrin.blogspot.com

John Cramer

John Cramer is a Professor Emeritus of Physics at the University of Washington in Seattle. John is an experimental physicist who approaches theoretical physics from an experimentalist viewpoint. He has published more than 200 physics research papers in peer-reviewed physics journals. John has written "hard" science fiction and is the author of two published novels: *Twistor* and *Einstein's Bridge*. John is also a science writer, and writes a bimonthly science-fact column, "The Alternate View," for *Analog Science Fiction and Fact* magazine. John has written more than 160 columns, all of which are available on the web.

www.npl.washington.edu/av

faculty.washington.edu/jcramer/TI/tiqm_1986.pdf

Ian Crawford

Ian Crawford is an astronomer turned planetary scientist, and is currently Professor of Planetary Science and Astrobiology at Birkbeck College, University of London, UK (www.bbk.ac.uk/geology). He is a Fellow, and currently the Senior Secretary, of the UK's Royal Astronomical Society (www.ras.org.uk/), and a Fellow of the British Interplanetary Society. He is a strong advocate for the renewed human exploration of the Moon, the development of a spacefaring infrastructure within the Solar System, and the eventual attainment of an interstellar spaceflight capability. In the latter

capacity he leads the science and astronomical target modules of the Project Icarus starship study (www.icarusinterstellar.org/). A more detailed summary of Professor Crawford's research interests, and list of publications, can be found at:

www.bbk.ac.uk/geology/our-staff/ian-crawford

Adam Crowl

Adam Crowl was born in Bendigo, Victoria, Australia in 1970. His first memory of TV is watching the BBC documentary on the Viking landings (1976) and (black & white) episodes of *Space 1999* and *Star Trek*. At age nine he learnt of a star-probe named "Daedalus," and was given a little book, *Road to the Stars* by Iain Nicholson, which opened his eyes to serious interstellar travel research. Since then Adam earned a B. Sc at the University of Queensland; raised a family; has retaught himself mathematics and physics while semi-completing an engineering/computing degree; written essays on SETI for the late Chris Boyce, on fusion propulsion for "Centauri Dreams, gas-mining Uranus" for *Discovery News*, and joined Project Icarus. Currently he is Team Leader for Project Icarus's Main Propulsion Module.

crowlspace.com

icarusinterstellar.org

Paul Davies

Paul Davies is a British-born theoretical physicist, cosmologist, astrobiologist, and best-selling author. He is Regents' Professor and Director of Beyond: Center for Fundamental Concepts in Science, at Arizona State University, where he is also co-Director of the Cosmology Initiative and Principal Investigator of the Center for the Convergence of Physical Science and Cancer Biology. His most recent is *The Eerie Silence: are we alone in the universe?* In 1995 he was awarded

the Templeton Prize for his work on the deeper meaning of science. He was also awarded the Faraday Prize by The Royal Society, the Kelvin Medal by the UK Institute of Physics, the 2011 Robinson Cosmology Prize, and many book awards, as well as three honorary degrees. In June 2007 he was named a Member of the Order of Australia in the Queen's birthday honors list and in December 2011 he was presented with the Bicentenary Medal of Chile. The asteroid 1992 OG was renamed (6870) Pauldavies in recognition of his work on cosmic impacts.

cosmos.asu.edu, beyond.asu.edu, cosmology@asu.edu, cancer-insights.asu.edu

Don Dixon

Don Dixon is a Fellow and founding member of the International Association of Astronomical Artists (IAAA). His artwork has been featured on the covers of *Scientific American, Astronomy, Sky and Telescope, Bild der Wissenschaft,* and dozens of books, ranging from physics compendiums to science fiction novels. His painting "Red Mars," cover painting for the first book in Kim Stanley Robinson's award-winning trilogy, rode the Phoenix spacecraft to a successful landing in the arctic region of the Red Planet on May 29, 2008, as part of Visions of Mars, The Planetary Society's digital "Martian library." Since 1991 he has served as Art Director at the Griffith Observatory in Los Angeles. www.cosmographica.com

Freeman J. Dyson

Freeman J. Dyson was born in 1923 in Crowthorne, England. He received a BA in mathematics from the University of Cambridge in 1945, and came to the United States in 1947 as a Commonwealth Fellow at Cornell University. He settled in the USA permanently in 1951, became a professor of physics at the Institute for Advanced Study at Princeton in 1953, and retired as Professor Emeritus in 1994. Professor Dyson began his career as a mathematician but then turned to the exciting new developments in physics in the 1940s, particularly the theory of quantized fields. He wrote two papers on the foundations of quantum electrodynamics that have had a lasting influence on many branches of modern physics. He went on to work in condensed-matter physics, statistical mechanics, nuclear engineering, climate studies, astrophysics, and biology. Beyond his professional work in physics, Freeman Dyson has a keen awareness of the human side of science and of the human consequences of technology. His books for the general public include *Disturbing the Universe, Weapons and Hope, Infinite in All Direction,* and *A Many-colored Glass.* In 2000 he was awarded the Templeton Prize for Progress in Religion. www.sns.ias.edu/~dyson

Stephen Hawking, CH, CBE, FRS, FRSA

Stephen Hawking is a British theoretical physicist, cosmologist, and author. His significant scientific works to date have been a collaboration with Roger Penrose on theorems on gravitational singularities in the framework of general relativity, and the theoretical prediction that black holes should emit radiation, often called Hawking radiation.

Joe Haldeman

Joe Haldeman is the youngest writer to be named a Grand Master by the Science Fiction and Fantasy Writers of America, and has earned steady awards over his forty-three-year career: his novels *The Forever War* and *Forever Peace* both made clean sweeps of the Hugo and Nebula Awards, and he has won four more Hugos and Nebulas for other novels and shorter works. Three times he's won the Rhysling Award for best science fiction poem of the year. In 2012 he was inducted into the Science Fiction Hall of Fame. The final novel in a trilogy, *Earthbound,* is out (after *Marsbound* in 2008 and *Starbound* in 2009). Ridley Scott has bought the movie rights to *The Forever War*. Joe's next novel is *Work Done For Hire*, appearing soon. When he's not writing or teaching—a professor at M.I.T., he has taught every fall semester since 1983—he paints and bicycles and spends as much time as he can out under the stars as an amateur astronomer. He's been married for forty-seven years to Mary Gay Potter Haldeman. www.home.earthlink.net/~haldeman

Nancy Kress

Nancy Kress is the author of thirty-two books, including twenty-five novels, four collections of short stories, and three books on writing. Her work has won four Nebulas, two Hugos, a Sturgeon, and the John W. Campbell Memorial Award. Most recent works are *After The Fall, Before The Fall, During The Fall* (Tachyon, 2012), a novel of apocalypse, and *FLASH POINT* (Viking, 2012), a YA novel. She is perhaps best known for the Sleepless trilogy, beginning with *Beggars In Spain*, which is about people genetically engineered to not need sleep. In addition to writing, Kress often teaches at various venues around the country and abroad; in 2008 she was the Picador visiting lecturer at the University of Leipzig. Kress lives in Seattle with her husband, writer Jack Skillingstead, and Cosette, the world's most spoiled toy poodle. nancykress.blogspot

Geoffrey Landis

Geoffrey Landis works for NASA on planetary exploration, interstellar propulsion, solar power and photovoltaics. He holds patents for solar cells and photovoltaic devices. Supported by his scientific background Landis also writes hard science fiction. He has won a Nebula Award, two Hugo Awards, and a Locus Award, as well as two Rhysling Awards for his poetry.

Jon Lomberg

Jon Lomberg is one of the world's leading artists inspired by science. He is an Emmy Award-winning television art director for *Cosmos*, a muralist for the National Air and Space Museum, and an award-winning science reporter for the Canadian Broadcasting Corporation. He designed the astronomical animation for the Warner Brothers film *Contact*, based on Carl Sagan's novel, named on the 100 Best Movie Openings of all time. In 1998, on the occasion of his fiftieth birthday, Jon was singularly honored by the International Astronautical Union, which officially designated an asteroid near Mars, previously called 6446 1990QL, as Asteroid Lomberg, in recognition of his many achievements in the field of science communication. The artist has produced some of the most unusual, durable, and far-flung images ever created by the human species. In 1977 he was Design Director for NASA's famous *Voyager* Golden Record, sent beyond the solar system on robot spacecraft. His cover art for that project, predicted to last for more than a thousand million years, may be the longest-lived piece of human art. His art is on four NASA spacecraft now on Mars: *Spirit, Opportunity, Phoenix* and *Curiosity*.
Facebook: Jon Lomberg Artist
www.jonlomberg.com, www.galaxygarden.net

Richard A. Lovett

Richard A. Lovett obtained a degree in astrophysics from Michigan State University before changing fields to pursue the intersection of law and economics. He then graduated second in his law school class from the University of Michigan, added a Ph.D. in economics, and was a law professor at the University of Minnesota. Then he took up writing. As of late 2012, he has sold more than 3,000 nonfiction articles and forty-eight science fiction stories. Most of his fiction has appeared in *Analog Science Fiction & Fact*, but his stories have also appeared in *Cosmos, Nature, Apex & Abyss, Running Times, Wisconsin Magazine*, and *Marathon & Beyond*. They have also been translated into Russian, Polish, Portuguese, and Finnish. His first science fiction book, just published, is *Phantom Sense and Other Stories*. Lovett is one of the most prolific contributors in the history of *Analog* magazine. In addition to science fiction, he's written science articles, writer profiles, and a popular series of how-to articles on short-story writing, for a total of 118 appearances.
www.richardalovett.com
www.facebook.com/pages/Richard-A-Lovett/104305422939478

Adrian Mann

Adrian Mann is a technical designer and illustrator. After studying math, physics, and art, he gained comprehensive experience working for many top-name design agencies in the UK, a wide variety of book and magazine illustration projects, and also for world-class management consultancies for clients such as Rolls-Royce and BAe. His illustrations of the Daedalus Starship are some of the most accurate to have been produced and he will be turning this skill to its successor, the Project Icarus vehicle, once designs begin to emerge. His illustrations have appeared in many books, magazines, and TV programs. Adrian also provides illustrations for other aerospace projects, especially unbuilt UK aircraft and missiles from the 1950s and 60s. Adrian has worked with Reaction Engines Ltd since their early days, and produces all the images and animations for their revolutionary SKYLON spaceplane and SABRE engine. Adrian and his wife Katinka currently live in rural Hungary.

Tom Peters

Tom Peters has served as a technical illustrator for a NASA contractor, a freelance illustrator, and a staff artist for multiple computer software companies. He is accomplished as both a 3D modeler, digital painter, and graphic artist. As a freelance illustrator, Tom has provided cover paintings for both volumes of Sharon Lee and Steve Miller's *Liaden Universe Companion* collections, as well as two of their chapbooks. He has also worked with SF author Allen Steele on the visual and functional design of the spacecraft in Steele's novel *Spindrift*. Tom now resides in the far-west suburbs of Chicago with his wife Diane and four art-directing cats. He works as a freelance artist, and as an instructor, teaching the fundamentals of Adobe Photoshop, Illustrator, and digital graphics production.

Martin Rees

Martin Rees is the author of many research papers, mainly on astrophysics and cosmology, as well as numerous general articles and eight books, some on science and some on politics and policy, including *Before the Beginning, Gravity's Fatal Attraction, Our Final Hour, 'ust Six Numbers, Our Cosmic Habitat* and (most recently) *From Here to Infinity: a Vision for the Future of Science.* His scientific research has been on black holes, "extreme" cosmic phenomena, galaxy formation, the early universe, and the concept of the multiverse. He has also been involved in more political issues, especially those involving space, and the role of "existential risks" to the planet stemming from excessive pressure on the environment, and the misuse of powerful new technologies.
www.ast.cam.ac.uk/~mjr/

Peter Schwartz

Peter Schwartz is Senior Vice President for Global Government Relations and Strategic Planning for Salesforce.com. In these roles he directs policy and politics throughout the world and manages the organization's ongoing strategic conversation. Prior to joining Salesforce he was cofounder and chairman of Global Business Network, a partner of the Monitor Group, a family of professional services firms devoted to enhancing client competitiveness. An internationally renowned futurist and business strategist, Peter specializes in scenario planning, working with corporations, governments, and institutions to create alternative perspectives of the future and develop robust strategies for a changing and uncertain world. From 1982 to 1986, Peter headed scenario planning for the Royal Dutch/Shell Group of Companies in London.

Allen Steele

Before becoming a science fiction author, Allen Steele was a journalist who'd worked for newspapers and magazines in Massachusetts, New Hampshire, Missouri, and his home state of Tennessee. But SF was his first love, so he eventually ditched journalism and instead began producing that which made him want to be a writer in the first place. Since then, Steele has published eighteen novels and nearly a hundred short stories. His work has received numerous awards, including three Hugos, and has been translated worldwide, mainly in languages he can't read. He serves on the Board of Advisors for the Space Frontier Foundation and the Science Fiction and Fantasy Writers of America. He also belongs to Sigma, a group of SF writers who frequently serve as unpaid consultants on matters regarding technology and security. A lifelong space buff, Steele has witnessed numerous space shuttle launches from Kennedy Space Center and has flown NASA's shuttle cockpit simulator at the Johnson Space Center. His most recent novel is *Apollo's Outcasts*. www.allensteele.com

Neal Stephenson

Neal Stephenson is the author of the three-volume historical epic *The Baroque Cycle* (*Quicksilver, The Confusion,* and *The System of the World*) and the novels *Cryptonomicon, The Diamond Age, Snow Crash,* and *Zodiac*. In 2011 he began the Hieroglyph Project to envision large ambitious future projects. He lives in Seattle, Washington.

Rick Sternbach

Rick Sternbach has been a space and science fiction artist since the early 1970s, often combining both interests in a project. His clients include NASA, *Sky and Telescope*, Data Products, Random House, *Smithsonian, Analog, Astronomy*, The Planetary Society, and Time-Life Books. He is a founding member and Fellow of the International Association of Astronomical Artists (IAAA), which was formed in 1981. He has written and illustrated articles on orbital transfer vehicles and interstellar flight for *Science Digest*. Beginning in the late 1970s Rick added film and television illustration and special effects to his background, with productions such as *Star Trek: The Motion Picture, The Last Starfighter, Future Flight*, and *Cosmos*, for which he and other members of the art team received an Emmy award, the first for visual effects. Rick also twice received the coveted Hugo award for best professional science fiction artist, in 1977 and 1978.

Dr. Robert Zubrin

Dr. Robert Zubrin is president of Pioneer Astronautics, a Senior Fellow with the Center for Security Policy, the president of the Mars Society, and the author of *The Case for Mars*, and *Energy Victory*. His newest book, *Merchants of Despair: Radical Environmentalists, Criminal Pseudo-Scientists, and the Fatal Cult of Antihumanism*, from Encounter Books.

Interstellar Organizations

The British Interplanetary Society

The British Interplanetary Society was founded in 1933 and is the oldest space organization in the world still in its original form. Its purpose is to promote the advancement of knowledge and the spread of education and particularly to promote the advancement and dissemination of knowledge relating to the science, engineering and technology of Astronautics and to support and engage in research studies and to disseminate the useful results. It this does

through organizing meetings, lectures and symposia, holding and promoting exhibitions and communication the results of global efforts through its world renowned publications *Spaceflight* magazine, *Odyssey, Space Chronicles* and the *Journal of the British Interplanetary Society*. www.bis-space.com

Tau Zero

Tau Zero, the oldest exclusively interstellar community is an international network of researchers, educators, writers, and artists who collaborate to solve the challenges of star flight and improve life on Earth in the process. They share that progress openly to improve public understanding and to inspire future pioneers. Specific goals: find today's pioneering work, share news of that progress (principally via the our "Centauri Dreams" news forum website, make more progress and inspire tomorrow's pioneers. www.tauzero.aero
www.centauri-dreams

Icarus Interstellar

The mission of **Icarus Interstellar** is to realize interstellar flight before the year 2100. It seeks to accomplish that objective by researching and developing the science and the technologies that will make interstellar flight a reality, igniting the public's interest, and engaging with all those prepared to invest in interstellar exploration. It is an umbrella organization with a number of separate projects under it.

www.icarusinterstellar.org

The Institute for Interstellar Studies

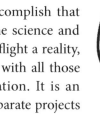

The **Institute for Interstellar Studies**™ is a global organization with a mission to foster and promote education, knowledge and technical capabilities which lead to designs, technologies or enterprise that will enable the construction and launch of interstellar spacecraft. It aspires towards an optimistic future for humans on Earth and in space. It has a bold vision to be an organization, which is central to catalyzing the conditions in society over the next century to enable robotic and human exploration of the frontier beyond our solar system and to other stars, as part of a long-term enduring strategy and towards a sustainable space-based economy.

www.i4is.org

100 Year Starship

100 Year Starship. "We exist to make the capability of human travel beyond our solar system to another star a reality within the next 100 years. We unreservedly dedicate ourselves to identifying and pushing the

radical leaps in knowledge and technology needed to achieve interstellar flight while pioneering and transforming breakthrough applications to enhance the quality of life for all on earth each step of the way. We actively seek to include the broadest swath of people and human experience in understanding, shaping and implementing this global aspiration."

100yss.org.html

ARTISTS

Graphics were provided by the following:

FURTHER READING

Non-Fiction

Interstellar Travel and Multi-generational Space Ships edited by Yoji Kondo, Fred Bruhweiler, John Moore and Charles Sheffield, Apogee Books, 2003

Going Interstellar edited by Jack McDevitt & Les Johnson, Baen Books, 2012

Skylife edited by Gregory Benford & George Zebrowski, Houghton Mifflin Harcourt, 2000

Interstellar Migration and the Human Experience edited by Ben Finney & Eric Jones, Unif Calif. Press 1985

Collapse: How societies choose to fail or succeed, J. Diamond, Penguin Group USA, 2005.

Project Orion, George Dyson, Henry Holt, 2002

Pale Blue Dot, Carl Sagan, Headline, 1995.

The High Frontier: Human Colonies in Space, G.K. O'Neill, William Morrow, New York, 1977.

Entering space: creating a spacefaring civilization, R. Zubrin, JP Tarcher, 2000

The Starflight Handbook: A Pioneer's Guide to Interstellar Travel, E. Mallove and G. Matloff, Wiley & Sons, 1989

Colonies in Space, T. A. Heppenheimer, Warner books, 1978

Mining the Sky, John Lewis, Addison Wesley, 1996

Deep-Space Probes, Gregory Matloff, Springer-Praxis, 2000

Solar Sailing, Colin McInnes, Springer-Praxis,1999

Deep Space Propulsion, Kelvin Long, Springer, 2012

Centauri Dreams-Imagining and Planing Interstellar Exploration, Paul Gilster, Copernicus Books, 2004

Future Spacecraft Propulsion Systems, Paul Czyst & Claudio Bruno, Springer-Praxis, 2006

Mirror Matter-Pioneering Antimatter Physics, Robert L. Forward & Joel Davis, John Wiley & Sons, 1988

Indistinguishable From Magic, Robert L. Forward, Baen Books, 1995

The Eerie Silence, Renewing Our Search for Alien Intelligence, Paul Davies, Houghton Mifflin Harcourt, 2010

The Milky Way-An Insiders Guide, William Waller, Princeton Univ. Press, 2013

Fiction

Voyage, Stephen Baxter. An epic saga of America's might-have-been. If President Kennedy had lived, we could have sent a manned mission to Mars in the 1980s.

Coyote, Allen Steele. Gallant misfits, led by a spaceship captain named Robert E. Lee, steal a starship. They flee a declining Earth rife with dictatorship and technophobia to found a new society on a new world. Good tech descriptions. First of a very popular series.

Time for the Stars, Robert A. Heinlein. Relativity and starships. Identical twins are enlisted to be the human radios that will keep the ships in contact with Earth, but one of them has to stay behind while the other explores the depths of space. Einstein intervenes.

Earth, David Brin. Published in 1991, this epic predicted many social and tech trends we've seen unfold. A small black hole escapes from the lab that made it, and Earth itself could be hollowed out. Wracked by gravity lasers from core to pole, EARTH explores whether humanity and freedom can survive. Call in the scientists!

Hull Zero Three, Greg Bear. Interstellar planet hunting on an enormous damaged starship. On the long voyage strange things have come to live in its vast corridors.

The Highest Frontier, Joan Slonczewski. Going to college in a space habitat in orbit. An intriguing biotech future. Global climate change and advanced social change amid a different future culture. Plus, you're in space.

A Fire Upon The Deep, Vernor Vinge. Supersmart entities rule in part of our galaxy, the Transcend. We're among the limited minds, though not as bad as the Unthinking Depths, where only simple animals can function. Add conflicts and stir.

Red Mars, Green Mars, Blue Mars, Kim Stanley Robinson. A trilogy about founding a Mars colony. It evolves political strains that differ on how to alter the Mars environment and govern the first society independent of Earth.

The Eight Worlds Universe, John Varley, *The Ophiuchi Hotline, Steel Beach, The Golden Globe, Red Thunder,* numerous short stories. Mankind has to expanded into space because we have been evicted from Earth.

Tau Zero, Poul Anderson, Doubleday. Relativity projects a starship that can't stop into the far future.

The Moon is a Harsh Mistress, Robert Heinlein, Doubleday. Aspects of founding colonies, in this case the moon. What a space frontier colony might look like.

Roscheworld, Robert L. Forward. Laser-driven Sailship expedition to Barnard's Star finds adventure

The Mote in God's Eye, Larry Niven & Jerry Pournelle. Classic tale of humans encountering an alien Sailship.

COPYRIGHT NOTICE

INDEX

Made in the USA
Charleston, SC
11 July 2013